PRINT SAMPLE

Size: 6 x 9
Pages: 564
Text stock: 60# Ta
Cover stock: Pe
Cover coat   nation

Sales   tive:
                        81

The Snakes of Thailand and Their Husbandry

The Snakes of Thailand and Their Husbandry

Merel J. Cox

Krieger Publishing Company
Malabar, Florida
1991

Original Edition 1991

Printed and Published by
**KRIEGER PUBLISHING COMPANY
KRIEGER DRIVE
MALABAR, FLORIDA 32950**

Copyright © 1991 by Krieger Publishing Company

All rights reserved. No part of this book may be reproduced in any form or by any means, electronic or mechanical, including information storage and retrieval systems without permission in writing from the publisher.
No liability is assumed with respect to the use of the information contained herein.
Printed in the United States of America.

Library of Congress Cataloging-in-Publication Data
Cox, Merel J., 1932-
    The snakes of Thailand and their husbandry / Merel J. Cox
    p.   cm.
    Includes bibliographical references.
    ISBN 0-89464-437-8
    1. Snake culture. 2. Snakes--Thailand. 3. Captive snakes.
I. Title.
SF515.5.S64C68 1990
639.3'96'09593--dc20   90-4069

10 9 8 7 6 5 4 3 2

Dedication

This book is dedicated to a person who knew nothing about snakes. In fact, she didn't even like them and grimaced every time she saw my children touching one. However, Prapa Peanchana, or "Maeo", liked people and people liked her. My sister-in-law and friend, Maeo, was a rare individual. She was always pleasant and cheerful and she seemed to take pleasure in being with and helping people. I cannot recall ever hearing anyone criticize her - there simply wasn't anything to criticize. Everyone liked Maeo. She was struck down and killed by a cowardly hit-and-run motorist on January 7, 1989. My wife, children, and I miss her very much. The world is a slightly less pleasant place without her.

The Year of the Snake

ปีมะเส็ง

"Now the serpent was more subtil than any beast of the field which the LORD God had made. And he said unto the woman, Yea, hath God said, Ye shall not eat of every tree of the garden? And the woman said unto the serpent, We may eat of the fruit of the trees of the garden: But of the fruit of the tree which is in the midst of the garden, God hath said, Ye shall not eat of it, neither shall ye touch it, lest ye die. And the serpent said unto the woman, Ye shall not surely die: For God doth know that in the day ye eat thereof, then your eyes shall be opened, and ye shall be as gods, knowing good and evil.........."

The Holy Bible
Genesis, Chapter 3, verses 1 to 5.

"And the LORD God said unto the woman, What is this that thou hast done? And the woman said, The serpent beguiled me, and I did eat. And the LORD God said unto the serpent, Because thou has done this, thou art cursed above all cattle, and above every beast in the field; upon thy belly shalt thou go, and dust shalt thou eat all the days of thy life: And I will put enmity between thee and the woman, and between thy seed and her seed; it shall bruise thy head, and thou shalt bruise his heel."

The Holy Bible
Genesis, Chapter 3, verses 13 to 15.

With these words from the sacred book of the Christian faith the serpent is condemned. The subtle, beguiling snake is found guilty of encouraging the first woman to disobey God and is therefore responsible for the original sin. As punishment for this treachery, the snake is the most cursed of all animals and, furthermore, God decreed a state of enmity between man and snake for all time. Within the Christian faith the serpent must be regarded as treacherous and morally dangerous and Christians must "bruise" its head for it will surely "bruise" the heel of man.

Not all of the people of the earth look upon snakes with such disdain. As recently as the late 19th century, pythons were widely worshipped in Africa and to molest or kill a python, even by accident, was punished by death. In pre-Vedic India, snakes were village gods and guardians. A huge snake named Sesha plays an important positive role in Hindu legend. Buddhist legend says that Lord Buddha was befriended and assisted by a cobra and the native people of the Americas, as well as the ancient Greeks and

Romans, worshipped snakes, at least to some extent. In both China and Southeast Asia, snakes have also been thought to have some desirable characteristics and have been objects of veneration.

The Kingdom of Thailand uses an official calendar which commences with the year 543 B.C., the year of the birth of the Lord Buddha. Thus, the year 1989 is equivalent to the Thai year of 2532 (1989 plus 543). In Thailand, in addition to other parts of Asia, a twelve-year cycle is widely known, commonly followed, and of great significance. Each twelve year period bears the name of an animal.

Gregorian Calendar	Twelve-year Cycle	Buddhist Ecclesiastical Calendar
1978	The Horse	2521
1979	The Goat	2522
1980	The Monkey	2523
1981	The Cock	2524
1982	The Dog	2525
1983	The Pig	2526
1984	The Rat	2527
1985	The Ox	2528
1986	The Tiger	2529
1987	The Rabbit	2530
1988	The Dragon	2531
1989	The Snake	2532

Most Asians are fully aware of the animal year in which they were born, as it has significance and influence in many ways. The completion of a twelve-year cycle is an important event and the completion of the fifth, twelve-year cycle is especially significant. The birthday of the King of Thailand is an important national holiday. In 1988, His Majesty the King reached his 60th birthday, the completion of his fifth, twelve-year cycle, and this birthday was celebrated with particular joy and happiness throughout the country.

The animal year of a person's birth is extremely significant as it is believed that it exerts a strong influence upon a person's personality, economic success, and marital happiness. Consequently, in matrimony some individuals are considered compatible and others are thought incompatible, depending on the animal year of birth. For example, persons born in the Year of the Monkey are advised to avoid those born in the Year of the Tiger and are urged to seek those born in the Year of the Dragon or Rat.

The people born in the Year of the Snake are considered wise and intense with a tendency toward physical beauty, but also vain and quick tempered. They are industrious, diligent, and persistent in business and enjoy good fortune. Their work is usually excellent. Although they enjoy good fortune in general, their fortunes often fluctuate and they do experience periods of misfortune. They sometimes fall short of their goals. They are not compatible with the Pig and are advised their best potential for happiness and success lies with the Year of the Cock or the Ox.

It is auspicious that this work was completed in the Year of the Snake, as the event occurred by chance. This coincidence will doubtless result in interesting and informative reading for those who open the pages and enjoy the content.

Foreword

Whether snakes are admired, despised or simply exploited, they have captured the fascination of people since at least the beginnings of recorded history. This book introduces the reader to the snakes of Thailand. Here is a descriptive account of the many species of snakes that are known to occur within the country, accompanied with photographs and simple aids to identification. These provide a useful, current guide for biologists or other persons who might seek basic information about Thailand's herpetofauna. The ensuing pages also contain practical information for amateur herpetologists who wish to maintain snakes in captivity, breed them, or in other ways become involved with these fascinating vertebrates.

Thailand is fortunate to support a rich and varied biota, of which snakes are conspicuous and important elements. Inspection of the distributions of major lineages of living snakes reveals that southern Asia exceeds other geographic regions in the number of such taxa that are present. This part of the world also has a large number of snakes that are either endemic, primitive, or morphologically specialized. Consequently, the snake fauna of Thailand is not only diverse, but also interesting and significant in a number of scientific contexts. Because of the presence of pythons and many venomous species, these reptiles hold a place of commercial importance as well.

But the biology of snakes residing in Thailand is poorly known, as is true for many parts of Asia and the tropics in general. The early works of Ed Taylor and others brought scientific attention to the herpetofauna of Thailand; yet present-day knowledge of elemental subjects such as systematics, reproduction, feeding habits, and geographic distribution is still rudimentary, and there is virtually no information concerning the ecology, life history, behavior or physiology of most species. Such limited information that exists has been derived largely from studies of similar or related species in surrounding parts of Asia (rather than Thailand itself), and is scattered in various scientific publications. Many of these are antiquated and are not widely available. Thus, there is a need for a general, up-to-date publication that can be used as a practical guide to the snake fauna of Thailand. This book provides such a resource.

Two indispensable elements have made this book possible. First, a dedicated individual who has had long standing interests in herpetology and seven years of experience in Thailand compiled the information presented in these pages. Equally significant is the simple fact that Thailand has been blessed with the variety of interesting creatures that are the subjects of this book. These have kindled the fire of curiosity and have provided a constant source of inspiration for the author. The reader may wish to reflect on a sentiment expressed by E. O. Wilson in his recent essay on *Biophilia*: "Life of any kind is infinitely more interesting than almost any conceivable variety of inanimate matter." Merel Cox and I share in the hope that this volume will become for some a *vade mecum* and will help numerous persons to discover Thailand's treasures of living resources. Perhaps now and then a photograph or a passage will move young people to swell the ranks of emerging biologists who are destined to play important roles in shaping Thailand's future.

Harvey B. Lillywhite
Gainesville, Florida

Preface

The purpose for this volume is twofold. The first purpose is to offer suggestions for the improved care of captive snakes. In the past, too many of us have been required to learn the husbandry of snakes by trial and error and, consequently, have suffered unnecessarily high mortality rates within our collections. This current study presents a summary of successful husbandry techniques acquired through considerable experience and research. Some general comments applicable to all collections are offered in addition to specific suggestions for the care of snakes which are native to Thailand. Probably the greatest single difficulty in maintaining tropical snakes in mid-latitude areas are the ensuing respiratory infections resulting from an inadequate understanding of the regimes of temperature and humidity under which tropical snakes live. The same misunderstanding of tropical climates may explain why attempts to breed particular tropical species have sometimes failed. To assist in the solution of some of these problems, comprehensive climatic data for all areas of Thailand have been collated and included within the Appendixes. Suggestions for the application of these data to the husbandry and breeding of Thai snakes are also included. Specific suggestions have also been provided in respect to housing, feeding, and the general care of specific Thai species.

The second purpose of this book is to present a comprehensive list of the snakes of Thailand with their characteristics and habits. The excellent work undertaken by Edward H. Taylor in Thailand significantly contributed to the field of herpetology and it will remain as a valuable and important source of information. Since Taylor's publications, however, our knowledge of the herpetology of Thailand has notably increased, taxonomic revisions have altered the classification of some Thai herpetofauna, and questions have been raised concerning the status of others. This volume attempts to expand upon Taylor's work, present commonly accepted current classifications, correct some errors occuring in current literature, report the presence of possible new species and subspecies, and help in clarifying the status of the genera *Naja* and *Trimeresurus* in Thailand.

Additions to the herpetofauna of Thailand.

Cantoria violacea
Dendrelaphis striatus
Dryocalamus subannulatus
Elaphe porphyracea nigrofasciata
Fordonia leucobalia
Hydrophis lapemoides
Hydrophis spiralis
Lepturophis borneensis
Naja sumatrana
Oligodon dorsalis
Trimeresurus hageni
Trimeresurus macrops, and
Xenelaphis hexagonotus

Reported range extensions within Thailand.

Ahaetulla nasuta
Boiga cyanea
Boiga dendrophila melanota
Boiga jaspidea
Dendrelaphis cyanochloris
Dendrelaphis formosus
Dryophiops rubescens
Elaphe flavolineata
Laticauda laticaudata
Naja sp. Black and White
 Spitting Cobra
Praescutata viperina
Rhabdophis nigrocinctus
Trimeresurus (Tropidolaemus) wagleri

Boiga saengsomi is offered as a new species and *Trimeresurus puniceus wiroti* as a new subspecies.

Two subspecies of sea snakes, *Hydrophis fasciatus fasciatus* and *Hydrophis torquatus torquatus*, are included as probable Thai specimens. Their present confirmed ranges make it probable that they will ultimately be found in Thai territorial waters. A comment regarding their photographs is necessary. The photographs were kindly supplied by Harold K. Voris as *Hydrophis fasciatus* and *Hydrophis torquatus*. Based upon the site of their capture, however, I have included them as *Hydrophis fasciatus fasciatus* and *Hydrophis torquatus torquatus*. Any error is mine, not that of Harold K. Voris.

Some species and subspecies listed as being native to Thailand in previous and contemporary literature have not been included here. They have been excluded as evidence does not exist to prove their presence in Thailand or they have proven to be synonymous with other species. These exclusions are listed on the opposite page.

Appendix 10 provides a translation and analysis of the Thai common names of snakes. It is interesting to note that Thai common names do often differ from the English common name.

The reader must be clear that this text is surely not the last word on the snakes of Thailand. Thailand is a large country that has received scant attention from the international herpetological community. Also, few Thai citizens have chosen to make a career of herpetology and few foreign herpetologists have spent extensive periods of time studying the herpetofauna in the Kingdom. Many border areas are also inaccessible for study. Doubtless future research will reveal that some specimens currently

Excluded Species and Subspecies

(syn. = synonymous with)

Species/Subspecies	Comments
Calamaria leucocephala	syn. *Calamaria lumbricoidea*
Calamaria siamensis	syn. *Calamaria pavimentata*
Calamaria uniformis	syn. *Calamaria pavimentata*
Calliophis maculiceps malcolmi	syn. *Calliophis maculiceps smithi*
Elaphe floweri	insufficient evidence
Gonyosoma jansenii	insufficient evidence
Gonyosoma jansenii elegans	insufficient evidence
Hydrophis mamillaris	insufficient evidence
Oligodon quadrilineatus	syn. *Oligodon taeniatus*
Trimeresurus erythrurus	none found in Thailand
Trimeresurus wiroti	syn. *Trimeresurus puniceus wiroti*
Typhlops muelleri	syn. *Typhlops diardi muelleri*
Typhlops siamensis	only one specimen found in "Siam," locality unknown

excluded from this study do, in fact, reside in Thailand. The existence of additional species in Thailand is also likely to be proven. Range extensions from adjacent countries seem likely, especially in areas of sparse population. The presence of additional species of sea snakes in Thai waters also appears likely. Results of future research are also likely to alter the status of genera *Naja* and *Trimeresurus*, in addition to other genera, in Thailand.

This current study has been based upon the personal experience of the author in association with a comprehensive survey of the available literature. Much personal experience has been gained in the field, through private collection, and also while serving the Siam Farm & Zoological Co., Ltd., in Bangkok.

Correspondence relating to this volume would be welcomed and should be addressed to:
Merel J. Cox
695/17 Pracharaj Road
Soi Homhual
Bangkok 10800, Thailand.

Acknowledgements

I am grateful to a number of people for their assistance. Jeff Kleinman and Gordon Wyllie of the A. U. A. Language Center in Bangkok were always willing to help, and frequently did. Thanks are also extended to Abraham Binstock, formerly of the A. U. A. Language Center, for undertaking the tedious job of proof reading. The staffs of both Siam Farm and the Pata Zoo were helpful in many ways. Dr. Porntep Ratanakorn, Veterinarian of the Pata Zoo, reviewed the veterinary aspects of this book and offered many useful suggestions. Wirot Nutphand provided me with a number of specimens, both living and preserved, for study. A special word of thanks is conveyed to Captain Prasert Suthon for his assistance in the preparation of several excellent drawings that are included in this text.

Mr. M. W. F. Tweedie most generously granted permission to use drawings from his excellent book, *The Snakes of Malaya*. My gratitude is also extended to Dr. William Dunson, of The Pennsylvania State University, and Dr. Harold K. Voris, of the Chicago Field Museum of Natural History, for supplying most of the photographs of sea snakes which appear in this book. I also wish to thank Dr. Lim Boo Liat, of Kuala Lumpur, Malaysia, who kindly supplied me with photographs of species that I was unable to photograph in Thailand.

Very special gratitude is extended to three gentlemen whose help and encouragement were instrumental in the completion of this in-depth study. Dr. Roger Conant, Adjunct Professor, The University of New Mexico, U.S.A., Dr. Hobart Smith, The University of Colorado at Boulder, U.S.A., and Mike Goode, Curator of Reptiles and Amphibians, The Columbus Zoo, U.S.A., were kind enough to take time from their very busy schedules to supply information pertinent to this study and offer help and encouragement in many forms. Whenever I requested advice or copies of literature not available in Thailand, each of these gentlemen was quick to respond. I am very much in debt to each of them.

Two gentlemen from Great Britain also deserve special mention. Dr. David Warrell, of the University of Oxford, has extensive knowledge of the venomous snakes of Southeast Asia, especially the genus *Trimeresurus*. Wolfgang Wüster, of the University of Aberdeen, has thoroughly researched genus *Naja*, as it exists in Southeast Asia. Both men have been generous with the results of their research and have supplied me with valuable information, as well as copies of useful literature. Their kindness and generosity have been most helpful to me.

I am profoundly grateful to Jarujin Nabhitabahata, Curator of Amphibians and Reptiles, Ecological Research Department, Thailand Institute for Scientific and Technological Research. Not only did he give me free use of his preserved specimens, laboratory, and library, but the reader will also note that a number of the photographs that appear in this book are his. Furthermore, this very knowledgeable herpetologist offered considerable and valuable information regarding the ecology and distribution of several Thai species. His help and advice have been a major contribution to this study.

Finally, Merel J. Cox is very grateful for the patience and forebearance of his family. My wife, Raynoo, not only assisted in numerous general ways, but also contributed several excellent drawings. Her forebearance extended to enduring the presence of serpentine "house guests" I brought home for further study. She even photographed several of them for inclusion in this book. My daughters, Becky, Irene, and Cathy, were very patient while their father spent much of his time writing rather than playing with them.

I hasten to add that any errors or mistaken judgements which may appear are mine alone. I am most grateful for help and advice received, but I alone am reponsible for the final decisions and judgements reflected in this book.

Contents

Foreword xi
Preface xv
Acknowledgements xxi
List of Figures xxxi
List of Tables xxxii
List of Plates xxxiii

PART 1
INTRODUCTION AND PRELIMINARIES

General Introduction 3

Geography of Thailand 5
Physical Regions 5
The North 5
The Central Region 8
The Northeast 9
The Southeast 10
The West 10
The South 11
Photoperiod 12

Implications for the Care of Thai Snakes 13
Photoperiod 14
Relative Humidity 14
Temperature 14

The Thai People and Their Snakes 16
Snakebites in Thailand 17

Some Characteristics of Snakes 20
Scalation 20
Locomotion 25
The Senses of Hearing, Smell, and Sight 27
Thermal Sensitivity 28
Food 28

Growth and Longevity 30
Reproduction 31
Habitat 34
Predators and Means of Defense 34

Husbandry in General 37
Laws Governing the Keeping of Snakes 38
Herpetological Societies 39
Housing 40
Cleanliness 42
Feeding 45
Water 47
Breeding 48
Record Keeping 52

Medical Aspects of Husbandry 53
External Problems 54
Poor Sheds 54
Parasites (External and Internal) 55
Fasts 58
Wounds 64

Diseases 65
Respiratory Infections 65
Mouth Rot 66
Scale Rot 67
Corneo-spectacular Swelling 67

Antibiotics and Their Use 68

PART 2
THE SNAKES OF THAILAND

Introduction 75

Infraorder Scolecophidia 81
Family Typhlopidae 82
Genus Ramphotyphlops 85
Genus Typhlops 87

Infraorder Alethinophidia 92
Superfamily Acrochordoidea 93
Family Acrochordidae 93
Genus Acrochordus 93

Superfamily Anilioidea 97
Family Uropeltidae 97
Genus Cylindrophis 98

Family Xenopeltidae 100
Genus Xenopeltis 100

Superfamily Booidea 102
Family Pythonidae 102
Genus Python 104

Superfamily Colubroidea 110
Family Colubridae 110
Subfamily Calamariinae 111
Genus Calamaria 111
Genus Pseudorabdion 115

Subfamily Colubrinae 117
   Genus Ahaetulla 117
   Genus Boiga 121
   Genus Chrysopelea 145
   Genus Dendrelaphis 149
   Genus Dryophiops 154
   Genus Elaphe 156
   Genus Gonyosoma 164
   Genus Psammodynastes 165
   Genus Psammophis 167
   Genus Ptyas 168
   Genus Xenelaphis 180

Subfamily Homalopsinae 181
   Genus Bitia 183
   Genus Cantoria 184
   Genus Cerberus 186
   Genus Enhydris 188
   Genus Erpeton 192
   Genus Fordonia 195
   Genus Gerarda 196
   Genus Homalopsis 198

Subfamily Lycodontinae 199
   Genus Dinodon 199
   Genus Dryocalamus 200
   Genus Lepturophis 204
   Genus Liopeltis 206
   Genus Lycodon 209
   Genus Oligodon 213

Subfamily Natricinae 233
   Genus Amphiesma 233
   Genus Opisthotropis 238
   Genus Parahelicops 240
   Genus Rhabdophis 241
   Genus Sinonatrix 246
   Genus Xenochrophis 250

Subfamily Pareatinae 254
    Genus Aplopeltura 255
    Genus Pareas 257

Subfamily Pseudoxenodontinae 262
    Genus Macropisthodon 262
    Genus Plagiopholis 264
    Genus Pseudoxenodon 266

Subfamily Sibynopheinae 276
    Genus Sibynophis 276

Subfamily Xenoderminae 280
    Genus Xenodermus 280

Family Elapidae 282
Subfamily Bungarinae 282
    Genus Bungarus 282
    Genus Naja 288
    Genus Ophiophagus 301

Subfamily Hydropheinae 305
    Genus Acalyptophis 315
    Genus Aipysurus 316
    Genus Astrotia 318
    Genus Enhydrina 319
    Genus Hydrophis 321
    Genus Kerilia 335
    Genus Kolpophis 336
    Genus Lapemis 337
    Genus Microcephalophis 339
    Genus Pelamis 340
    Genus Praescutata 342
    Genus Thalassophis 344

Subfamily Laticaudinae 345
    Genus Laticauda 346

Subfamily Maticorinae 349
Genus Calliophis 349
Genus Maticora 363

Family Viperidae 366
Subfamily Crotalinae 367
Genus Calloselasma 367
Genus Trimeresurus 371

Subfamily Viperinae 406
Genus Vipera 406

Appendixes

1. World Herpetological Societies and Organizations 411
2. Climatic Data. The North (1951-1980) 427
3. Climatic Data. The Central Region (1951-1980) 433
4. Climatic Data. The Northeast (1951-1980) 439
5. Climatic Data. The Southeast (1951-1980) 445
6. Climatic Data. The West (1951-1980) 451
7. Climatic Data. The South (1951-1980) 457
8. Water Surface Conditions Off the Coasts of Thailand 463
9. Protected Snakes of Thailand 465
10. Analysis of Common Thai Snake Names 467

Bibliography 501

Index 513

List of Figures

		Page
1.	Head Scales in General	21
2.	Head Scales of Genus Trimeresurus	22
3.	Body Scales in General	23
4.	Mouth Gag	63
5.	Diagram of the Mouth	63
6.	The Bones of the Head of an Idealized Snake	78
7.	Types of Dentition	80
8.	*Ramphotyphlops braminus*	86
9.	*Acrochordus javanicus*	96
10.	*Cylindrophis rufus rufus*	99
11.	*Xenopeltis unicolor*	101
12.	Heads of Reed Snakes: *Calamaria lumbricoidea* and *Calamaria schlegeli schlegeli*	113
13.	*Pseudorabdion longiceps*	116
14.	*Ahaetulla prasina*	121
15.	*Boiga dendrophila melanota*	134
16.	Heads of Racers: *Elaphe flavolineata, Elaphe radiata,* and *Elaphe porphyracea porphyracea*	158
17.	*Psammodynastes pulverulentus*	166
18.	*Ptyas korros*	170
19.	Heads of Homalopsine Snakes: *Enhydris plumbea, Homalopsis buccata,* and *Cerberus rynchops*	187
20.	*Lycodon subcinctus*	212
21.	*Amphiesma inas*	236
22.	*Rhabdophis subminiatus subminiatus*	245
23.	Present division of Genus *Natrix*	247
24.	*Xenochrophis piscator*	252
25.	Heads of Pareine Snakes	254
26.	Subfamily *Pareatinae* Criteria	255
27.	*Sibynophis collaris*	277
28.	*Bungarus fasciatus*	285
29.	Heads of Cobras: *Naja naja* and *Ophiophagus hannah*	289
30.	*Enhydrina schistosa*	320
31.	*Hydrophis cyanocinctus*	324

	Page
32. *Laticauda colubrina*	347
33. *Calloselasma rhodostoma*	369
34. *Trimeresurus albolabris albolabris*	374
35. *Trimeresurus (Ovophis) monticola meridionalis*	382
36. *Trimeresurus popeorum popeorum*	385

Figures 1, 3, 8 - 22, 24, 25, 28 - 33, and 35 have been reproduced by kind permission of Mr. M. W. F. Tweedie, author of The Snakes of Malaya where the drawings were first published in 1983.

Figures 2, 5, 6, and 7 were drawn by Captain Prasert Suthon, Royal Thai Air Force. Figures 4, 27, 34, and 36 were drawn by Raynoo Cox.

List of Tables

	Page
1. Provinces of Thailand	6
2. Photoperiods of Thailand	13
3. Reported Bites and Confirmed Deaths from Snakebites. Thailand (1974 - 1985)	18
4. Antibiotic Dosages	71
5. Reclassification List	77
6. Characteristics of *Boiga cynodon* and *Boiga ocellata*	141
7. Characteristics of *Trimeresurus hageni* and *Trimeresurus sumatranus*	376
8. Characteristics of *Trimeresurus albolabris albolabris* and *Trimeresurus macrops*	380
9. Characteristics of *Trimeresurus puniceus puniceus* and *Trimeresurus puniceus wirtoi*	389

Maps

1. Map of Thailand	7

List of Plates

		Page
1. *Ramphotyphlops braminus*, งูดินธรรมดา		123
2. *Typhlops lineatus*, งูดินลายขีด		123
3. *Acrochordus granulatus*, งูผ้าขี้ริ้ว		123
4. *Acrochordus javanicus*, งูงวงช้าง		124
5. *Acrochordus javanicus*, งูงวงช้าง		124
6. *Cylindrophis rufus rufus*, งูก้นขบ, งูสองหัว		124
7. *Cylindrophis rufus rufus*, งูก้นขบ, งูสองหัว. Brown variety		125
8. *Xenopeltis unicolor*, งูแสงอาทิตย์		125
9. *Xenopeltis unicolor*, งูแสงอาทิตย์. A Juvenile		125
10. *Python curtus brongersmai*, งูหลามปากเป็ด		126

11. *Python molurus bivittatus*, งูหลาม.
    Albino and normal specimen breeding 126
12. *Python molurus bivittatus*, งูหลาม. An albino 126
13. *Python molurus bivittatus*, งูหลาม. A color mutation 127
14. *Python molurus bivittatus*, งูหลาม. A color mutation 127
15. *Python molurus bivittatus*, งูหลาม.
    A mutation with a normal specimen 127
16. *Python reticulatus*, งูเหลือม 128
17. *Python reticulatus*, งูเหลือม. Female laying eggs 128
18. *Python reticulatus*, งูเหลือม.
    Hatchling emerging from egg 128
19. *Ahaetulla mycterizans*, งูเขียวหัวจิ้งจกยมลาย 129
20. *Ahaetulla nasuta*, งูเขียวปากแหนบ 129
21. *Ahaetulla prasina*, งูเขียวหัวจิ้งจก, งูง่วงกลางดง 129
22. *Ahaetulla prasina*, งูเขียวหัวจิ้งจก, งูง่วงกลางดง. Yellow phase 130
23. *Boiga cyanea*, งูเขียวดง, งูเขียวบอน 130
24. *Boiga cynodon*, งูแส้หางม้า, งูกะปิ, งูกินไข่ 130
25. *Boiga cynodon*, งูแส้หางม้า, งูกะปิ, งูกินไข่. Melanistic form 171
26. *Boiga cynodon*, งูแส้หางม้า, งูกะปิ, งูกินไข่. Juvenile 171
27. *Boiga dendrophila melanota*, งูปล้องทอง 171
28. *Boiga drapiezii*, งูดงคาทอง. Green phase 172
29. *Boiga drapiezii*, งูดงคาทอง. Brown phase 172
30. *Boiga jaspidea*, งูกะ 172
31. *Boiga multomaculata*, งูแม่ตะงาวรังนก 173
32. *Boiga nigriceps*, งูต้องไฟ 173
33. *Boiga saengsomi*, งูเขียวดงลาย 173
34. *Chrysopelea ornata*, งูเขียวพระอินทร์, งูเขียวดอกหมาก 174
35. *Chrysopelea paradisi*, งูเขียวร่อน, งูเขียวบิน 174
36. *Chrysopelea pelias*, งูดอกหมากแดง 174
37. *Dendrelaphis caudolineatus*, งูสายม่านแดงหลังลาย 175
38. *Dendrelaphis formosus*, งูสายม่านหลังทอง 175
39. *Dendrelaphis p. pictus*, งูสายม่านพระอินทร์, งูสายม่านธรรมดา 175
40. *Dendrelaphis striatus*, งูสายม่านลายเฉียง 176
41. *Dendrelaphis subocularis*, งูสายม่านเกล็ดใต้ตาใหญ่ 176
42. *Dryophiops rubescens*, งูสายน้ำผึ้ง, งูเถา 176
43. *Elaphe flavolineata*, งูทางมะพร้าวดำ, งูหลุนซุน 177
44. *Elaphe porphyracea porphyracea*, งูทางมะพร้าวแดง, งูบ้องไฟ 177
45. *Elaphe radiata*, งูทางมะพร้าวธรรมดา 177

List of Plates   xxxv

46. *Elaphe taeniura ridleyi*, งูกาบหมากหางนิล, งูไป้   178
47. *Elaphe taeniura taeniura*, งูกาบหมากดำ   178
48. *Gonyosoma oxycephalum*, งูเขียวกาบหมาก   178
49. *Psammodynastes pulverulentus*, งูหมอก   219
50. *Psammodynates pulverulentus*, งูหมอก.
     Reddish color phase   219
51. *Psammophis condarnarus indochinensis*, งูม่านทอง   219
52. *Ptyas korros*, งูสิงธรรมดา   220
53. *Ptyas mucosus*, งูสิงหางลาย   220
54. *Cerberus rynchops*, งูปากกว้างน้ำเค็ม   220
55. *Enhydris bocourti*, งูไซ   221
56. *Enhydris enhydris*, งูสายรุ้งธรรมดา   221
57. *Enhydris jagori*, งูสายรุ้งลาย   221
58. *Enhydris plumbea*, งูปลิง   222
59. *Erpeton tentaculatum*, งูกระด้าง   222
60. *Erpeton tentaculatum*, งูกระด้าง.
     Head showing appendages   222
61. *Homalopsis buccata*, งูหัวกระโหลก, งูเหลียมอ้อ   223
62. *Homalopsis buccata*, งูหัวกระโหลก, งูเหลียมอ้อ. Juvenile   223
63. *Homalopsis buccata*, งูหัวกระโหลก, งูเหลียมอ้อ.
     Head of an albino   223
64. *Dryocalamus subannulatus*, งูปล้องฉนวนมลายู.
     Striped form   224
65. *Dryocalamus subannulatus*, งูปล้องฉนวนมลายู. Banded form   224
66. *Lepturophis borneensis*, งูปล้องฉนวนบอร์เนียว   224
67. *Lycodon capucinus*, งูสร้อยเหลือง   225
68. *Lycodon laoensis*, งูปล้องฉนวนลายเหลือง, งูปล้องฉนวนลาว   225
69. *Lycodon subcinctus*, งูปล้องฉนวนบ้าน   225
70. *Lycodon subcinctus*, งูปล้องฉนวนบ้าน. Juvenile   226
71. *Oligodon barroni*, งูปี่แก้วหัวลายหัวใจ   226
72. *Oligodon cyclurus smithi*, งูปี่แก้วธรรมดาลายจาง   226
73. *Oligodon inornatus*, งูปี่แก้วสีจาง   267
74. *Oligodon joynsoni*, งูปี่แก้วใหญ่   267
75. *Oligodon taeniatus*, งูงอด   267
76. *Amphiesma deschauenseei*, งูลายสาบท้องสามขีด   268
77. *Amphiesma stolata chinensis*, งูลายสาบดอกหญ้า   268
78. *Rhabdophis chrysargus*, งูลายสาบจุดดำขาว   268
79. *Rhabdophis subminiatus subminiatus*, งูลายสาบคอแดง   269

80. *Rhabdophis subminiatus subminiatus*,
    งูลายสาบคอแดง. Juvenile ... 269
81. *Sinonatrix trianguligera*, งูลายสอลายสามเหลี่ยม ... 269
82. *Sinonatrix trianguligera*, งูลายสอลายสามเหลี่ยม. Juvenile ... 270
83. *Xenochrophis flavipunctata*, งูลายสอธรรมดา ... 270
84. *Xenochrophis piscator*, งูลายสอใหญ่ ... 270
85. *Aplopeltura boa*, งูกินทากหัวโหนก, งูบอ ... 271
86. *Pareas carinatus*, งูกินทากเกล็ดสัน ... 271
87. *Pareas malaccanus*, งูกินทากมลายู ... 271
88. *Pareas margaritophorus*, งูกินทากจุดขาว ... 272
89. *Macropisthodon flaviceps*, งูรังแหหัวแดง ... 272
90. *Macropisthodon rhodomelas*, งูรังแหหลังศร ... 272
91. *Plagiopholis nuchalis*, งูหัวศร, งูหัวลายลูกศร ... 273
92. *Sibynophis melanocephalus*, งูคอขั้นปลายหัวดำ ... 273
93. *Sibynophis triangularis*, งูคอขั้นหัวลายสามเหลี่ยม ... 273
94. *Bungarus candidus*, งูทับสมิงคลา ... 274
95. *Bungarus candidus*, งูทับสมิงคลา. Juvenile ... 274
96. *Bungarus fasciatus*, งูสามเหลี่ยม ... 274
97. *Bungarus flaviceps*, งูสามเหลี่ยมหัวหางแดง ... 307
98. *Naja kaouthia*, งูเห่าหม้อ ... 307
99. *Naja kaouthia*, งูเห่าหม้อ. Defensive stance ... 307
100 - 103.
    Naja kaouthia, งูเห่าหม้อ. Atypical hood patterns ... 308
104 - 107
    Naja kaouthia, งูเห่าหม้อ. Atypical hood patterns ... 309
108. *Naja kaouthia*, งูเห่าหม้อ. Atypical hood pattern ... 310
109. *Naja kaouthia*, งูเห่าหม้อ. Dorsal view of albino ... 310
110. *Naja kaouthia*, งูเห่าหม้อ. Ventral view of albino ... 310
111. *Naja kaouthia*, งูเห่าหม้อ. White cobra ... 311
112. *Naja sp.* งูเห่าสีนวล. Suphan Cobra ... 311
113. *Naja sp.* งูเห่าด่างพ่นพิษ. Black and White Spitting Cobra. Dorsal view ... 311
114. *Naja sp.* งูเห่าด่างพ่นพิษ. Black and White Spitting Cobra. Ventral view ... 312
115. *Naja sp.* งูเห่าดำพ่นพิษ. Black Spitting Cobra. Dorsal view ... 312
116. *Naja sp.* งูเห่าดำพ่นพิษ. Black Spitting Cobra. Ventral view ... 312

List of Plates xxxvii

117. *Naja sp.* งูเห่าอิสานพ่นพิษ. Isan Spitting Cobra.
 Dorsal view 313
118. *Naja sp.* งูเห่าอิสานพ่นพิษ. Isan Spitting Cobra.
 Ventral view 313
119. *Naja sumatrana*, งูเห่าทองพ่นพิษ. Dorsal view of hood 313
120. *Naja sumatrana*, งูเห่าทองพ่นพิษ. Ventral view of hood 313
121. *Ophiophagus hannah*, งูจงอาง 314
122. *Acalyptophis peronii*, งูทากลายท้องขาว 314
123. *Acalyptophis peronii*, งูทากลายท้องขาว. View of head 314
124. *Aipysurus eydouxii*, งูทะเลจุดขาว 355
125. *Astrotia stokesii*, งูทากลาย 355
126. *Astrotia stokesii*, งูทากลาย. Melanistic form 355
127. *Enhydrina schistosa*, งูคออ่อนปากจะงอย, งูคออ่อนหัวโต 356
128. *Hydrophis brookii*, งูแสมรังท้องเหลือง 356
129. *Hydrophis caerulescens*, งูแสมรังลายเยื้อง, งูแสมรังเกล็ดหยาบ 356
130. *Hydrophis cyanocinctus*, งูแสมรังเหลืองลายคราม, งูแสมรังลายฟ้า 357
131. *Hydrophis fasciatus fasciatus*, งูแสมรังลายแถบ 357
132. *Hydrophis ornatus ornatus*, งูแสมรังหางขาว 358
133. *Hydrophis spiralis*, งูแสมรังเหลือง 358
134. *Hydrophis torquatus torquatus*, งูแสมรังเทาหัวดำเหลือง 358
135. *Lapemis hardwickii*, งูอ้ายงั่ว 359
136. *Lapemis hardwickii*, งูอ้ายงั่ว. Male sex organs 359
137. *Microcephalophis gracilis gracilis*, งูคออ่อนหัวเข็ม 360
138. *Pelamis platurus*, งูชายธงหลังดำ 360
139. *Pelamis platurus*, งูชายธงหลังดำ. Various patterns 360
140. *Pelamis platurus*, งูชายธงหลังดำ. Mating 361
141. *Praescutata viperina*, งูชายธงท้องขาว 361
142. *Laticauda colubrina*, งูสมิงทะเลปากเหลือง 362
143. *Calliophis gracilis*, งูปล้องหวายเทา 362
144. *Cailiophis maculiceps maculiceps*, งูปล้องหวายหัวดำ 362
145. *Maticora bivirgata flaviceps*, งูพริกท้องแดง 395
146. *Maticora intestinalis lineata*, งูพริกสีน้ำตาล 395
147. *Calloselasma rhodostoma*, งูกะปะ. From the South 396
148. *Calloselasma rhodostoma*, งูกะปะ. From the Northeast 396
149. *Calloselasma rhodostoma*, งูกะปะ.
 Hatchlings leaving the eggs 397
150. *Trimeresurus albolabris albolabris*,
 งูเขียวหางไหม้ท้องเหลือง. Shedding 397

151. *Trimeresurus albolabris albolabris,* งูเขียวหางไหม้ท้องเหลือง 398
152. *Trimeresurus hageni,* งูเขียวหางไหม้สุมาตราหัวเขียว 398
153. *Trimeresurus kanburiensis,* งูหางแฮ่มกาญจน์ 398
154. *Trimeresurus macrops,* งูเขียวหางไหม้ตาโต 399
155. *Trimeresurus (Ovophis) monticola meridionalis,* งูหางแฮ่มภูเขา. Specimen from the North 399
156. *Trimeresurus (Ovophis) monticola meridionalis,* งูหางแฮ่มภูเขา. Specimen from the South 399
157. *Trimeresurus popeorum popeorum,* งูเขียวหางไหม้ท้องเขียว 400
158. *Trimeresurus puniceus puniceus,* งูปาล์มแดง 400
159. *Trimeresurus puniceus wiroti,* งูปาล์ม 400
160. *Trimeresurus purpureomaculatus purpureomaculatus,* งูพังกา, งูเขียวม่วงหางไหม้ 401
161. *Trimeresurus (Tropidolaemus) wagleri,* งูเขียวตุ๊กแก, งูกะปะเลือ 402
162. *Trimeresurus (Tropidolaemus) wagleri,* งูเขียวตุ๊กแก, งูกะปะเลือ. A juvenile 402
163. *Vipera russelli siamensis,* งูแมวเซา 402
164. *Vipera russelli siamensis and Vipera russelli russelli,* งูแมวเซาและงูแมวเซาอินเดีย 402

Plates are individually credited to each of the photographers. These include, the author, Merel J. Cox, Raynoo Cox, Dr. Lim Boo Liat, J. P. Kleinman, Suthigit Patramangorn, Jarujin Nabhitabhata, Dr. Roger Conant, Don Wells, Piboon Jintakune, Dr. William Dunson, Harold K. Voris, Corey Blanc, and Preecha Varavichit.

PART 1

Introduction and Preliminaries

General Introduction

The origin of snakes is unknown. Contemporary studies reveal snakes are more closely related to lizards than any other vertebrate, but fossil records have not yet revealed information regarding the origin of snakes. The difficulty is mainly due to the fact that fossil skeletons of snakes are rarely found intact. Individual vertebrae are common but are an inadequate tool for determining the origin of snakes.

The fossil record proves that a number of extinct forms inhabited the earth between approximately 36 million B.C. and 140 million B.C., or from the Eocene to the late Cretaceous Epoch. Available evidence indicates they were mostly secretive creatures that inhabited crevices or burrowed in loose soil. Both terrestrial and marine forms existed during that time and at the advent of the Oligocene Epoch snake fauna were rather varied although not to the present extent. Oligocene (22.5 million B.C. to 36 million B.C.) sediments have not yet yielded substantial fossil remains and it appears that that epoch was unfavorable for snakes and represents a period of regression in snake evolution. The Miocene Epoch (5.5 million B.C. to 22.5 million B.C.), however, was a period so rich in snake fauna that it has been referred to by some as the "Age of Snakes." The ancestors of many of our present species appeared including venomous *Elapidae* and *Viperidae* as well as *Acrochordoidea, Anilioidea,* and *Booidea. Scolecophidia* were widespread and it was then that *Colubroidea* surpassed *Booidea* in total numbers. By the Pleistocene Epoch (which began at approximately 1.8 million B.C.) most of today's extinct species were very uncommon and the composition of the world's snake population was similar to what it is today.

Regardless of their origin and evolution, snakes are both frightening and fascinating to most people. People are most probably frightened of them because some religions portray snakes in a very bad light. This may be part of the explanation for the many superstitions surrounding snakes, some of which give snakes impossible capabilities. Justifiable fear results from the knowledge that certain species of snakes have very toxic venom and constitute a real danger to mankind. It is unfortunate that fear, coupled with limited knowledge, has led to the destruction of countless numbers of harmless and beneficial snakes.

On the other hand, even those frightened by snakes are also fascinated by them. Reptile houses are usually filled with spectators. Although few would admit it, snakes are usually graceful and beautiful animals with several possessing brilliant colors. They are fascinating not only because they are limbless but also they have been compensated in many interesting ways and manage to survive in a very competitive world.

The Geography of Thailand

The term "geography" as used encompasses the physical and human environments in which the snakes of Thailand live. The physical environment of the six regions of Thailand are described and the successful husbandry of Thai snakes discussed. Various aspects of the man - snake relationship will be discussed within the human environment.

Physical Regions

Thailand is divided into six regions based upon physical and cultural features. These will be discussed in respect to the physical environments in which the snakes of Thailand live. Refer to: Map of Thailand, page 7.

THE NORTH

This is a region of mountains and valleys. Mountain peaks may exceed 2500 m in elevation. Most valleys are over 200 m above sea level. Valley floors are usually cultivated and highland areas are often cultivated by slash and burn agriculturalists. Much of this region, however, is forested.

Table 1.   Provinces of Thailand.

1. The North

1. Chiang Rai
2. Chiang Mai
3. Mae Hong Son
4. Lampang
5. Nan
6. Lamphun
7. Phrae
8. Uttaradit
9. Sukhothai
10. Tak
11. Phayao

2. The Central Region

12. Phitsanulok
13. Kamphaeng Phet
14. Phichit
15. Phetchabun
16. Nakhon Sawan
17. Chai Nat
18. Uthai Thani
19. Sing Buri
20. Lop Buri
21. Saraburi
22. Ang Thong
23. Phra Nakhon Si Ayutthaya (Ayutthaya)
24. Suphan Buri
25. Nakhon Pathom
26. Nonthaburi
27. Pathum Thani
28. Nakhon Nayok
29. Samut Songkhram
30. Samut Sakhon
31. Krung Thep Maha Nakhon (Bangkok)
32. Samut Prakan

3. The Northeast

33. Loei
34. Nong Khai
35. Udon Thani
36. Sakon Nakhon
37. Nakhon Phanom
38. Khon Kaen
39. Maha Sarakham
40. Kalasin
41. Mukdahan
42. Chaiyaphum
43. Roi Et
44. Yasothon
45. Ubon Ratchathani
46. Nakhon Ratchasima
47. Buri Ram
48. Surin
49. Si Sa Ket

5. The Southeast

50. Prachin Buri
51. Chachoengsao
52. Chon Buri
53. Rayong
54. Chanthaburi
55. Trat

6. The West

56. Kanchanaburi
57. Ratchaburi
58. Phetchaburi
59. Prachuap Khiri Khan

7. The South

60. Ranong
61. Chumphon
62. Phangnga
63. Surat Thani
64. Phuket
65. Krabi
66. Nakhon Si Thammarat
67. Trang
68. Phatthalung
69. Satun
70. Songkhla
71. Pattani
72. Yala
73. Narathiwat

Map of Thailand

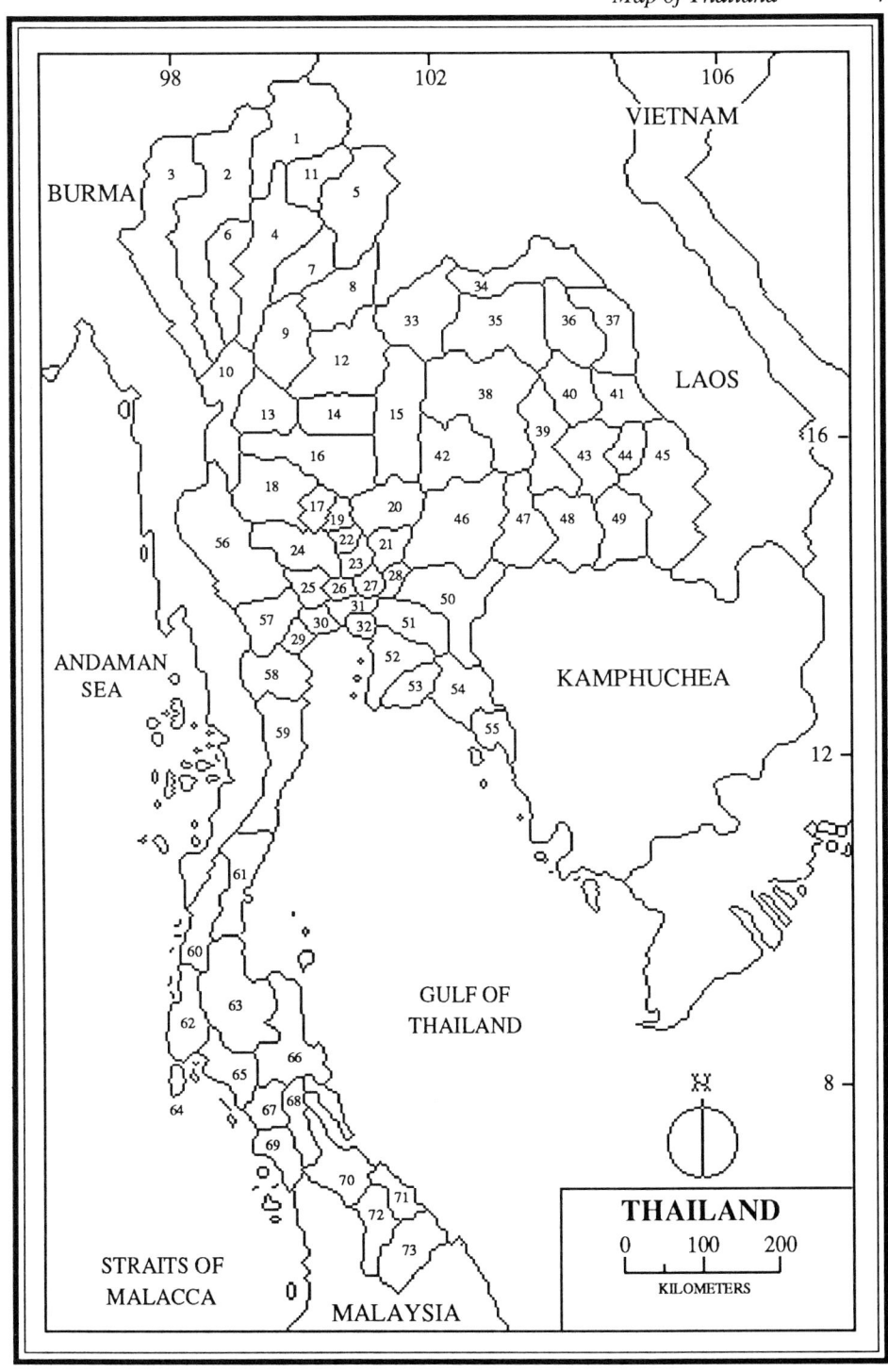

Because of the elevation and distance from the sea, the North experiences generally cooler temperatures than the other regions. Refer to: Appendix 2, pages 425 to 430. Average daytime peak temperatures are quite warm, but are slightly cooler than the rest of the country. The cold season is more pronounced than elsewhere in the country. During the cold season, the average daily low temperatures fall in the range of 12° C to 18° C, and the extreme lows between 3° C and 4° C. The North experiences the greatest diurnal temperature ranges. At Mae Sariang, Mae Hong Son Province, for example, the daily high during February averages 33.6°C and the average low is 12.9°C, resulting in a diurnal temperature range of 20.7° C.

Relative humidity is in the 80's during the rainy season (May - October) but drops into the 60's during the hot season (March-April). Relative humidity averages around 75 percent throughout the year.

Rainfall averages are between 1400 and 1500 mm in most parts of this region. Most rainfall occurs between May and October, with January, February and March being the driest months.

THE CENTRAL REGION

This is a large alluvial lowland with an elevation at sea level in the south and approximately 140 m in the north. This region, which includes Bangkok, is the most densely populated area of Thailand. During the past years most of the natural vegetation has been replaced by cultivation.

In the southern area, that nearest the Gulf of Thailand, the temperatures are mild. Average maximum daytime temperatures fall below 35° C during the hottest months of the year, with average lows during the cold season above 20° C. The coldest temperature ever recorded in Bangkok was 9.9° C. The Bangkok diurnal temperature range is 11.3° C for January. Refer to: Appendix 3, pages 431 to 436.

The average highs and lows in the northern area are more extreme than those for the southern area. Pitsanulok is a good example; average maximum daytime temperature is 37.4° C for April, the average nighttime low is 17.7° C for January, and the diurnal temperature range is 13.9° C during January. These temperatures are more extreme than those of Bangkok.

The average annual mean relative humidity is 78 percent in the south and 66 percent in the north. The seasonality and amount of rainfall of the Central Region is comparable to that of the North.

THE NORTHEAST

The Northeast, also known as the Khorat Plateau, is a plateau that gently tilts to the north and east, towards the Mekong River. It is flat to gently rolling country with elevations mostly between 100 and 200 m. The western and southern margins, however, range from over 200 m to 1000 m above sea level. The Northeast is a region of sparse forests and grasslands with areas of heavy cultivation.

This region, like the North, is quite far from marine influences. It is not shielded by mountain ranges, and is often invaded by cool air masses from the Asian mainland during the cold season. Consequently, the cold season in the Northeast is very similar to that of the North. Temperatures during the day are warm but become fairly cold at night. The lowest temperature ever recorded in

Thailand, 0.1°C, was recorded at Loei, in the Northeast. The hot season in this region is usually the hottest in all of Thailand, with extreme maximum temperatures often above 40° C. The Northeast receives the least amount of precipitation of any region in Thailand, therefore the relative humidity is the lowest in the nation. Refer to: Appendix 4, pages 437 to 442.

THE SOUTHEAST

This is essentially a coastal plain, although elevations do increase to the north and east. The level areas are heavily farmed and the rolling country to the north and east is covered with forests and plantations. The Southeast is the smallest region in Thailand.

The Southeast is under marine influences that tend to make climatic conditions similar to those of the southern Central Region. Temperatures in the hill country are slightly more variable. Mean relative humidity ranges from 70 to 80 percent. Refer to: Appendix 5, pages 443 to 448.

THE WEST

This region is composed of mountains and valleys. It differs from the North in several ways, however. The maximum elevations are generally lower than those of the North. Furthermore, the southern portion of this region borders the Gulf of Thailand and is therefore under marine influences. The mountain slopes are sometimes heavily forested and the valleys and coastal plains often farmed.

Climatically, there are really two regions. The northern part is more mountainous and farther removed from marine influences. Therefore, the cold season is cooler than in the southern portion of this region. Also, the average nighttime temperatures are lower and the diurnal temperature range is greater in the northern portion of this region. The moderating presence of the Gulf of Thailand gives the southern portion a climate free of extreme temperature ranges.

The relative humidity also varies from north to south. In the north the mean relative humidity is in the upper 60's; in the south it is in the upper 70's. This is another consequence of the southern half of the region being located near a large, warm body of water. Refer to: Appendix 6, pages 449 to 454. Rainfall is greater in the south than in the north and probably averages over 1100 mm throughout the region. May through October are the wettest months.

THE SOUTH

The South is peninsular, being flanked by the Gulf of Thailand to the east and by the Andaman Sea and the Straits of Malacca to the west. Three mountain ranges play a prominent role in the topography of this region. The Phuket Range extends from the north in a south - southwesterly direction and out to sea in the vicinity of 8°N, forming the island of Phuket and a number of other offshore islands. Some of the peaks of this range are nearly 1900 m above sea level with general elevations of an average of 1000 meters. The second major mountain range, the Nakhon Si Thammarat Range, first appears in the Gulf of Thailand in the form of Samui Island and other offshore islands. It then crosses the peninsula trending in a north - south direction and enters the Straits of Malacca, forming offshore islands, at the Malaysian border. Elevations in the Nakhon Si Thammarat Range are comparable to those of the Phuket

Range. The Semgalasiri Range, a northern extension of a mountain range which extends through much of peninsular Malaysia, is restricted to the southern provinces of Narathiwat, Yala, Pattani, and Songkhla. Its maximum elevation is slightly under 1500 meters. The land lying between each of these ranges is rolling to rugged, with some isolated mountain peaks reaching elevations of almost 800 meters. Of the two coastal plains, the eastern is the wider and more continuous. The western is narrower and more discontinuous, especially in the north. The slopes of the mountain ranges and isolated mountains are forested, the lower elevations cultivated. The rolling country is most often used to produce plantation crops, whereas the most level, easily irrigated land produces rice.

Temperatures in this region are the mildest and most stable in all of Thailand due to the adjacent warm waters. Extreme high temperatures are the lowest of all regions and even during the coldest month the diurnal temperature range is under 10° C in all parts. The coldest temperature ever recorded in this region was 12.1° C in Chumphon. Rainfall is heavy and the relative humidity remains high throughout the year. Refer to: Appendix 7, pages 455 to 460.

Photoperiod

Thailand is a low-latitude country, therefore the number of daylight hours is less variable than areas of higher latitude. The photoperiod for Thailand is displayed in Table 2.

As the table indicates daylight hours reach a maximum during June and a minimum during December. Daylight hours vary by only two hours and twenty minutes during the year.

Table 2. Photoperiods of Thailand.

Date	Sunrise	Sunset	Daylight hours
Dec. 21	0637	1755	11 hr., 18 min.
Mar. 21	0622	1829	12 hr., 7 min.
Jun. 21	0551	1849	13 hr., 38 min.
Sep. 21	0609	1815	12 hr., 6 min.

Implications for the Care of Thai Snakes

It is not possible to replicate the natural environment in which snakes live, but the closer the approximations the greater the chances of successfully keeping a snake in captivity. An animal at ease with its surroundings will breed more easily and satisfactorally. With this in mind, we will examine aspects of natural environment in Thailand and offer suggestions for the husbandry of Thai snakes. The ideal situation is to determine the locality in Thailand from which a snake came and then approximate the humidity, photoperiod, and daily temperature conditions.

Most of Thailand experiences seasons based upon changes in rainfall, temperature, and humidity. The rainy season generally extends from May to October. These are the wettest months when the relative humidity is the highest, and when the daily temperature range, the duirnal range, is the least. The cold season lasts from November to February, the coolest months and also when the daily temperature range is the greatest. March and April are the hottest months. At this time of year, temperatures are highest and relative humidity is lowest. The South is the exception to this general pattern. The South is peninsular and flanked by bodies of warm water. Therefore, seasonality is not as pronounced as it is in other parts of Thailand. Relative humidity and temperatures vary little from month to month, and the daily temperature range is the least of any region in Thailand. Both rainfall and relative humidity remain at high levels throughout the year.

PHOTOPERIOD

The length of daylight is not important if a person is not interested in breeding snakes. It may, or, may not be a factor if one wishes to breed them. Some Thai species have been bred in captivity when no attempt was made to adjust the photoperiod to that of the natural environment. If a person wishes to breed a species which has not yet bred in captivity, it might be wise to simulate the photoperiod of the snake's natural environment. The data for Thailand are provided in Table 2, page 13.

RELATIVE HUMIDITY

Relative humidity is high throughout the year with the lowest monthly average above 50 percent. It is, therefore, suitable to keep Thai snakes in an environment of high relative humidity. Appropriate percentages are shown in Appendixes 2 to 7. If the relative humidity is kept at an appropriate level, shedding problems will also be minimized.

TEMPERATURE

This is the most important factor of successful husbandry and, most likely, successful breeding. A study of the data presented in Appendixes 2 to 7 will reveal that Thailand does not experience prolonged periods of cold. The difference between the average daytime high temperatures of the warmest and coldest months falls within the range of 4° C to 6° C. This does not mean that Thailand is a land of constantly high temperatures, however. The daily range is a key factor in determining seasonality. During the cold season, the daily range can be extreme. In the North at Mae Sariang, Mae Hong Son Province, for instance, the difference between the average daytime high and the average nighttime low is 20.7° C in February. This means that during a 24-hour period in the cold season, a snake may experience temperatures of 30° C or higher, and 10° C or lower. Although some species may avoid the extremely cold temperatures by retreating underground many can survive a few hours of cold temperature, as this is part of their natural environment. However, they are vulnerable to health problems if exposed to prolonged periods of cold, or even cool, temperatures. Note that snakes native to the South do not experience the extreme highs and lows that occur in other regions. Snakes from southern Thailand should be maintained within the temperature ranges indicated by Appendix 7.

It is clear that if one wishes to keep snakes from Thailand successfully, or snakes from other parts of the world with a similar climate, they must be kept warm. A temperature of 30° C will keep them comfortable. Overnight cooling, however, is a part of their natural environment and should occur. Prolonged periods of cold temperature, however, must be avoided or health problems will certainly follow. Diseases which seem to be a common problem among captive snakes are uncommon among wild specimens. If captive specimens are kept in a temperature regime similar to that of their natural environment, health problems will probably be substantially reduced.

A breeding program will have a greater chance of success if the temperature regime of the natural environment is approximated. Indeed, such a regime might be critical for those species which have not yet bred in captivity. Although there is evidence that some species breed throughout the year, most Thai snakes breed during the cold season (November-February), when nighttime temperatures are at their lowest and the daily temperature range is the greatest. Most eggs and young begin appearing in March. Healthy, stress-free Thai snakes may be induced to breed if they are exposed to nighttime temperatures within the ranges shown in Appendixes 2 to 7. Determine the region from where the specimen has come and follow the temperature guidelines provided in the appendixes. Daytime temperatures, however, must be raised to an appropriately high level and the hours of daylight adjusted seasonally, as displayed in Table 2, page 13.

The Thai People and Their Snakes

The attitudes of the people of Thailand toward snakes are generally similar to that of most other people of the world. Most Thai are unable to distinguish between the venomous and the non-venomous, and assume that any snake they encounter is venomous although generally they seem more reluctant than Westerners to kill a snake. Most Thai simply avoid any snake they encounter and continue on with their business. Perhaps this is the influence of Buddhism. On the other hand, since there is a large market for snakes and snake products in Thailand, many are caught and marketed. Large numbers of live snakes are also exported. Before the Thai Government restricted the export of reptile leather, many were killed for their skins. Even today many are killed for the domestic leather market. A number are also killed and eaten as the blood and certain other parts of the snake are thought to have desirable effects on the human body. There is a growing awareness that many types of snakes prey primarily upon rodents and are, therefore, beneficial. In recent years, the Government of Thailand, for conservation purposes, has banned the export of several species of snake. Refer to: Appendix 9, page 463. Unfortunately, the snake product industries are generally unregulated, and continue to take a heavy toll upon the snake population.

Snakebites in Thailand

Thailand is a tropical country, therefore animals do not hibernate. Venomous snakes are active throughout the year and pose a threat to anyone who may contact them. The majority of the Thai labor force is engaged in agriculture and therefore likely to be in contact with venomous snakes. The majority of bites occur in areas of heavy cultivation. In 1985 recorded bites were as follows, the South (51%) and Central Region (47%), 98% of the national total. Persons between the ages of 15 and 34, the bulk of the agricultural labor force, received 49 percent of all reported bites. A substantial number of bites were inflicted between April and December, when agricultural activity is at its maximum. Fewer bites occur in January, February, and March, when agricultural activity is less intense.

It is most likely that there are more bites and perhaps more deaths than these figures indicate. Most farmers know that bites from certain snakes, those of genus *Trimeresurus* for instance, though painful, are not usually fatal. They are likely to endure the pain rather than the inconvenience and expense of going to a hospital for treatment. Furthermore, not all of the provinces report statistics on snakebites.

Table 3. Reported Bites and Confirmed Deaths From Snakebites.

Year	Bites	Deaths	Year	Bites	Deaths
1974	32	0	1980	1411	16
1975	57	1	1981	1809	13
1976	229	4	1982	2407	18
1977	1090	15	1983	2776	4
1978	1085	12	1984	2673	6
1979	1527	14	1985	3377	8

Nationwide, the two genera *Calloselasma* and *Trimeresurus* account for the largest number of reported bites, having inflicted, 629 and 229 respectively, of the identifiable bites (85.8% of the total) in 1985. The large number of bites inflicted by *Calloselasma rhodostoma* and members of *Trimeresurus*, mainly *Trimeresurus albolabris albolabris* and *Trimeresurus macrops*, is attributed to two main factors. These snakes are abundant throughout the major agricultural regions of Thailand and usually rely on camouflage to avoid detection. An unsuspecting person might inadvertently approach too close, inviting a defensive strike. Members of *Naja* are also abundant in heavily farmed areas, but their first line of defense is to flee so there are fewer confrontations and bites. *Vipera russelli siamensis* is present and locally abundant in heavily farmed areas, however, its range is discontinuous and it is absent from many agricultural areas. In 1985, cobras were proven to have inflicted 120 bites and *Vipera russelli siamensis* 101 bites.

In approximately 66 percent of the reported cases the type of snake was unknown. In 1985, for example, 3337 cases were reported. Of these the originating snake was identified in only 1085 cases, as it had either escaped or been killed and not brought for identification. *Calloselasma rhodostoma* accounted for 629 bites, genus *Trimeresurus* 229, genus *Naja* 120, *Vipera russelli siamensis* 101, genus *Bungarus* 4, and *Ophiophagus hannah* and a sea snake one each. The mortality rate is quite low for a nation whose herpetofauna includes a large number of very toxic species. This is a tribute to the quality of the treatment available at Thai medical facilities.

Some Characteristics Of Snakes

Snakes are limbless carnivores which possess a limited sense of hearing. This places them at a real disadvantage when capturing, overpowering, and eating prey. Furthermore, they are cold-blooded, a factor which exerts a strong influence on their biological functions. The following generalities relate to certain characteristics of these interesting animals.

Scalation

The body of every known snake is covered by scales arranged in fairly regular patterns rather than randomly. Furthermore, snake scales are not separate, removable parts of the skin, as those of fish, but are an integral part. Each scale is a thickened and hardened portion of the skin separated by thinner, more flexible skin portions. The flexible portions between the scales are seen when the snake is coiled or when its body has been distended after swallowing a large meal. The scales vary in size and shape, larger ones are known as plates, scutes, or shields. The smaller, more numerous dorsal body scales are not named. Snake scalation can best be discussed in four particular scale categories: head, dorsal body, ventral body, plus the anal plate and subcaudal (those on the ventral side of the tail).

Scalation

Figure 1. Head Scales in General. *Elaphe flavolineata*.
1. Side view. 2. Top view. 3. Under-side of head.

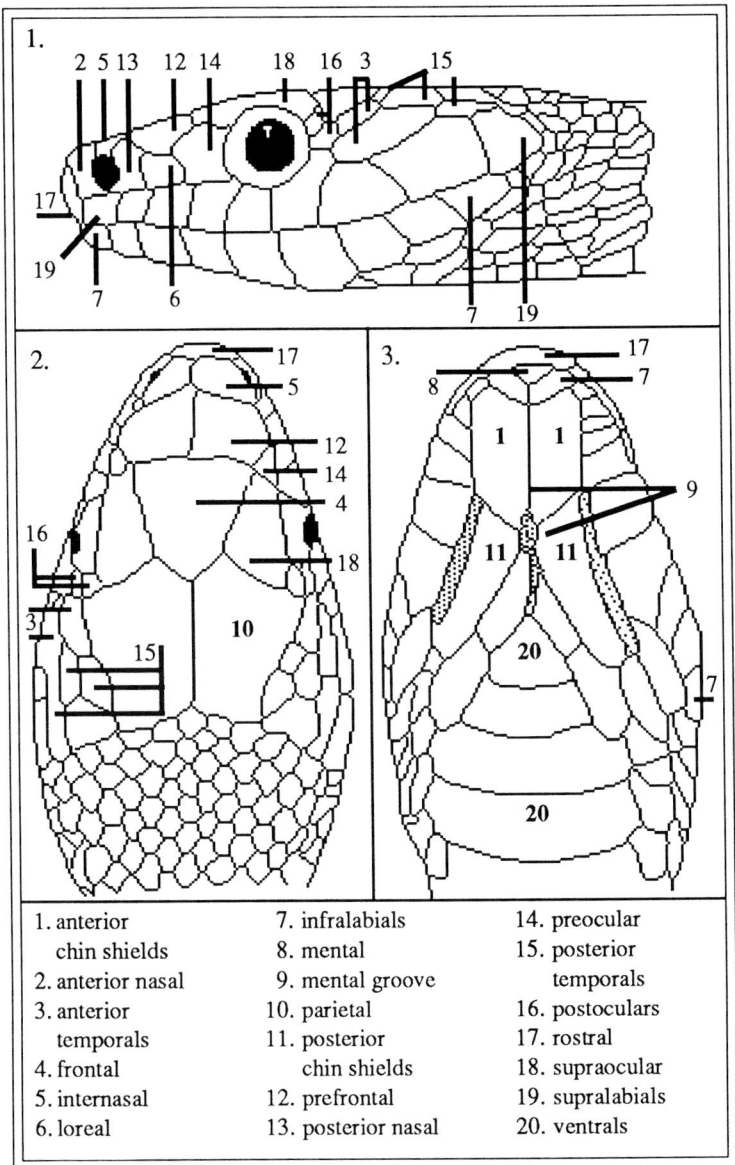

1. anterior chin shields	7. infralabials	14. preocular
2. anterior nasal	8. mental	15. posterior temporals
3. anterior temporals	9. mental groove	16. postoculars
4. frontal	10. parietal	17. rostral
5. internasal	11. posterior chin shields	18. supraocular
6. loreal	12. prefrontal	19. supralabials
	13. posterior nasal	20. ventrals

In most kinds of snakes, both non-venomous and venomous, the dorsal surface of the head is covered by a symmetric arrangement of large scales, or plates. Figure 1 displays the typical head scalation of most snakes and names of each of the common head plates. In some species, for example Homalopsine snakes (Figure 19) some of the larger head plates may be greatly reduced in size or entirely replaced by considerably smaller head scales. In some genera the majority of the plates are replaced by smaller scales (Figure 2) as in members of genus *Trimeresurus*.

Figure 2. Head Scales of Genus *Trimeresurus*.

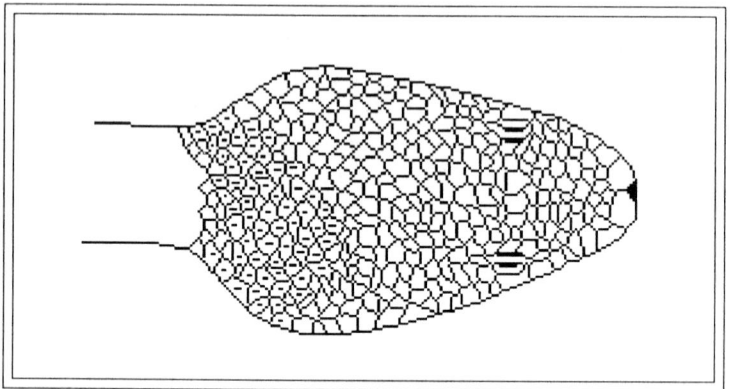

The dorsal body scales usually overlap (are imbricate) and are arranged in an oblique, longitudinal series of rows (Figure 3.4). There are usually an odd number of dorsal scale rows, but, on a few species such as *Ptyas carinatus*, an even number of scale rows may be present. Body scales vary in shape; for example, they may be nearly as broad as long, as on *Ptyas*, or long, narrow, and pointed as on *Dendrelaphis*. Furthermore, the scales may be smooth or keeled (Figures 3.5 and 3.6). Some scales bear one apical

Figure 3. Body Scales in General.

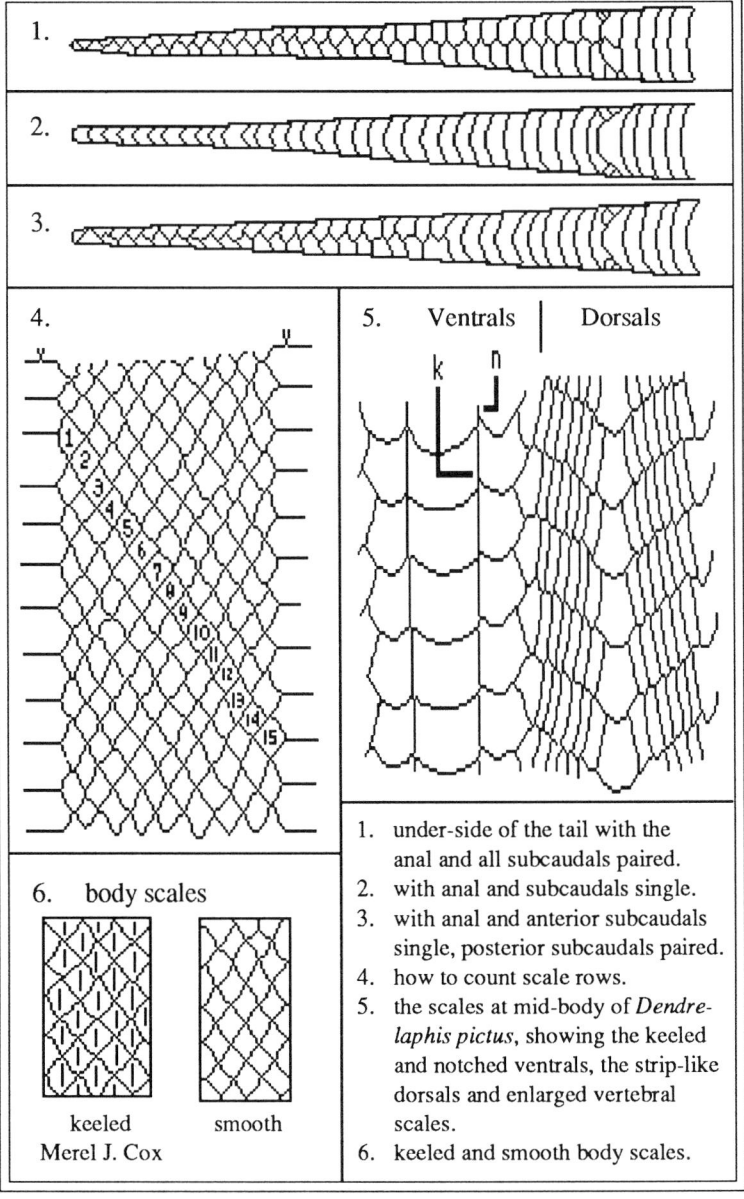

1. under-side of the tail with the anal and all subcaudals paired.
2. with anal and subcaudals single.
3. with anal and anterior subcaudals single, posterior subcaudals paired.
4. how to count scale rows.
5. the scales at mid-body of *Dendrelaphis pictus*, showing the keeled and notched ventrals, the strip-like dorsals and enlarged vertebral scales.
6. keeled and smooth body scales.

6. body scales
keeled   smooth
Merel J. Cox

pit, as does *Dendrelaphis*, others, such as *Elaphe*, have two. Apical pits are tiny pores at the apex of individual scales and may be either obvious or very difficult to see. The first row or rows of dorsal scales nearest the ventral surface are often larger than the others and in some genera, such as *Boiga* or *Dendrelaphis*, the vertebral series of scales may be enlarged (Figure 3.5). The number of scale rows around the body is usually at its maximum at mid-body and the least number around the neck or before the anus. On most sea snakes, however, the posterior portion, rather than the middle, is the thickest part of the body where the maximum scale row count is also found.

Ventral scales are enlarged, transverse plates located on the belly. They are narrow at the neck, attain maximum width at mid-body, and narrow again as they reach the vent. Their contribution to the locomotion of terrestrial species is obvious and in the case of some arboreal forms, for example, *Dendrelaphis* and *Lepturophis*, they are keeled and notched to facilitate rapid, sometimes vertical, movement through shrubs and trees (Figure 3.5). Ventral plates are not present on all species, however. Ventral plates, or ventrals, are not required for species which do not need them for locomotion. Consequently, they are absent on some burrowing forms, for example, *Typhlops*, and either absent, or, greatly reduced on many marine forms, such as *Acrochordus* and *Hydrophis*. In some cases where ventrals are absent the number of dorsal scale rows is very high as they cover the entire body. The dorsal scale row count may reach as high as 93 on genus *Kolpophis*.

The anal plate, sometimes referred to as the anal shield, covers the anal opening and may be single or divided.

The subcaudal scales may be single, paired, or both single and paired. If both single and paired, those closest to the anal opening are usually single and the remainder paired (Figures 3.1, 3.2, 3.3).

Scale characteristics are consistent down to the subspecies level and, therefore, provide an important, highly visible taxonomic tool. Furthermore, microscopic study has revealed characteristics of scalation which have been used in the classification of snakes. The most obvious external characteristics, however, are often key diagnostic factors. For example, certain species of the genus *Trimeresurus* are distinguished by the fusion or lack of fusion of the first supralabial and the nasal, *Ophiophagus hannah* may be identified by the presence of occipital shields, and genus *Calamaria* by 13 dorsal scale rows with the anal entire. These are a few of the numerous cases where scalation is a convincing taxonomic tool. In addition, characteristics of scalation may indicate sexual dimorphism, especially by the number of ventral and subcaudal scales. Refer to the section: Reproduction, page 31.

Locomotion

Being limbless, snakes must rely solely upon their long, cylindrical bodies for locomotion. On a smooth surface they have great difficulty in moving, but on a rough surface, such as the ground, they move with ease. This is done in two ways. Irregularities in their paths, such as stones, vegetation, etc., are used to push against propelling them forward. This requires great muscle control and coordination, but terrestrial snakes have become masters of this form of locomotion. A second method for moving over rough surfaces employs the ventral scales. The overlapping ventral scales make contact with rough spots, and by contraction of the abdominal muscles the snake moves forward with caterpillar-like movement.

Snakes are not fast. The fastest can only move at ten kilometers per hour, a speed surpassed by almost any man. They appear to move faster because when encountered by man, they quickly move into piles of stones, dense vegetation, or other obstacles. In the open, they are easily caught. Similarly, they capture their prey by moving slowly and stalking, or, by remaining coiled and motionless and then ambushing it. In most cases, they are too slow to outrun their prey.

Arboreal snakes move by using the two methods discussed above. In some Thai snakes, for example, in the genus *Chrysopelea*, the ventral scales are notched and keeled, providing a better grip on rough surfaces resulting in improved locomotion. Such specimens can climb nearly vertical surfaces and move over branches with astonishing ease and quickness. Members of genus *Chrysopelea* can also leap and glide short distances. They leap short distances between branches by tightly coiling their bodies and springing forward. Gliding is accomplished by laterally flattening the ribs and drawing in the ventral surface so that it becomes concave. This offers sufficient lift for gliding short distances from limb to limb or from a tree to the ground.

Any snake can swim. They do so with lateral undulations of the body. Sea snakes, most of which are exclusively marine, have vertically compressed tails, like the blades of oars, to give them greater propulsion through water.

The Senses of Hearing, Smell, and Sight

Snakes have neither an external ear nor an ear drum; therefore, they have little or no sensitivity to airborne vibrations. Snakes "hear" by being sensitive to vibrations conveyed through the surface on which they are lying. Although cobras appear to dance to the music of a snake charmer, they do not. They cannot hear the music but take a defensive position and merely move in response to the motions of the snake charmer.

The sense of smell is well developed in snakes. They have a normal sense of smell which utilizes the nostrils, but they also rely heavily upon an organ located in the roof of the mouth known as Jacobson's organ. The tongue collects particles from the air and surface and in turn passes them to the organ for analysis. This is why the tongue is in constant use when a snake is searching for food or is alarmed. Jacobson's organ is invaluable. Snakes rely upon it to detect prey in addition to the presence of their enemies. It is also essential to reproduction because without it males would have great difficulty in locating receptive females during the breeding season.

The ability of snakes to see varies greatly from genus to genus. In some burrowing forms, especially those in the genera *Ramphotyphlops* and *Typhlops*, sight has degenerated and these snakes are virtually blind. Many of the colubrids have excellent eyesight over relatively short distances. They detect motion, focus on objects, and some also have a limited ability to distinguish colors. A snake does not have eyelids. Each eye is covered by a transparent scale which is shed periodically, along with the other body scales. The eyes are always open.

Thermal Sensitivity

Some snakes, such as those members of the Superfamily *Booidae* and Subfamily *Crotalinae*, have a means of detecting heat emitted by warm-blooded animals such as rodents and birds. In the genus *Python* the heat-sensitive tissues are located in pits in the rostral and/or labials. In the Subfamily *Crotalinae*, they are located in loreal pits. The ability to seek prey at night or in the total dark is a great advantage to such serpents. It has been determined that thermal receptors within the pit itself are able to detect temperature fluctuations as small as $0.002°C$. The effectiveness of these receptors, however, diminishes as distance from the thermal receptors increases and most probably becomes nil at a range of 300 mm from the pit. Thermal receptors can detect objects closely positioned above or below the head, or items placed at right angles to it. They are important for locating prey and self defense.

Food

All snakes are carnivores. Prey ranges from minute insect larvae to fairly large mammals, depending upon the snake species. Some species prey exclusively upon cold-blooded prey, others only upon warm-blooded, while many feed upon both.

Since snakes have neither arms nor legs to help them capture and eat food, their limbless bodies have developed special adaptations for these purposes. Because they are usually slower than their prey, they stalk cautiously or wait in motionless ambush. At the right moment, they either strike or seize prey with their mouth. The method of killing prey varies according to the species. Some merely seize prey and swallow it alive. Others, sometimes referred to as semi-constrictors, pin down the prey with their bodies and then swallow it. Constrictors seize prey with their mouths, encircle it with coils, and then squeeze until it dies of suffocation before swallowing. Mildly venomous snakes bite their prey and chew, working their venom into the prey until it becomes numb and ceases to resist. Some of the mildly venomous snakes also employ constriction to overpower their prey. Cobras and their relatives usually bite and hold their prey until the venom eventually kills it. Vipers most often strike, envenomate, and then release their prey. The prey is swallowed after death.

Most carnivores dismember their prey in order to eat it. They do so by using enlarged teeth and claws to tear food into pieces small enough to swallow. Snakes have neither teeth designed for tearing nor clawed limbs with which to do this, so they must swallow prey whole. Most snakes can swallow an animal larger than their own heads as their mouths open to nearly 180 degrees. In addition, most jawbones are very flexible. The lower jawbones are not fused at the chin but attached by elastic ligaments which permit greater flexibility. Thus, the jawbones can work independently to move prey from the mouth into the throat. The sharp recurved teeth also help to hold the prey in place as it enters the throat. To facilitate this process of swallowing, prey is usually, but not always, swallowed head first. Mucus glands, located in the jaws, secrete large amounts of mucus that lubricates the prey and eases passage to the digestive tract. It may take one hour or more to swallow large prey. During this process, breathing is maintained by thrusting the windpipe, or glottis, outside of the mouth.

Once the meal has been swallowed the snake will retreat to a safe and comfortable place and wait for approximately one week until the meal has been digested. Then the quest for food will be renewed.

Growth and Longevity

When provided with adequate food and water, captive snakes grow quite rapidly. The fastest rate of growth occurs when snakes are juveniles and adolescents with slow growth at maturity. Some contend that growth continues until death; if this is true, the annual growth rate is very small. Most information on the growth rate has been by observation of captive specimens which have a regular supply of food and water, and this may not be an accurate reflection of the growth rate in the wild. Furthermore, measuring live snakes is difficult. Weighing them is relatively easy, but measuring their length accurately is not, as it is very difficult to control them.

Longevity in the wild is difficult to determine and we can only guess the actual age. Longevity records for captive specimens are helpful in some cases. For those which adjust well to captivity, longevity is impressive and might reflect the situation in the wild, excluding the intervention of predators. A boa constrictor, for instance, has lived for more than 40 years in captivity. For those species which do not adjust well to captivity, captive longevity records are misleading. For instance, available records indicate that *Maticora bivirgata flaviceps* have not lived longer than a few months in captivity. Obviously, this species lives much longer in the wild.

Reproduction

Sexual maturity is reached by the third year in most species, although it might be later in some of the larger *Booidea* or earlier in smaller species. Unfortunately, data on the seasonality and frequency of reproduction in the wild for tropical snakes is meager. My observations are, that most matings in Thailand occur in the cold season, November to February, and that most eggs, hatchlings, and newborn appear between April and August. I have, however, observed both eggs and newborn in early March and in late November. I dispute the general assumption that because temperatures vary relatively little in the tropics, temperature influences on reproduction cycles are minimal.

The climatic data in the appendixes display a pronounced cold season in most parts of Thailand except for the South. It is true that daily maximum temperatures remain high in November, December, and January, but daily low temperatures are quite low in many parts of Thailand and this probably exerts a strong influence on the reproductive cycles of many, if not most, species. Clearly, though, some reproduce aseasonally or have an extended breeding season. *Homalopsis buccata*, for instance, has been proven to reproduce throughout the year and *Cerberus rynchops* has been known to have a breeding season of six to seven months. It seems likely that further research will discover other species which have either extended reproductive seasons or reproduce aseasonally.

The frequency of breeding of most snakes remains unknown. It has been shown that the proportion of breeding females in a number of temperate species varies from thirty to one hundred percent per year; a breeding female is defined as one which contains shelled eggs, developing offspring, or enlarged follicles. Unfortunately, data regarding the proportion of breeding females in tropical areas is very meager. The only data that I am aware of indicate the percentages to be 7 percent among *Acrochordus arafura*, a non-Thai species, and 75 percent each among *Acrochordus granulatus* and *Cerberus rynchops*. It is, of course impossible to draw definite conclusions regarding the frequency of reproduction of Thai snakes from such inadequate data. At this point we do not know if Thai and other tropical wild species reproduce yearly, every other year, or even every third year. It may depend on a number of factors, such as the availability of food. A limited food supply might result in poor physical conditioning and, consequently, an inability to reproduce. Another factor might be genetic. It is not yet known if the frequency of snake reproduction is genetically determined.

There is very little external variation by which male and female snakes can be distinguished. In a few species there are sexual differences in coloration, in some the body or head scales may be more heavily keeled in one sex, and in some cases males may have pronounced cloacal spurs, but most species lack such easily observed evidence of sexual dimorphism. Nevertheless, a few generalizations can be made. In most, but not all species, mature females are heavier and longer than mature males. Therefore, females usually have more ventral plates than males. On the other hand, male reproductive organs are located in the base of the tail, posterior to the anal opening. Consequently, males usually have longer tails and more subcaudal scales than females. Because of the presence of the male copulatory organs inside the base of the tail, in some, but by no means all, species the base of the tail is wider and tapers more abruptly in males. These subtle differences are for those who wish to know if a snake is male or female, but not of much use or interest to a snake seeking a relationship with a member of the opposite sex. When a female is ready to breed, she secretes fluid from an anal gland which leaves a scent, clearly marking her trail. A foraging, interested male detects the odor, his Jacobson's organ analyzes it, and his sense of smell enables him to find a mate.

After copulation, either young or eggs eventually appear. The female, however, may retain and store some of the living sperm to be used for future fertilization. Females of some species have retained viable sperm for four years; thus, copulation is not necessary for each brood. For all practical purposes, there are two methods of parturition. The species may be viviparous, where the young receive nourishment and develop inside the mother and emerge as fully developed young animals ready to fend for themselves. Most snakes, however, are oviparous. Their young develop in eggs outside of the mother's body. Less than five or more than 100 eggs may be produced, depending upon the species. The incubation period is approximately 30 to 90 days, depending upon the species and air temperature surrounding the eggs.

Parental care is minimal. It is, in fact, non-existent in viviparous species. The young are on their own when they leave the mother's body. Among oviparous species parental care is minimal to the extent that females lay their eggs in secluded places. These may include cavities in the ground, under decayed vegetation, bark or moss. In most cases, the eggs are then abandoned. In some cases, genus *Python* for instance, the eggs are incubated by the female.

In the case of *Ophiophagus hannah* a nest is built, the eggs are deposited and the female guards the nest until the eggs hatch. Upon hatching, however, the young are on their own.

Habitat

Snakes live in any geographic environment where temperatures are sufficiently warm to permit their bodies to function. Within any given environment, they may be primarily terrestrial, fossorial, arboreal, semi-aquatic, aquatic, or marine. However, primarily terrestrial species, such as cobras, will occasionally climb trees in search of food. By the same token, primarily arboreal species of the genus *Trimeresurus* will often descend to the ground for the same purpose and fossorial species are sometimes seen on the ground.

Every part of Thailand is inhabited by some kind of snake, including the surrounding seas. Most enjoy the warmth of the lowlands, but the range of some species is restricted to the cooler highlands. Many are seen near human habitation, and, surprisingly a large number of species appear within the congested urban centers.

Predators and Means of Defense

As is the case with all of the fauna of the world, man is the greatest enemy of snakes. Many kill snakes out of fear because they cannot distinguish between the poisonous and harmless ones, or simply because they feel that the only good snake is a dead snake. Economic reasons have also taken a heavy toll. A large market for snake skins has always existed, especially in tropical areas, and this has been an important factor with the slaughter of many snakes, the dangerous as well as the harmless. Fortunately, in recent years enlightened attitudes have evolved and more people and governments have become aware of the beneficial role these animals play in the balance of nature. They exercise a major control over the rodent population of the world and carry few, if any, diseases harmful to mankind. Today the trend is to suppress the trade in reptile skins and extend protection to some species of harmless snakes in addition to some venomous forms.

Although man is the most devastating of predators, he is not the only creature that preys upon snakes. Because of their small size, most baby snakes are very vulnerable to a wide range of predators including toads and birds. Even venomous adults may become a meal for a hawk, monitor, or mongoose. Many serpents fall prey to fellow snakes, such as the King Cobra or krait.

Many Thai snakes rely upon color as a defense against predators. In most cases body colors tend to make the snake blend into its environment and make it difficult to see. If it remains quiet and motionless, *Vipera russelli siamensis* is very difficult to detect when against a background of soil and dry vegetation. Similarly, many arboreal forms, such as *Trimeresurus* and *Ahaetulla* are primarily green and difficult to see when among the leaves of bushes and trees. Some harmless species have coloration and markings similar to those of dangerous species. As an example, a juvenile *Lycodon subcinctus* is banded in a manner very similar to the dangerous *Bungarus candidus* and *Bungarus fasciatus*. No doubt more than one potential predator has been frightened by such mimicry. Refer to: Plate 70, page 226.

Behavior is another means of defense. Cobras do not usually rely upon protective coloration as a means of defense. If they cannot flee from a predator, they rely upon a menacing defense, as do many venomous and some non-venomous snakes. Loud hissing, accompanied by strikes, is employed by cobras and *Vipera russelli siamensis* to frighten off aggressors. Both the harmless *Cylindrophis rufus rufus* and the venomous *Maticora bivirgata flaviceps* raise their tails and expose the brightly colored ventral surface in an attempt to distract predators from their heads.

Husbandry in General

Some snakes do very well in captivity whereas some others survive for only short periods of time. The stress of capture and confinement results in some refusing food, being very vulnerable to disease, or inflicting damage to themselves by repeatedly striking against the glass of their cage or rubbing their rostrals in an attempt to escape. Some simply die for no apparent reason. Chances of successfully keeping any species in captivity are enhanced if correct husbandry techniques are employed.

Laws Governing the Keeping of Snakes

It is wise to become familiar with the appropriate laws before beginning or adding to a collection. The maintenance of snakes in captivity is regulated, directly or indirectly, at many local, national, and international political levels. The purpose of various regulations is to ensure public safety, conserve natural resources, and to protect species thought to be endangered.

Local laws, such as city or state laws, are generally concerned with public safety and regulate the keeping of potentially dangerous animals. These laws often make it illegal for private individuals to keep venomous snakes, pythons, or boas. The reason is obvious, of course. All too often there are reports of irresponsible persons who have kept venomous snakes, such as cobras, in unsafe cages in an apartment with subsequent escape posing a threat to the community. Such irresponsible persons, however, pose a greater threat to themselves. Recent reports indicate that in Great Britain, roughly 75 percent of the persons reporting bites from poisonous snakes were bitten **in their own homes**. Similarly in the United States of America, 71 percent of the persons treated for snake bites were amateur collectors. Large pythons and boas are a threat, especially to children, due to their size and strength. There are many species of beautiful, interesting, and harmless snakes which pose no threat and can be kept legally. Local laws, however, may prohibit catching and keeping of protected harmless species. It is always wise to become familiar with all applicable local laws **before** starting a collection.

National laws are usually focused on public safety and conservation. These laws often prevent the keeping and/or importing of animals considered a threat to the general public. Similarly, national laws may prohibit the capture and export of rodent-eating snakes, considered a valuable natural resource, and may prohibit the importation of species protected by other nations.

International laws are mainly concerned with conservation. The international animal trade is heavily regulated due to the worldwide decline in wildlife populations, which is as it should be. The most comprehensive and effective treaty regulating the international animal trade is the Convention on International Trade in Endangered Species of Flora and Fauna (CITES) also known as the Washington Convention. It is designed to protect animals and plants considered to be endangered or threatened. Many species of snakes are under its jurisdiction. Information relating to CITES may be obtained from: U.S. Fish and Wildlife Service, U.S. Department of the Interior, 18th and C Streets N.W., Washington, D.C. 20240, U.S.A.

Herpetological Societies

It is always useful to exchange information and experiences with other colleagues. Among snake enthusiasts this can be done by joining a local herpetology society. These societies offer an excellent forum to meet people who share your interest, make friends, exchange information, and perhaps exchange animals. You may inquire about a local herpetological society through your nearest zoo; many zoos also sponsor such groups.

Many nations have national and regional societies who publish newsletters, bulletins, and journals useful to all herpetologists. Refer to: Appendix 1, page 411.

Housing

The size of the quarters in which a snake is kept should suit the snake species. Captive specimens should have sufficient space to be able to stretch full length. It is better to have housing that is too large rather than too small. Specimens kept in cramped quarters have restricted mobility and are forced to remain coiled for extended periods of time which may result in a lack of muscle tone and obesity, thus shortening the life and breeding potential of the specimen. Avoid overcrowding. Too many specimens in a single cage will also reduce mobility, increase stress, and result in problems during feeding. Refer to: Feeding, page 45.

Aquariums often make suitable and adequate quarters for many species, especially for semi-aquatic and marine specimens. Aquariums have the added advantage of smooth surfaces, which minimize the chances of rostral damage to the snake who will continuously probe for an exit. It may be necessary to use wooden cages for large specimens, although they are not as desirable as aquariums. The primary disadvantage of wooden cages is that they are easily soiled and can harbor infectious organisms. If they must be used they should, firstly, be sanded smooth to remove all rough spots which may damage rostrals and then painted to facilitate cleaning. The floor should be given several applications of paint so excrement or spilled water will not be absorbed. Whether an aquarium or a wooden cage is used, it is desirable that the door or entrance be on the top. Aquariums can easily be fitted with a wire frame to cover the top. The frame should fit tightly and be equipped with a locking device as snakes are skilled escape artists. Access from the top of the cage makes removal of the snake easier, reduces the chance of being bitten by an unfriendly inhabitant, and makes cleaning easier. Caution should be taken with all wire surfaces as damage to the rostral will result from rubbing by the snake. If a snake is inclined to push against the wire, and most new arrivals will, cover the inside surface of the wire with a cloth to protect the snake.

Cages manufactured and designed specifically for reptiles are now available in some parts of the world. They are usually made of plastic or a similar stain resistant, non-absorbent substance, are seamless, and have the corners rounded for ease of cleaning. They provide excellent housing for many terrestrial species.

A cage can be as plain or as elaborate as the owner wishes. Minimum requirements are a water dish, a hiding box, and a substrate. A branch must be added for arboreal snakes. Many specimens thrive in this form of confinement. Cages can be decorated with real or artificial plants. Either will offer certain advantages. Although it is not possible to replicate an animal's natural environment in captivity, plants provide an attractive and natural look to a cage. Furthermore, plants offer hiding places to nervous specimens which reduces stress. Cages containing plants are particularly advantageous to arboreal species as plants give them places to climb and hide. In addition, many species of arboreal snakes obtain most of their water by rainwater run-off from leaves of trees or bushes in which they live. This is especially true of babies and juveniles. Rainfall can be simulated in captivity by frequently spraying the plants.

It may be necessary to provide cages with a form of lighting. Heat lamps or incandescent bulbs are often used for heating and lighting in cold climates. If they are to be used, care must be taken to shield them so the snake is unable to climb on them and burn itself. It is recommended that lighting units such as Gro-lux, True-lite, or Vita-lite be used. These units cannot be used to heat cages but, because they produce a reasonable facsimile of natural sunlight, they have a healthy effect on both reptiles and plants.

Cages for venomous snakes must include special safety features. The cages should be equipped with locks which should be locked at all times. Wire surfaces should have two layers of wire spaced far enough apart to prevent people from being envenomated by a strike against the wires. If spitting cobras are kept, surfaces of wires should be covered with cloth or similar substance. Finally, each cage should include a hiding box that can be closed and removed. This will enable the keeper to move venomous snakes into the box, close it, and then remove the box. The keeper can then clean the cage with complete safety.

Cleanliness

This is a very important factor in maintaining a thriving, healthy collection of animals. Fecal matter must be removed as quickly as possible, and fresh water must be available at all times. Water dishes should be cleaned daily. Both fecal matter and stagnant water offer an attractive environment for harmful bacteria and parasites which can pose a threat to any collection.

Fresh water should be provided daily or more often if the snake chooses to relieve itself in the water dish which they often do. Simply remove the dish, dispose of the foul water and thoroughly clean with hot water. Dip the water dish, as well as any tools and equipment used, into a disinfectant and then finally rinse thoroughly. A 70 percent alcohol solution or bleach is useful for this purpose. Add room-temperature water after thoroughly rinsing, but do not fill the dish to the rim. Allow sufficient space for the snake to enter the dish to soak without causing an overflow and wetting the floor. Snakes which must live on damp surfaces often develop fatal infections of their ventral surfaces. Water dishes recessed into the floor of the cage offer a more natural situation and minimize the possibility of the water dish being overturned. A similar, natural situation can be created by placing rocks or stones around the dish. The rough surface of the rocks and stones will also help the snake to shed.

Snakes should not be kept on a bare surface for reasons of cleanliness. It is preferable to use a substrate that will contain solid excrement, regurgitated food items and which also absorb liquid excrement. There are many choices available, all of which have advantages and disadvantages. Soil or sand are not recommended, except when used for burrowing species. Each looks natural but both have the disadvantage of retaining moisture and harboring parasites. Herpetologists often use pine bark or wood shavings. These substrates are attractive and do have a pleasant aroma. However, they retain moisture and are difficult to clean. Furthermore, fatalities may result as snakes, particularly smaller ones, often swallow pieces of wood with their food. Gravel has the disadvantage of being very hard to clean. Newspapers are a good substrate. They are, of course, unattractive and, unless the edges are firmly taped to the floor of the cage, snakes get under them, hide, and defecate on the floor of the cage. They do, however, have the advantage of being cheap, contain solid waste material and regurgitated food items, absorb fluids, and are easy to remove. Indoor/outdoor carpeting was suggested by a good friend, Rodney Hosterman, of Altoona, Pennsylvania, U.S.A., and it has proven to be a very satisfactory substrate for most species. It is made in a variety of colors, however, shades of green invariably look natural. Usually it is sufficiently rigid to prevent the inhabitant getting under when exactly cut to fit the floor. Its rough surface is an aid to a shedding snake and indoor/outdoor carpeting retains solid materials, but does not absorb liquid wastes. This deficiency is easily compensated for by placing some newspapers under the carpeting. Both are easily removed for cleaning. The carpeting easily washes clean and dries very quickly. Obtain two pieces to fit each cage and alternate their use between each cleaning. Regardless of the substrate used, objects which the snake can hide under should be placed on top of the surface.

Cleaning should be frequent. Ideally it should follow each defecation or regurgitation. The substrate should be removed and the cage thoroughly washed to remove all organic material and then the cage disinfected. It is important to remember that disinfectants are unable to penetrate organic matter; if any remain after cleaning, disinfectants will not be completely effective. A number of disinfectants are available and each has its proponents. Probably the cheapest and most readily available is household bleach. Usually sold in concentrations of five percent, it can be diluted by adding ten parts water to provide an effective disinfectant. After the cage has been washed clean of organic matter and disinfected, rinse thoroughly with clean water and let dry. If indoor/outdoor carpeting is used the same treatment is recommended. Make sure the cage and substrate are completely dry before returning the snake. A clean garbage can with a tight fitting lid may be used to hold snakes during the cleaning operation.

Coal tar derivatives should not be used to disinfect cages. They are toxic to all reptiles and may prove fatal.

Feeding

Snakes are exclusively carnivorous, feeding on prey ranging in size from small insects and their larvae to fairly large mammals, depending upon the size and the species of snake. Standard textbooks describe the natural prey of various species. The descriptions which follow in Part 2 of this volume include the natural prey of Thai snakes.

Before establishing a snake collection it is necessary to find a reliable source of food animals for the collection. Some collectors raise their own food animals, but if space and time make this impossible, food animals can be purchased from commercial sources. Fish, for example, can be obtained from local markets. Research institutions often sell surplus rats and mice at low prices, and poultry farms often sell culled chicks and ducklings. In all cases food animals should be of good quality and disease free.

The frequency of feeding depends primarily on the species and age of the snake. Species which feed upon insects and soft-bodied animals such as earthworms eat rather frequently, probably once a week or more often. Those which prey upon vertebrates eat less frequently, perhaps every two or three weeks. Juveniles eat more often than adults as they grow rapidly and require more energy. Juveniles which feed upon invertebrates should have food available to them at all times. Those which feed upon vertebrates seem to do well when fed once each week, commencing after their first shed. After they reach one year, they usually thrive when fed once every two or three weeks. Use common sense to decide the amount of food to be given to a snake. If too much is consumed, regurgitation may occur. Additionally, over-feeding causes obesity, shortens the life span, and reduces the breeding potential. It is good to establish a feeding routine that will maintain specimens in trim physical condition, neither too skinny nor too fat. The diet should be varied as much as possible. Snakes which feed upon warm-blooded prey can be maintained on a diet of rats and mice, but an occasional chick or duckling provides variety. Similarly, fish eaters should be given frogs occasionally. Do not feed toads to snakes. Some species are toxic and could prove fatal. If snakes are provided with a balanced diet, vitamin supplements are not necessary.

Avoid offering food items which are too big. Although snakes can often swallow large animals, they do not always retain them. Once in the stomach, prey swells before it is digested. If too big, the swelling of the ingested prey will cause discomfort and will often be regurgitated.

It is wise to train snakes to eat dead animals. This is done easily, except for those snakes that eat invertebrates. Those with diets consisting of cold-blooded prey will usually accept dead animals readily. This is often the case with varieties which prey upon warm-blooded animals. A number of benefits are derived from feeding freshly killed animals to snakes. Dead animals can be purchased in bulk at low prices, frozen, and stored until needed which is often more preferable than maintaining a stock of live animals.

In addition, live prey can injure or even kill snakes. Many herpetologists have left a snake overnight with a live rat or mouse and returned in the morning to find the rodent alive and well but the snake dead or badly chewed.

Many people prefer to use forceps to offer food to members of their collection. This has the advantage of accelerating the feeding process and enables the keeper to determine the quantity of food that each member of the collection has eaten. The process involves using a pair of long forceps to avoid accidental bites during feeding. The rear portion of the dead animal is held by the forceps with the head and front portion held in front of the snake. If the snake does not show sufficient interest, the animal can be moved slightly to simulate life-like movements. This will usually result in the snake grabbing the animal.

It is wise to separate cagemates during feeding times. This is especially true if live animals are being offered. Aggressive feeders may seize the same prey and, unless separated by the keeper, the weaker snake may be killed and become a meal for the stronger. Errant strikes by vipers at fast moving prey have resulted in damage to cagemates caused by fangs. Even when feeding with forceps, a struggle may result unless cagemates have been separated. To separate two or more snakes trying to eat the same animal is extremely difficult, especially if they are constrictors.

Water

Clean, fresh water should be available at all times. Some snakes are unable to find elevated water dishes and will dehydrate if not "led" to water. If you believe a specimen is not drinking water, gently lift it on a snake hook and place its head in the water dish. Most likely it will then drink. A recessed water dish, as described earlier, may make it easier for snakes to find water.

It is particularly important that snakes which are fasting drink regularly. If a snake is fasting and you suspect it is not drinking, follow the above procedure.

Breeding

In the past 15 years experienced amateurs have made substantial contributions to captive snake breeding. This is significant as we live in an age when wild animal populations are rapidly declining, a result of man's encroachment. The diminishing numbers of wild animals has resulted in tighter controls for international trading in animals. It now seems likely that in future the majority of the non-native species maintained in captivity will originate from those previously born in captivity. Therefore, the breeding of captive wild animals should be an ultimate goal of responsible keepers.

The successful captive breeding of snakes is the true measure of a keeper's skills as success indicates that the animals have been maintained in a clean, disease and stress-free environment. Also that the nutritional needs have been met and that the snakes are in excellent condition.

A successful breeding program begins, of course, by acquiring a pair of snakes of your choice. In most species, it is extremely difficult to determine the sex by visual examination. It is generally true that in any two snakes of the same species and length, the tail of the male is longer and tapers more abruptly than that of the female. Usually the tail of the male has more subcaudal scales. However, it is not usual to get snakes of the same length, and these differences are not obvious in every species. The most suitable way to determine the sex of a snake is by probing. This involves gently inserting a blunt probe into the cloaca, on either side of the mid-line, and gently pushing it into the tail. If the animal is a male, the probe will move a considerable distance into the tail. The sex organs of males are carried in their tails and allow deeper penetration by the probe. Refer to: Plate 136, page 359.

After determining the sex, house males and females separately and condition with a balanced diet. Ensure they do not get fat as it is healthier for them to be slim. Study carefully the natural environment and habitat of the specimens and attempt to replicate them as far as possible. In some cases this may require altering the photoperiod or briefly raising or lowering temperatures. If attempting to breed Thai snakes, consult the large range of data and information provided in this book and use the guidelines accordingly.

Once the animals are correctly conditioned and the environment adjusted, introduce the male into the female's cage and allow them to remain undisturbed. When observing courtship behavior, ensure that the snakes are not disturbed. The behavior normally consists of the male crawling over the female and, often, both animals crawl around the cage together, the male keeping his head just behind that of the female. This crawling is often accompanied by abrupt twitching and jerking motions of both snakes. Whether terrestrial, arboreal, or marine, snakes copulate by entwining their bodies and bringing the cloacal openings into contact. See Plate 11, page 126 and Plate 140, page 361. The hemipenis is then inserted into the female, where fertilization takes place. Some herpetologists keep males and females housed together until the female appears to be pregnant. Others consider it necessary to remove the male after courtship and cessation of mating. Successful breeding has resulted from both methods.

If all has gone well the female will eat normally and gradually her body increases in girth. Do not be alarmed if she refuses some meals; females often fast before delivery. Eventually, either, a clutch of eggs or a number of live young will appear, depending upon the species. Some of the live young will be born in a membrane and may require assistance to get out. Once the young have safely arrived, place them in separate quarters. The mother has done all that she will do.

In most cases, a clutch of eggs requires considerable attention from the keeper. Some snakes, for example various species of *Python*, incubate their eggs. In such cases, it may be possible to leave the eggs with the female and allow her to incubate them. Most collectors, however, prefer to remove the eggs and incubate them by other means as this gives a greater opportunity to control the temperature and humidity and to guard against fungus and other problems. For most oviparous species there is no choice. The keeper must incubate the eggs. This can be achieved in a number of ways. Incubators, large or small, can be purchased and simple but effective ones can be built. The most important consideration is to maintain the developing eggs in a temperature range of 25° C to 31° C depending upon the species. Additionally it is important to maintain a humid, but not wet, environment. This is usually done by burying the eggs in a moisture-retaining substrate such as vermiculite or similar substance. During this period, it is necessary to closely observe the eggs to avoid mold. If mold appears wipe it away with a soft cloth dampened with a weak solution of bleach. Clear plastic bags are an effective means to incubate eggs. The bag is to be perforated; moist, but not wet, vermiculite is added. The eggs are partially buried in the moist vermiculite. Close the bag and place in a sufficiently warm temperature to ensure hatching. Condensed droplets of water in the form of a fine, partial layer should be present on the inside of the bag. This ensures that the eggs are not becoming dehydrated. The eggs are to be regularly observed to ensure they remain healthy. A dripping water layer covering the walls of the bag indicates excessive moisture. If all has gone well, the young will emerge from the eggs within 30 to 90 days, depending upon the temperature maintained and the species. The young do not usually require assistance to get out of the egg, but they will immediately require clean, warm quarters.

If the first breeding attempt fails, try again. Review the procedures followed and make adjustments that further research and common sense suggest. There are a number of explanations for failure with some being natural. Perhaps, one, or both of the snakes were sexually immature or perhaps the species was one in which the female only conceives in alternate years. Keep trying; the rewards are worth it.

The rearing of young rarely presents problems and is a great deal of fun. Occasionally, the young may be too small to eat the known prey of adults. Furthermore, some species may have an entirely different diet as juveniles. Offer hatchlings or newborn a variety of small prey suitable for them to comfortably swallow; this may also include insects. If suitable small prey is not available, they might be induced to eat parts of natural prey, such as mouse or lizard tails, or frog legs. In most cases, however, this is not a problem. They do not eat until after their first shed, which usually occurs near to the tenth day after birth. After commencing to eat, feed weekly for the first year. Most captive-born and raised snakes become quite tame and make excellent pets.

Record Keeping

Herpetologists, amateur or professional, should maintain records of feeding, shedding, breeding and any other interesting activity noticed in their collection. Our present knowledge of husbandry is not exact and we can all learn from others. Feeding records of animals in your collection are essential to maintain correct feeding schedules and diet. Newborn snakes offer a wonderful opportunity to record data according to type, and quantity of food consumed and rate of growth; that is, length and weight gains. Such carefully compiled data is lacking for many species.

Medical Aspects of Husbandry

Ailments, infections and injuries are always a problem in any animal collection, and snakes are not exempted. Although medical problems are perhaps inevitable, there are measures which can be taken to minimize them. For instance, all new additions to a collection should be quarantined for a period of 90 days to avoid the risk of a potentially ill animal entering a healthy collection. When possible take the opportunity to examine a snake before purchase. Examine specimens closely to detect if any of the symptoms described in the following paragraphs are present. Finally, many ailments and injuries can be substantially reduced if the suggestions offered in the sections describing housing and cleanliness are carefully followed.

External Problems

External problems are normally related to shedding, parasites, fasts, and wounds. Each one poses a threat to the health of a snake in its own right, and their presence will also make the snake vulnerable to secondary infections. Consequently, they should be rapidly corrected and treated.

POOR SHEDS

Snakes occasionally have trouble in shedding their skins. This is especially true during winter in high latitude areas when humidity is extremely low. If high humidity is maintained, the cage equipped with a suitably sized water receptacle for the snake to soak, and rough surfaces such as stones are provided, most sheds will be normal. Several days before shedding commences the coloration of a snake becomes dull and the scales covering the eyes, or eye caps, become opaque. The eyes then become clear and shedding follows. If, however, normal shedding does not occur after the eyes have become clear, or, if shedding commences and is incomplete, the snake may require some assistance. Firstly, place the snake into a container filled with moist sphagnum moss or inside a wet sack. This usually helps the snake accomplish the shed overnight. In the second instance, place the animal in a container filled with tepid water to allow soaking for a few hours. Subsequently remove the snake and, using fingers, peal the dead skin from the body. In both instances, examine the shed skin to ensure that the eye caps have been shed. If not, and they are allowed to accumulate over the eyes, the specimen could develop serious eye problems. Unshed eye caps can usually be removed by touching them with the sticky side of a piece of adhesive tape. If this fails, use a finger nail to locate the edge of the eye cap, and gently lift it free. Avoid the use of anything metallic as the risk of injury to the eye is too great. Refer to: Plate 150, page 397.

PARASITES (EXTERNAL AND INTERNAL)

Mites and ticks often present an external problem posing a real threat to a collection, as they may be carriers of diseases. They feed on the blood of their hosts and, if left unchecked, they can debilitate a snake, cause it to refuse food, and, ultimately, kill it. Loss of blood to these parasites causes the host to become anemic. Anemia lowers the host's natural resistance or immunity to infections and diseases that eventually result in the snake's death. Mites and ticks multiply rapidly and migrate from cage to cage infecting an entire collection quickly. Mites are very small black or brown arthropods that lodge between scales, particularly those on the head. Ticks attach their head to the skin between the scales and drain a considerable quantity of blood in proportion to their size. Usually they are noticed only when their body swells with the host's blood. Fortunately, they are rather easy to eliminate. The easiest and the most effective way is to use a Shell pest strip, obtainable from most hardware stores. Cut a small piece and place it on the wire top of the cage or aquarium. Do not place above water surfaces. This method will kill ticks and mites on both the inside and outside of the cage. Shell pest strip is most effective within a confined area, therefore, place the infected cage in a closet or confined space during treatment. Remove the pest strip after two or three days.

Dead mites fall from the snake, but ticks will probably need removal with forceps. Remove them carefully to extract both the head and body. If the head remains, an infection may result. Repeat one week later to eliminate newly hatched parasites. If Shell pest strip or similar item is unavailable, effective water-soluble ectoparasite insecticides exist in powder form, Asuntol being a good example. The powder is dissolved in water according to the instructions provided. This solution can be sprayed on the snake or the snake dipped into the solution. After approximately 20 minutes rinse the snake with tepid, clean water to remove the dead parasites. Again, be sure that the heads of ticks are removed. Repeat as and when necessary.

If Shell pest strip or another suitable ectoparasite insecticide is unavailable, mites and ticks can be suffocated. For ticks, this is done by covering each one with a drop of cod liver oil, olive oil, or other similar oil. Mites are eliminated by placing the infected snake in a container, perhaps a clean garbage can with a secure lid, filled with tepid water to a depth that will cover the body but allow for the head to be above the water level. The mites will move on to the head of the snake and then they can be suffocated with an application of cod liver oil, olive oil, or similar substance. In both cases, the cages will require decontamination. Aquariums can be easily cleaned with ammonia and then rinsed to rid parasites and their eggs. Wooden cages are a problem as parasites can find many places to hide. Remove the water dish and cover the surfaces, each corner and crevice, with a flea powder that does **not** contain DDT as it is toxic to reptiles. Ectoparasite solutions can also be used to decontaminate cages. Wash, rinse and dry the cage thoroughly before replacing stock.

Strict quarantine procedures and thorough cleanliness are the keys to controlling external parasites. Cages should be frequently cleaned and all wooden surfaces should be sealed with a sealant or sound covering of paint.

Snakes often play host to internal parasites, much like any other animal. These include protozoans in addition to various worms i.e., roundworms and tapeworms. Their presence is usually detected in one of two ways. The bodies of worms or the segments of tapeworms may be seen in the excrement. Other symptoms include weight loss although feeding, regurgitation, or frequent excretion of white to yellow soft, pungent material. The presence of internal parasites may, or, may not be a problem. Many animals host internal parasites with no apparent ill effect. If you observe evidence of infestation but the snake appears healthy, it is probably suitable to let the condition go untreated. To treat internal parasites, a form of poison needs to be introduced into the body of the snake. Since this is always risky, it is wiser to avoid introduction of such medication, unless you consider a real danger of losing the animal exists. Two things should be done, however. First, the substrate and the cage should be cleaned and disinfected after each and every defecation. This will kill eggs and worms, prevent the infection of other animals, and perhaps reduce the re-infestation of the host. Secondly, it may be helpful to raise the temperature in the cage to 35°C - 37°C for one or two days. Studies indicate that some diseases and infestations of internal parasites can be successfully treated by this method. This treatment is known as thermotherapy.

If the infected snake shows signs of physical decline and you fear it will die, then it is time to attempt treatment. If you are fortunate to know a veterinarian with knowledge of reptiles and their treatment, by all means consult him or her. If you are not so fortunate, try to identify the type of worm you wish to eliminate. Bodies of worms and tapeworm segments found in excrement will help to identify the parasite, but it is also necessary to examine stool specimens for eggs. The eggs of worms cannot be seen with the naked eye, therefore a microscope will be required for detection. Take a fresh stool sample to a veterinarian, have the worms identified and the proper treatment prescribed. If a veterinarian is not available and you have access to a microscope, you may attempt identification yourself. It is usually possible to find relevant reference books in the local library to assist in this. After the worms have been identified, purchase from a pet store an appropriate remedy. Determine the appropriate dosage. This is the quantity of medicine in milligrams per kilogram of the animal's body weight. When determined, the medicine is given by a plastic tube and syringe. Refer to pages 62 and 69. Repeat the treatment every two weeks until the worms cease being passed and no eggs appear in stool samples. During this time, the cleaning and thermotherapy previously discussed should be followed. If no other remedy is available, use thermotherapy. It may prove effective.

FASTS

Occasionally a collector might acquire a specimen that refuses to eat. Snakes may refuse to eat for a variety of reasons, some are:
- incorrect food is being offered and/or feeding times are wrong
- cage temperature incorrect, too high or too low

- animal is female and gravid
- preparing to shed the skin
- nervousness
- infestation of parasites, either external or internal
- sickness.

The first situation is corrected by offering a variety of food, warm- and cold- blooded, both alive and dead. Vary feeding times. Offer during the day and night.

The second situation is easily corrected. Study temperatures of the area from where the specimen came and try to replicate them. Consult a local library or, an experienced herpetologist. Most snakes maintain body temperatures between 24^0 C and $30°$ C. Digestion ceases below 15^0 C or $10°$ C, and often at higher temperatures in tropical snakes.

The third reason is also easily resolved. Simply wait for two months. If the specimen is female and gravid, she will have given birth or laid eggs during that time. Gravid females resume feeding after giving birth or laying eggs.

In the fourth case, shedding, merely wait until the snake sheds its skin and then offer it food which it will probably accept eagerly.

The fifth possibility, nervousness, is a more difficult problem to resolve. As mentioned earlier, include a hiding place in the cage, as this provides a snake the chance to hide from what it perceives in the form of a person, as the menacing presence of an enormous predator. It also might help to cover all glass surfaces of the cage with a cloth so the snake is unable to see outside of the cage. Do not, however, keep the snake in darkness. Provide internal lighting and simulate the normal diurnal photoperiod. In addition, the snake should not be handled or disturbed in any way.

The latter two cases, parasites, viral or bacterial infections require direct treatment of the snake. Parasites were discussed earlier. Viral and bacterial infections are discussed on pages 65 to 68.

Fasting is quite common among snakes. It can be of a short duration, as in the case of skin shedding or delivery of young, or it can be lengthy, perhaps a year or more. If a member of your collection is fasting provide plenty of clean water, watch it closely for signs of physical deterioration, and offer a variety of food at weekly intervals. There are two reasons for varying the type of food offered. If the snake is new to your collection and the feeding habits are unknown, it may have a particular preference; for example, it may readily eat hamsters but refuse mice. When dead food is refused, offer a variety of live food but ensure that this live prey does not injure the snake. A clever rat or mouse often senses that a fasting snake is not a threat and may harm the snake. When varying the offerings, you may chance upon the food the snake prefers. Secondly, even among long-term captives that suddenly fast, offering a change of diet may encourage an end to the fast. If the snake refuses all food offered over a long period of time, action must be taken. If the specimen was collected locally and the season is favorable release it in the area where it was caught. If this is not the case, it may be necessary to forcibly feed the snake.

Force feeding is a last resort and should be done with great care. It usually requires a second person to hold the snake in the correct position. If solid food is being used, it should be small so that it will easily pass down the throat, despite any resistance offered by the snake. Whole animals, such as small mice for example, are preferable to pieces of meat as they offer a balanced diet. The dead animal should be lubricated with water or egg white, held in forceps, worked into the mouth and gently forced into the throat. Once the food animal has entered the throat, the forceps should be removed, the mouth held shut, and the food animal gently massaged deeper into the throat with the free hand. When several centimeters into the throat, release the snake as the swallowing process may then continue voluntarily. If this occurs it is likely that the snake will continue to eat voluntarily. If the meal is regurgitated, the process is to be repeated and the meal worked further into the stomach. If a snake regurgitates each forced solid meal, then force feed a liquefied meal as this is easier to undertake and some nutrition will remain even if the snake regurgitates. A basic mixture includes finely ground meat, an egg, and liquid vitamins. Milk can be added to create more fluid. Place the mixture into a caulking gun or large syringe and force it through a plastic tube into the stomach. Lubricate the plastic tube with water, or similar liquid to ease passage through the esophagus. Great care must be exercised in each step of the operation as the snake can be easily injured, perhaps fatally.

    Select the tube carefully ensuring that the diameter is suitable and without rough spots or edges which may injure soft tissue. Baby feeding tubes or stomach tubes with rounded tips and holes on the sides are the most suitable. The tube tip should be soft, yet flexible and sturdy so it does not kink when passing through the esophagus. Secondly, ensure that the tube is being passed into the esophagus and not into the trachea. If foreign matter is passed into the trachea, the snake will undoubtedly die. Approximate the stomach-to-mouth distance, one third of the body length measured back from the snout, and mark this on the tube. Use a device to keep the mouth open, otherwise the snake will bite the tube and prevent its passage. An effective mouth gag is made from the case of a disposable syringe. Cut off the end, make a hole for the tube to pass, smooth the edges, and insert into the mouth. Look into the mouth through the mouth gag. You will see the trachea opening and glottis on the floor of the mouth and the esophagus opening at the back of the mouth. Refer: Fig. 5.

Figure 4. A Mouth Gag Fabricated From a Disposable Syringe.

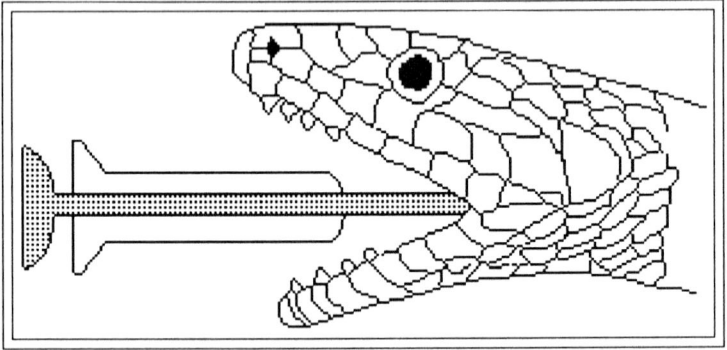

Pass the tube through the mouth gag, along the roof of the mouth back into the esophagus and gently work into the stomach area. Then release the liquefied mixture. After releasing the food, remove the syringe or caulking gun and hold the tube above the snake to ensure that all of the liquid flows down the tube into the stomach. Thirdly, prepare to remove the tube. Before removing the tube, plug the end, perhaps with your thumb. This will prevent any foreign matter remaining in the tube from entering the trachea as the tube exits the body. Next, remove the mouth gag and hold the head of the snake up for a short period of time. Then return the snake to its quarters.

Whichever method of force feeding is employed, continue to offer natural food, both alive and dead, between force feedings. Eventually the snake may begin eating

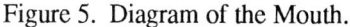
Figure 5. Diagram of the Mouth.

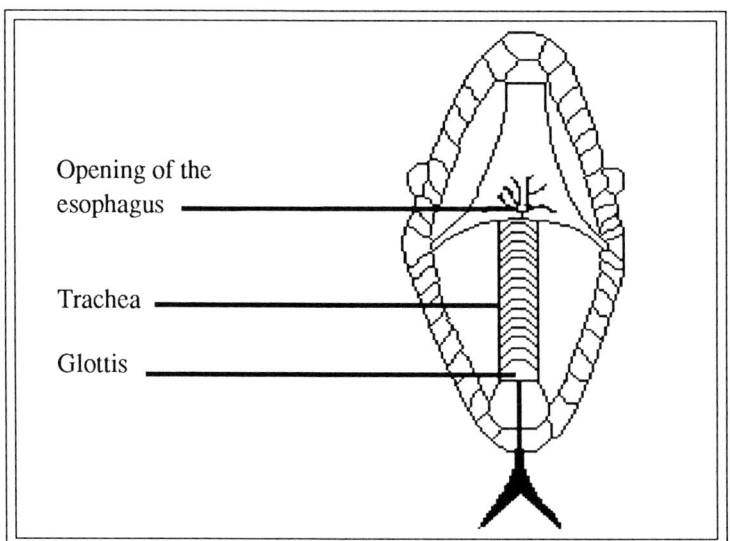

voluntarily. Place either a dead or live food animal in the cage with the snake and remove it in the morning if not eaten. However, do not leave a live rat or mouse with a reluctant feeder for more than a couple of hours as they may injure the snake.

Occasionally baby snakes are so small that natural food cannot be found and it is necessary to force feed them until they are grown and able to eat on their own. In this case, follow the steps listed above with great care as babies are delicate and easily injured.

Inexperienced amateurs should **not** keep poisonous snakes. If this advice is not heeded, please heed the following. Do **not** attempt to force feed venomous snakes. It is a very dangerous operation.

WOUNDS

The most common wound is the rubbed rostral that results from snakes probing with their noses to escape from a cage. This is common among newly acquired specimens. It is usually avoided by providing housing of comfortable dimensions. Housing should be free of rough spots and the interior sides of wire surfaces covered with cloth to protect the animal. When the snake is settled and ceases searching for escape the cloth may be removed. Preventive measures are important as damage to the rostral can be irreversible. See Plate 114, page 312.

Lighting used inside of cages which generates high temperatures must be shielded to prevent specimens from burning themselves as unshielded lighting can quickly inflict severe burns.

Snakes occasionally injure each other when feeding within close quarters. Good feeders become excited and tend to grab anything that moves. Members of *Viperidae* often injure each other. The problem is not caused by envenomation from the strike but rather from the physical damage inflicted by the long fangs. If a fang were to strike an eye, it is likely to be permanently damaged. Constrictors are able to kill cage mates during struggles for food and large specimens may accidentally swallow smaller ones when contesting for the same food. All of these problems are avoided by separating snakes during feeding.

Rodents, particularly rats, have large incisors and their bites can inflict severe wounds, even on poisonous species. It is wise to kill rodents before feeding.

The wounds mentioned above will usually heal without treatment. It is wise, however, to apply a general antibiotic to wounds to prevent secondary infections. Gentian Violet or iodine are good for this purpose.

Diseases

Snakes are vulnerable to a number of diseases which may prove fatal if not treated. Unfortunately, there are few veterinarians trained to treat reptiles and even fewer who are willing to treat snakes. If you live in an area with a veterinarian capable of, and willing to treat snakes, consult him as soon as you notice the symptoms subsequently described. Ask the nearest zoo to recommend a qualified veterinarian. Treating a snake yourself should be the last resort, but if required, carefully conduct the following recommended procedures. In all cases, isolate the diseased animal to minimize the risk of contaminating other members of a collection. **Do not** try to administer antibiotics to poisonous snakes. The risk is too great. Treat them with thermotherapy only.

RESPIRATORY INFECTIONS

Respiratory infections are a common problem, especially among tropical snakes. They are usually caused by prolonged exposure to cold or even cool temperatures. For that reason snakes imported directly from the tropics should be quarantined and watched closely after their arrival into mid-latitude areas. Symptoms include gaping, labored breathing, wheezing, coughing, and discharges of mucus. When infections become severe, snakes will often keep their heads elevated and their mouths open to facilitate breathing. Begin treatment when any of these symptoms are noticed. Daily injection of chloramphenicol in association with injections of gentamicin every three days is recommended. If no improvement is seen after one week, other drugs, such as ampicillin, tetracycline, or tylosin should be used. Concurrent with this treatment or, as an alternative if drugs are not available, thermotherapy may be used. Raise and maintain the temperature in the quarters at approximately 35°C - 37°C until the symptoms disappear. There are reports that respiratory illnesses have been cured by thermotherapy alone.

MOUTH ROT

Mouth rot, canker, or necrotizing stomatitis is a disorder of the mouth leading to death if not treated. Symptoms are swelling and discoloration of the gums, they bleed easily, and the swelling reaches the point where the mouth cannot be closed. If unchecked, the gum tissue dies and the teeth become loose and may fall out. It is highly contagious and treatment should begin as soon as any of these symptoms are observed. The mouth should be washed with water, the dead tissue removed, and Polysporin applied to the infected surfaces. This should be done daily. If improvement is not observed within a week, begin treating with amikacin, ampicillin, or gentamicin. Injections should be given according to the instructions on page 70, and the treatment should be continued for several days beyond apparent recovery.

SCALE ROT

Scale rot, or necrotizing dermatitis, results from keeping snakes on a damp substrate. It first appears as brownish-red discolorations on the ventral surface. If neglected, the discoloration becomes blisters filled with fluid which eventually burst. The scales and skin decline and die, and the disease may enter the stomach cavity. Death is sure to follow. As soon as any of these symptoms are detected, the snake should be moved to warm, dry quarters. If the disease is discovered when in the early stages, treatment with Polysporin ointment will probably cure it. The medication should be rubbed into the skin between the scales. If the disease is advanced, treatment with the drugs amikacin, chloramphenicol, or gentamicin is recommended, following the procedures used for treating respiratory infections. The animal should recover in about two weeks.

CORNEO - SPECTACULAR SWELLING

Corneo-spectacular swelling, or panophthalmitis, is an eye disease which may lead to blindness. A duct leading to the eye becomes infected and the infection spreads to the eye. The disease may first be detected when the eye becomes opaque, as occurs before shedding, or by the eye swelling. Eventually the eye becomes completely opaque, very swollen, and appears to be filled with fluid. As soon as any of these symptoms are observed, begin applying

Polysporin daily and commence treatment with drugs such as amikacin, chloramphenicol, gentamicin, or tylosin. For determination of the correct dosages it is always wise to seek professional help.

It is recognized that snakes carry bacteria which produce diseases that are common in captivity, yet snakes rarely have them in the wild. A natural resistance exists in the wild which, for various reasons, does not correctly function when they are in captivity. Stress may be the most important factor in causing the natural resistance to be inadequate. Stress, of course, takes many forms. It includes exposing tropical snakes to prolonged cold and denying them a normal 24-hour temperature range. This aspect will be discussed in the section relating to Thai snakes. It also includes denying a hiding place, as well as excessive handling. Natural environments cannot be created in captivity. By definition, it is impossible. However, following good husbandry techniques, carefully considering the various aspects of stress and taking steps to reduce stress will create an improved environment for snakes and one in which problems of disease will be significantly reduced.

Antibiotics and Their Use

Reptiles should always be treated by a qualified veterinarian, if available. When a veterinarian is not available, it may be necessary for the collector to undertake the treatment. The following information is useful for such emergencies.

Firstly, venomous snakes should **not** be treated by the inexperienced amateur.

Antibiotics are supplied as powders or as concentrated liquids, therefore they require reconstitution or dilution.

The ratio of drug to volume of water is normally expressed as milligrams/cubic centimeter, or mg/cc. Drugs which are prepared in powder form can be reconstituted to a required strength. Therefore, if a drug is required in the ratio of 10 mg/cc, dissolve 100 mg of the antibiotic in 10 cc of sterile water. Most liquid antibiotics are highly concentrated and not suitable for direct use so they require dilution. To dilute 50 mg/cc of an antibiotic to the concentration of 10 mg/cc, add 8 cc of sterile water to 2 cc of the concentrate.

Dosages are given in milligrams of drug per kilogram of body weight per interval between treatment, or mg/kg/interval between treatment. Calculating the correct dosage is relatively simple.

Example.

To treat a 1.3 kg snake with a 50 mg/cc concentration of amikacin at the recommended dosage of 2.5 mg/kg/every 72 hours.

Multiply the weight of the snake by the weight of the recommended dosage/cc.

[1.3 (kg) x 2.5 (mg/kg) = 3.25]

Then divide the product by the weight of drug in solution, that is, 3.25 ÷ 50 = 0.065. The correct dosage is therefore 0.065 cc of the concentrated solution to be given every 72 hours.

However, it is impractical for this dosage to be given due to the minimal volume. This is a common problem when treating snakes as most species are limited in body weight. The most satisfactory method is to increase the volume by additional dilution.

For the example shown, the drug is diluted using a ratio of 1:10 for the dosage to become 0.65 cc. This amount can be injected by using a Tuberculin syringe, that is, a one milliliter syringe with a fine scale.

> **To prepare a practical dosage for the 1.3 kg snake:**
> Aspirate 1 ml of the 50 mg/cc amikacin and thoroughly mix with 9 ml of distilled water. This provides a concentration of 5 mg/cc. If the weight of the snake is multiplied by the weight of the recommended dosage we obtain 3.25.
>
> $$[1.3 \text{ (kg)} \times 2.5 \text{ (mg/kg)} = 3.25]$$
>
> Next, divide 3.25 by the weight of drug in solution to achieve 0.65.
>
> $$(3.25 \div 5 = 0.65)$$
>
> Ten milliliters of this dosage would be injected, using a Tuberculin syringe with a 27-gauge needle, every 72 hours. This general procedure will often need to be followed to prepare easily administered dosages.

The drug should be given by means of a subcutaneous injection. Since the snake is unlikely to cooperate, it will be necessary for someone to hold it. After ensuring that air bubbles are not present, align the syringe parallel to the body, point towards the head, and insert into the skin at the edge of a scute. The needle should pass through the skin into the space between skin and muscle. If the needle is correctly positioned a small bump will then form on the surface of the skin as the dosage is pressed into the subcutaneous area. If this does not happen, withdraw the needle and try again.

Table 4. Recommended Dosages of the Antibiotics Mentioned.

Antibiotic	Dosage
Amikacin	2.5 mg/kg/every 72 hours
Ampicillin	50-75 mg/kg/day
Chloramphenicol	50-75 mg/kg/day
Gentamicin	2.5 mg/kg/every 72 hours
Tetracycline	50-75 mg/kg/every 72 hours
Tylosin	50-75 mg/kg/day

Intestinal diseases and most worms are treated by oral administration of drugs. If the snake is still eating, although ill, it may be possible to apply antibiotics in capsule form. Place the capsule of an appropriate antibiotic in the mouth of dead food animals and feed to the snake. If the capsule does not fit into the mouth of the food animal, make an incision in the skin and place the medication under the skin. It the snake accepts dead food, this is a convenient and safe way to administer a drug.

If the snake is not eating, another method of administration is required. Pass a soft plastic tube through the mouth into the stomach, using water as a lubricant. Great care must be exercised, however, as the snake could be easily injured. Carefully follow the procedures recommended within the discussion in respect to force feeding, page 60. Remember that the potential for injury is great, therefore, pass the tube gently and carefully into and out of the esophagus. Ensure the tube does not enter the trachea to avoid introducing foreign matter into the trachea.

PART 2

The Snakes of Thailand

Introduction

There are 175 species and subspecies of snakes native to Thailand, depending upon the acceptance of proposed new species and subspecies. Furthermore, some species not previously found in Thailand have been included in this current listing. Refer to: Page xvii. Of the 175 snakes presented and listed in this volume, 62 are nonvenomous (none of the teeth are markedly enlarged and/or Duvernoy's gland is absent), 50 are mildly venomous (Duvernoy's gland **and** enlarged rear maxillary teeth), 56 are very venomous (anterior fangs and venom gland), with insufficient data to classify seven species. These seven, however, are certainly no more toxic than mildly venomous. The distinctions between mildly venomous and venomous are variable as the envenomation by a rear-fanged *Colubroidae* might prove very toxic to one individual but produce insignificant affects on another. Many variables affect venom toxicity, the health and age of the victim and the quantity of the venom injected are some examples. Within Appendix 10, Analysis of Common Thai Snake Names, the venom toxicity of each snake is clearly specified. The 175 species and subspecies of Thailand are classified within 4 superfamilies, 8 families, 15 subfamilies, and 67 genera.

Thailand is rich in snake fauna with body sizes ranging from the small blind snakes to the large *Python reticulatus*. A surprising number of species are still found in densely populated Bangkok. Snakes represent a significant natural resource in this agricultural nation as rodents, which must compete with man for food and also carry potentially lethal diseases, comprise the bulk of the diet of many Thai snakes. Destruction of their natural habitat by man has resulted in the decline of a number of species with the leather trade taking a heavy toll on others. Fortunately, snakes are becoming more appreciated as a natural resource, and some species are now protected and their export prohibited.

Classification has presented a problem throughout this study. Superfamily *Colubroidae* in particular, is undergoing taxonomic review and numerous reclassifications have recently been made. The classifications used in this book are the most recent and widely accepted. Refer to: Table 5, page 77.

Snakes usually have distinctive taxonomic characteristics and those of Thai snakes are described in this volume. Scalation characteristics are always described. Less frequently, but where appropriate, characteristics of dentition and skull structure are explained. Terminology used in describing scalation, dentition, and skull structure can be clarified by reference to Figures 1, 2, 3, 6, and 7.

Table 5. Reclassification List.

Formerly	Currently
Agkistrodon rhodostoma	Calloselasma rhodostoma
Ahaetulla ahaetulla	Dendrelaphis pictus pictus
Ahaetulla caudolineata	Dendrelaphis caudolineatus
Ahaetulla formosa	Dendrelaphis formosus
Calamaria leucocephala	Calamaria schlegeli schlegeli
Calamaria vermiformis	Calamaria lumbricoidea
Dryophis mycterizans	Ahaetulla mycterizans
Dryophis nasuta	Ahaetulla nasuta
Dryophis prasinus	Ahaetulla prasina
Natrix inas	Amphiesma inas
Natrix percarinata	Sinonatrix percarinata
Natrix piscator	Xenochrophis piscator
Natrix trianguligera	Sinonatrix trianguligera
Typhlops albiceps	Ramphotyphlops albiceps
Typhlops braminus	Ramphotyphlops braminus

Unanimous agreement among taxonomists has not been reached on snake classification. The reader will doubtless find references to other classification systems used in literature which vary from the following system. The one chosen here, and used throughout has the general support of the herpetological community. Detailed discussions and explanations of the criteria used in establishing levels of classification lay beyond the scope of this book. Only superficial mention is made of certain criteria, especially the higher levels of classification. Readers who seek more in-depth discussion are referred to:

- *Classification of the Serpents: A Critical Review*, by Herndon G. Dowling
- *Osteology of the Reptiles*, by Alfred S. Romer
- *Snakes: Ecology and Evolutionary Biology*. Chap.1, Systematics, by Samuel B. McDowell.

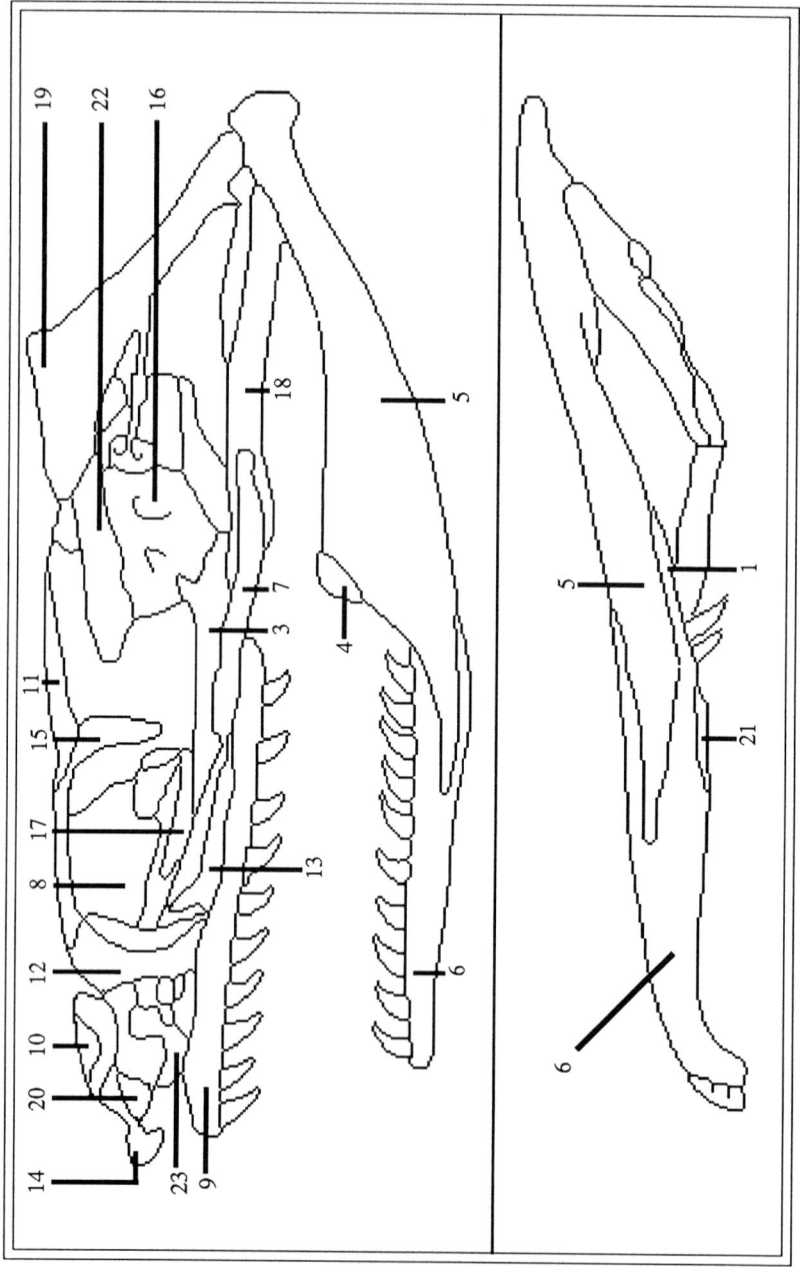

Figure 6. The Bones of the Head of an Idealized Snake.

Figure 6. The Bones of the Head of an Idealized Snake.

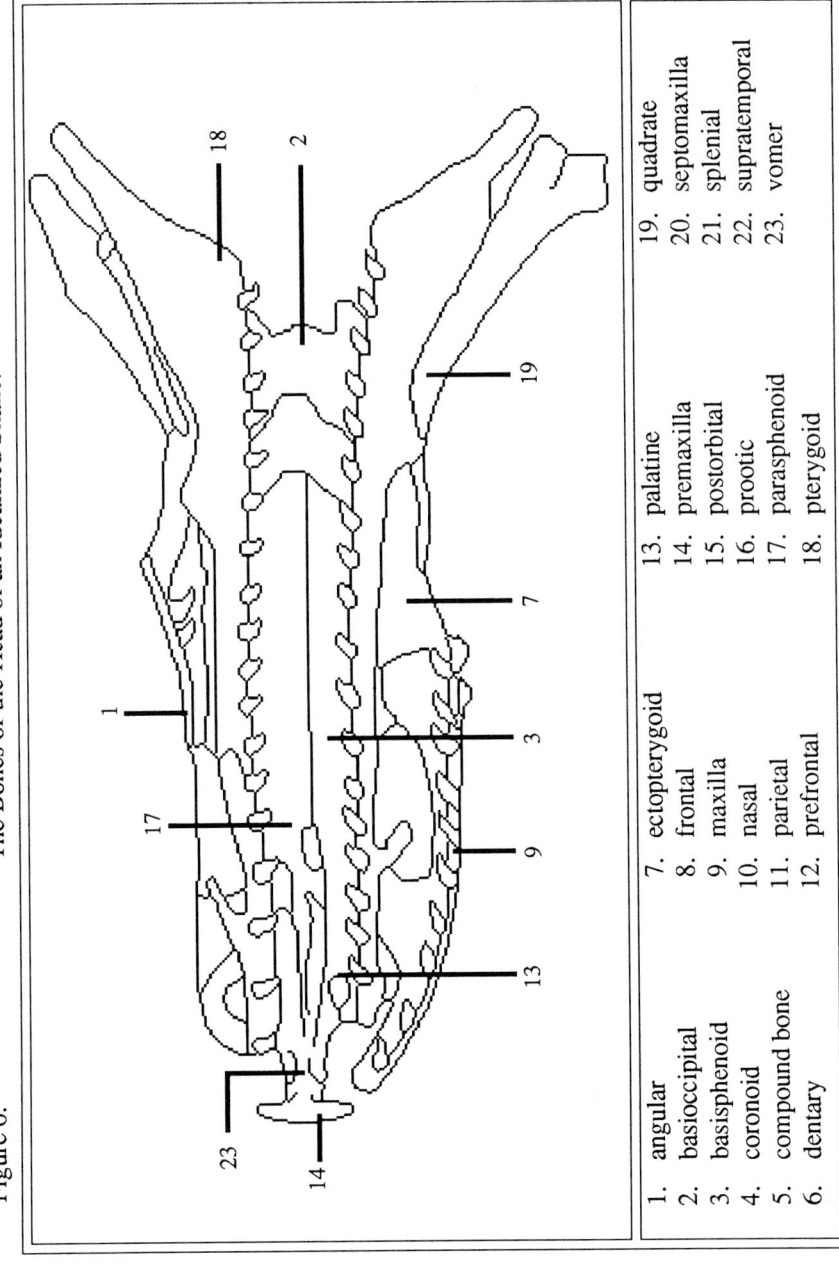

1. angular
2. basioccipital
3. basisphenoid
4. coronoid
5. compound bone
6. dentary
7. ectopterygoid
8. frontal
9. maxilla
10. nasal
11. parietal
12. prefrontal
13. palatine
14. premaxilla
15. postorbital
16. prootic
17. parasphenoid
18. pterygoid
19. quadrate
20. septomaxilla
21. splenial
22. supratemporal
23. vomer

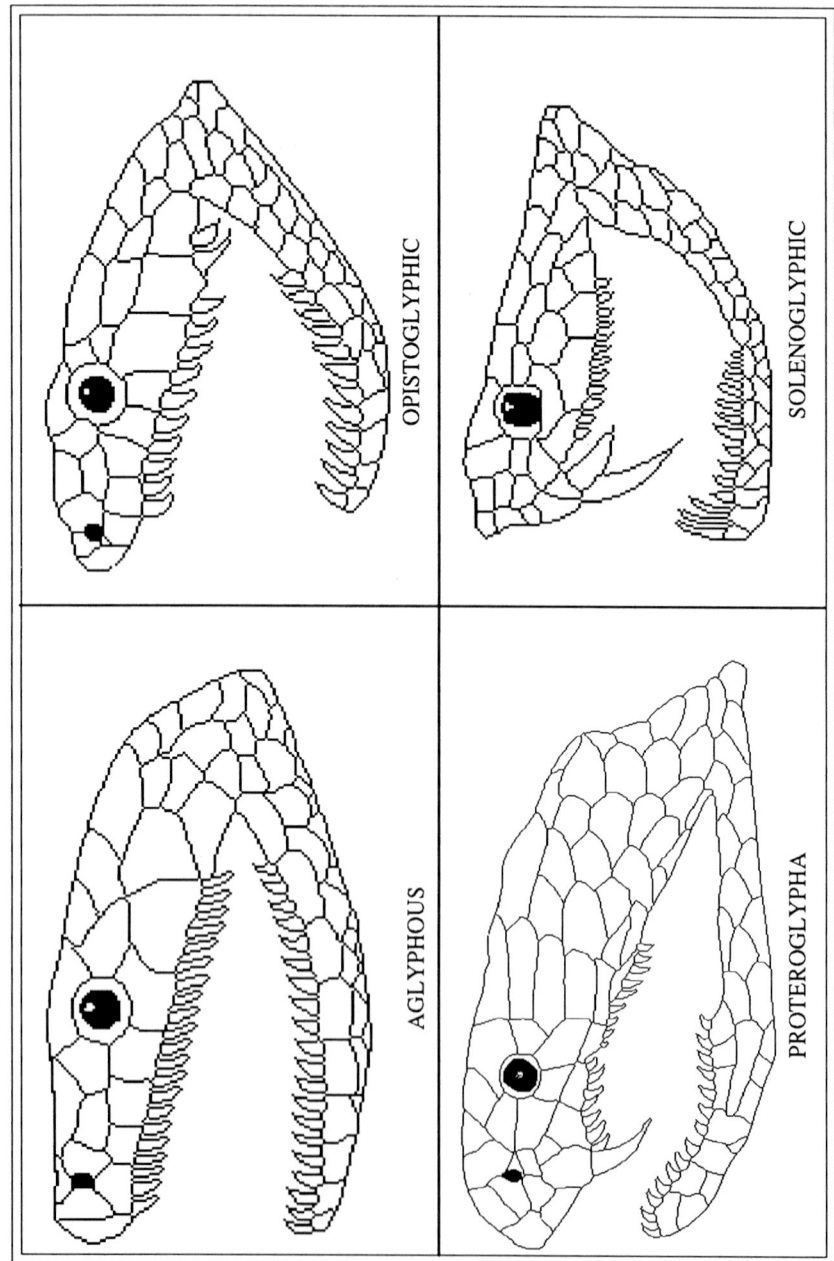

Figure 7. Types of Dentition.

INFRAORDER SCOLECOPHIDIA

Members of this infraorder are characterized by a consolidation of the skeletal structure of the head which facilitates their lives as burrowers. Their ability to see is extremely reduced, hence, they are called blind snakes. They are considered to be primitive, in evolutionary terms, possessing the following "primitive" features:

- jaw bones are short and united with the braincase, reducing flexibility and ability to swallow

- glands, probably sebaceous, along bases of smaller head scales

- palatine and pterygoid bones without teeth

- contact or a suture exists between the premaxillary and maxillary bones of the upper jaw

- coronoid bone is present in the lower jaw

- rudimentary pelvic elements and appendage present

- left lung is relatively large

- eye covered by a large scale not conforming to the shape of the eye

- ventral scales not fully enlarged to form "true" ventral scales.

The families included within this infraorder are those of *Anomalepididae, Leptotyphlopidae*, and *Typhlopidae*. Only Family Typhlopidae is found in Thailand.

FAMILY TYPHLOPIDAE

The members of this family are known as Weak-jawed Blind Snakes. Although they are usually visible, the eyes are covered by an ocular shield and are often difficult to see. They are burrowing snakes, or fossorial, and are well adapted for such a life. They are generally small and worm-like. The rostrals are large and the heads indistinguishable from the cylindrical bodies. Tails are short, blunt, and often equipped with a short spine. The head and body scales are smooth, overlap, and polished. Body scales are uniform without ventral or subcaudal scales. Thus, with a consolidated skull structure, well protected eyes, a smooth and cylindrical body, and a spine for added push, these animals are well equipped for a subterranean existence. Other characteristics are:

- maxillary bones of the upper jaw are toothed and movable on the palatine bones

- premaxillary, palatine, and pterygoid bones of the skull are toothless; the dentary bone of the lower jaw is either toothless or with a single tooth

- supratemporal bone of the skull is absent

- ectopterygoid bone of the upper jaw is present in some genera, but presumed to be fused with the palatine bone in others

- coronoid bone of the lower jaw is present

- pelvic elements present in some species

- four supralabials

- large tracheal lung.

Weak-jawed Blind Snakes are usually found, in loose soil, under rocks or fallen trees, and in, or, under decayed logs. They also frequent termite mounds and are often dug from the ground by farmers and gardeners. They are nocturnal and often seen on the ground surface during evening hours searching for food, particularly after rain. Food consists of termites and ants, in addition to their pupae.

Eggs of other insects and grubs, as well as burrowing insects and worms are also eaten. Although not thoroughly researched this family seems to be mostly oviparous, members probably laying less than ten eggs at a time. Females are larger than males, and one genus is known to be parthenogenetic. Blind Snakes are quite harmless, although when handled, they often emit a foul odor.

These snakes have not been popular with collectors due to their small size, secretive habits, and drab coloration. Furthermore, it is difficult to obtain food for them in areas with cold winters. If one chooses to add members of this family to a collection, their burrowing habits will need to be accommodated. An aquarium with a deep layer of firm, sandy soil will meet this requirement, especially if stones and pieces of bark and wood are strewn on the surface. A small water dish with the rim level to the surface should be placed into the soil. When soil shows signs of becoming dehydrated, spray lightly with water to moisten. A small vaporizer is excellent for this purpose. A few plants will create an attractive display. Food should be present in the aquarium at all times.

Typhlopidae is represented by two genera within Thailand.

Genus Ramphotyphlops

There are no obvious external differences by which either the genera *Ramphotyphlops* or *Typhlops* can be distinguished. *Ramphotyphlops* is identified by the structure of the hemipenis and the presence of paired retrocloacal sacs, which might serve as a place for the storage of sperm.

Although *Typhlina* is the senior taxon, it has been suppressed in favor of *Ramphotyphlops*.

Two members of the genus are present in Thailand.

Ramphotyphlops albiceps
White-headed Blind Snake   งูดินหัวขาว

The dorsal surface of this species is brown, the ventral surface light brown. The head, neck, tail and tongue are all white. The White-headed Blind Snake probably never exceeds 0.175 m in length.

The nasals are divided below the nostril but united above it. The supraoculars are small, and there is one preocular and one postocular. Of the four supralabials, the last is the largest and is in contact with the postocular. Suboculars are absent. There are 20 scale rows around the body and the end of the tail has a spine. This species produces 2 to 8 eggs at a time.

This species is reported to be rare in Thailand, but it has been found in the North, South, Southeast, and Central Regions. It has also been reported in Burma, Vietnam, southern China, peninsular Malaysia, and Hong Kong.

Ramphotyphlops braminus Plate 1
Common Blind Snake
(Flower Pot Snake)   งูดินธรรมดา

The Flower Pot Snake may reach an adult length of 0.17 meters. The rounded snout projects a little beyond the mouth, and the nostrils are placed laterally on the snout. The head is indistinguishable from the cylindrical body. The dorsal surface is black or dark brown and highly polished, the ventral surface is somewhat lighter. There is a spine at the end of the tail. This may be the only parthenogenetic species of snake in the world, the females producing fertile eggs without the assistance of a male.

Figure 8. Head Structure of *Ramphotyphlops braminus*.

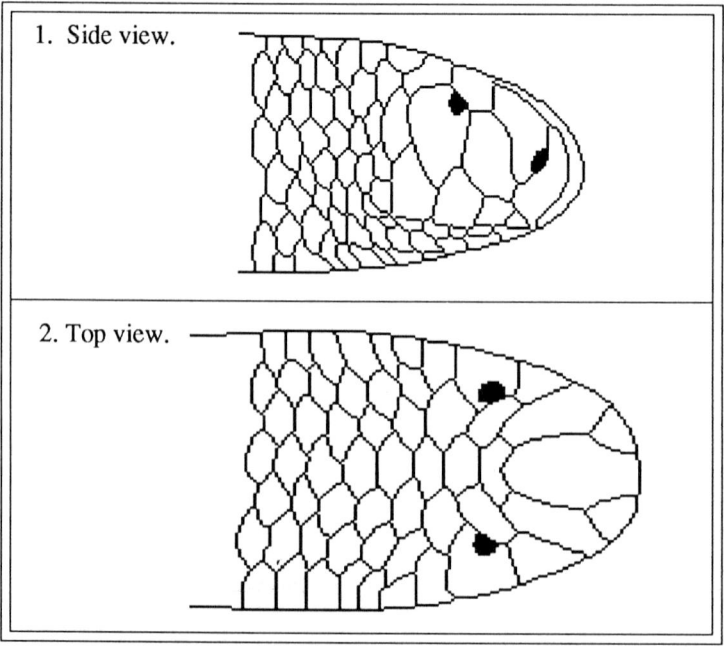

There is one preocular which is in contact with the second and third supralabials. The third and fourth of the four supralabials are in contact with the ocular scale, which obscures the eye. Twenty rows of smooth scales encircle the body and, of course, no ventrals.

The Flower Pot Snake is found in every part of Thailand although it is seldom seen due to its burrowing habits. It is most often discovered by persons digging in gardens or lifting stones or pieces of wood. Occasionally it can be found on the surface of the ground after a heavy rain. This species, no doubt, has the greatest range of any terrestrial snake because it has been accidentally introduced into many parts of the world in flower pots and the balled roots of plants. Its range extends from Africa through the Middle East, South and Southeast Asia, Taiwan, New Guinea, northern Australia, some of the Pacific Islands, and Mexico. It has also been recently discovered in the State of Florida, in the southeastern United States.

Genus Typhlops

In genus *Typhlops* the hemipenis is completely inverted when retracted, whereas in *Ramphotyphlops* it remains partially everted. *Typhlops* lacks the retrocloacal sacs found in *Ramphotyphlops*.

This genus is represented by six members in Thailand.

Typhlops diardi diardi
Indochinese Blind Snake   งูดินใหญ่อินโดจีน

This snake attains a relatively large adult length of approximately 0.43 meters. Its dorsal color is brown to blackish-brown; the ventral color is a lighter shade of brown. The tail ends with a small spine.

The snout is rounded and projects strongly. The upper portion of the rostral covers approximately 60 percent of the head width. The eyes are fairly distinct, although covered by an ocular shield which intrudes between the third and fourth supralabials. There are 24 to 26, rarely 28, scales around the body.

The Indochinese Blind Snake is widely distributed in Thailand north of roughly 16°N latitude. It is also found in northeastern India, Bangladesh, Burma, Laos, and northern Vietnam.

Typhlops diardi muelleri
Malayan Blind Snake   งูดินใหญ่มลายู

Typhlops diardi muelleri is also a relatively large blind snake, sometimes exceeding 0.4 m in length. Its dorsal surface is blackish-olive to brown and is patternless; the ventral surface is light yellow. The blunt tail ends in a small, downward directed spine. This species differs from most other blind snakes in being viviparous, producing three to eight living young per litter. The newborn average 0.1 m in length.

The nasal is divided below the nostril but undivided above. The eye is distinct but covered by an ocular shield. There is one large preocular, two small postoculars, and four supralabials. The postoculars are directly above the fourth supralabial. There are usually 24 to 26 scale rows around the body but rarely as few as 22. The scales are smooth and shiny.

Thai specimens have only been found south of 14° N latitude. The Malayan Blind Snake also exists in Burma, Kampuchea, southern Vietnam, peninsular Malaysia, and western Indonesia.

Typhlops floweri
Flower's Blind Snake   งูดินหัวเหลือง

This is a small snake, with an adult length of a little over 0.2 meters. It possesses the typical *Typhlops* body in that the head is not distinct from the neck, the body is cylindrical, and the tail is short. However, in this species the spine at the end of the tail is missing. Its color is dark brown or black. The tip of the snout is yellow, the rostral and nasals are light brown with tiny yellow specks, and the anal area is yellow.

The head is relatively flat and wide and the snout projects well beyond the mouth. The rostral is approximately 60 percent as wide as the head. The nasals are divided, there is one preocular, an ocular, two postoculars, and one subocular. There are 18 scale rows on the body.

This species is found in the Central Region around Bangkok. It has not been reported outside of Thailand.

Typhlops khoratensis
Khorat Blind Snake   งูดินโคราช

This is a short snake with the typical *Typhlops* body characteristics previously mentioned. It is almost entirely gray but the labials, chin, and throat are dull white. Adults are 0.1 to 0.128 m in length. A posterior spine is present.

The rostral is approximately 33 percent of the width of the head. The nasals are completely divided. A supraocular, preocular, ocular, and postocular are present. The eyes are barely visible. The supralabials and infralabials are relatively large. There are 20 rows of scales on the body.

The species is found in the North, Northeast, and Central Regions. The range may also continue into Laos.

Typhlops lineatus   Plate 2
Striped Blind Snake   งูดินลายขีด

One of the largest members of the genus, this snake attains a length of 0.48 meters. Its dorsal side is yellowish-brown with a series of dark lines from the neck to the tail. The ventral surface is creamy white and the head light brown with a few small black spots. The end of the tail has a spine.

The rounded snout projects over the mouth; the eyes are invisible. The rostral is quite broad, about 75 percent of the head width. The nasals divide below the nostrils but are fused above. There is an ocular, two postoculars, large supraoculars but no preocular. Of the four supralabials, three touch the ocular. There are 22 body scale rows.

Only two specimens have been found in the Kingdom, both in southern provinces; one in Pattani and the second in Narathiwat. Elsewhere it exists in Singapore, Sumatra,

Java and Borneo and is widespread, but uncommon, in peninsular Malaysia. It has been seen at elevations of 1425 meters. Plate 2 was photographed in Malaysia.

Typhlops trangensis
Trang Blind Snake   งูดินตรัง

This species is short and relatively thick bodied. The only specimen so far discovered was 0.155 meters and was possibly an immature specimen. The 13 uppermost body scales are gray to ultramarine, the lower 11 cream. The nasal is partially divided and the suture extends to the second supralabial. The nostril, nasals, oculars, preoculars, prefrontal, frontal, supraoculars, and parietals are covered with pits, or depressions, visible through the posterior portion of each scale. The eye is not visible. The preocular is roughly the size of the ocular and intrudes between the second and third supralabial. The body is encircled by 24 rows of scales, with a spine at the end of the tail.

This little known snake is extremely rare. The holotype was discovered by Edward H. Taylor under a rotting log on Khao Chong Forest Experimental Station, Trang Province.[1] Perhaps it is not surprising that this holotype is the only specimen recorded to date. Every member, by virtue of being fossorial, is secretive and seldom encountered. Doubtless additional specimens will be discovered in the future, possibly in northern peninsular Malaysia.

---

1. Dr. Taylor proposed this new species in the University of Kansas Science Bulletin, Vol. 43 (1962): 251-253. This holotype is cataloged as No. 35754, Edward H. Taylor-Hobart M. Smith Collection, Lawrence, Kansas.

INFRAORDER ALETHINOPHIDIA

This vast infraorder includes most of the world's snakes. Members are considered to be more advanced along the evolutionary scale than are the members of infraorder *Scolecophidia*. A minimum number of "primitive" characteristics are present, although some exist in the subfamilies of *Booidea, Calamariinae* and others. *Alethinophidia* share a flexibility, to varying degrees, of the bones and muscles of the skull. This flexibility enables opening of the mouth to a wide angle, permitting relatively large prey to be swallowed. Swallowing is accomplished, at least in part, by asymmetric movement of the lower jaws. This requires drawing back the left and right tooth-bearing bones in sequence so the prey is ultimately swallowed. The palatine and pterygoid bones of the skull are almost always toothed. The eye is covered by an ocular scale that closely conforms to the shape of the eye and is quite indistinguishable from it, unlike the large, obvious ocular scale of *Scolecophidia*.

Superfamily Acrochordoidea

This superfamily is distinguished from all others by modification of two bones in the skull. The tabular is large and disk-like and extends forward, approaching the postorbital bone. The prefrontal bone is greatly reduced in size.

FAMILY ACROCHORDIDAE

This family contains only three species, two of which exist in Thailand. Both are well adapted for life in the water. In fact, these snakes are almost helpless on land as they do not have ventrals or subcaudals. Their respiratory systems are equipped with valved nostrils which prevent water from entering the lungs. It does not seem that they have a venom producing capacity.

Both Thai members of this family belong to genus *Acrochordus*.

Genus Acrochordus

These are large, heavy-bodied reptiles whose primary prey is fish and eels. They bear living young which are miniature replicas of the adults. Females are larger than males.

Each species does fairly well in large aerated aquariums maintained at 27° C to 29° C. Some specimens may reach two meters thus requiring spacious and securely covered aquariums. They are nocturnal and tend to be lethargic during the day. Although not poisonous, they can give a nasty bite. Dry areas are not necessary in their housing.

Acrochordus granulatus   Plate 3
File Snake   งูผ้าขี้ริ้ว
(Granular Snake)

The head is blunt and indistinguishable from the neck with the nostrils at the top of the snout. The eyes are small and the pupils round, more closely spaced than those of the *Acrochordus javanicus* and with 8 to 11 small scales between them. The skin, covered with keeled scales, hangs in loose folds from the body. One prominent fold of skin extends along the center of the belly. The tail is short and somewhat compressed. Females tend to be larger than males, and the average length of the species is approximately one meter. It is a more attractive snake than *Acrochordus javanicus*. The body is crossed by a series of alternating white and black or dark brown bands. The dark bands are broad on the back but become narrow on the sides and may, or, may not extend across the belly. The head is covered with small granular scales and is dark with several small white spots. Females produce from 6 to 8 young per litter, the young averaging 0.22 meters.

    Acrochordus granulatus has a blood volume greater than that reported for many other reptiles (12.5%), including that of either of its congeners. Furthermore, the percentage of red blood corpuscles to the volume of whole blood is very high (mean 50.1%). The combination of these two factors gives this species an unusually large capacity for the storage and transport of oxygen, enabling it to remain submerged for long periods before surfacing to breathe. Captive specimens have remained submerged for 139 minutes. Longer periods of submergence are believed possible by accessory oxygen uptake through the skin. This capability is a great advantage to this totally marine, nocturnal snake, enabling it to reduce exposure to the high daytime water temperatures as well as exposure to daytime predators.

Supralabials number from 14 to 22 and the infralabials from 12 to 18. Scale rows number approximately 100 at mid-body. The scales are keeled or tubular and rough to the touch.

Found in salt water, it frequents coastal areas and estuaries. It has been found 10 km offshore and at depths of 20 m making it more marine than *Acrochordus javanicus*. Because of its habitat and markings, it is often confused with sea snakes.

File Snakes are found on both Thai coasts. They have been reported in Sri Lanka, Bangladesh, all of Southeast Asia, New Guinea, the northern coast of Australia, and on the island groups surrounding New Guinea as far east as the Solomons.

Acrochordus javanicus Figure 9
Elephant Trunk Snake Plates 4 and 5
(Wart Snake) งูงวงช้าง

This is a heavy-bodied snake with a head that is short, blunt and indistinct from the neck. The nostrils are at the end of the snout and directed forward. Refer to: Plate 5. The eyes are small with round pupils, widely spaced and with 18 to 22 small scales between the eyes. The head is covered with small scales. The skin is wrinkled and hangs in loose folds from the body. Body scales are small, keeled, and very rough to the touch. The basic color is olive-brown to gray-brown with a hint of a marbled, black pattern on the sides.

Figure 9. *Acrochordus javanicus.*

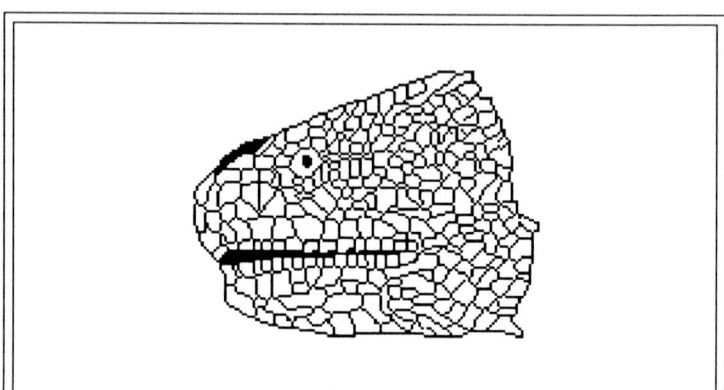

The ventral side is lighter in color and there is a slightly developed ridge along the center of the belly.

The body scales are larger than those on the head. Midbody scale rows number from 126 to 150. Supralabials are only slightly enlarged and number from 22 to 36. Infralabials vary between 31 and 34. Ventrals and subcaudals are absent.

Females may reach two meters in length, but the average for this species is probably less. The Elephant Trunk Snake is aquatic and prefers slow moving water, either fresh or brackish. It may be found in the mouths of rivers which empty into the sea, or in fresh water rivers and canals. Females can be prolific, producing from 18 to as many as 48 live young at a time. Captives do well on a diet of goldfish.

In Thailand, this species is found in the Central Region, the Southeast, and in the South. It is also found in India, and the remainder of Southeast Asia.

Superfamily Anilioidea

Anilioidea differs from all other *Alethinophidia* except *Acrochordoidea* in that the dentigerous (bearing teeth or dental structures) process of the palatine bone of the upper jaw continues forward to the vomer bone. This is true whether teeth are actually present or not. Pelvic vestiges could be present or totally absent. Similarly, the left lung may be more than half the length of the right lung and possibly vestigial, or absent. There is no tracheal lung.

Two families are present in Thailand, *Uropeltidae* and *Xenopeltidae*.

FAMILY UROPELTIDAE

The burrowing snakes of this family differ from those of *Typhlopidae* in several significant ways. They are larger, their eyes are visible, and there are differences in their dentistry and skeletal structures. For example, both the families have vestigial pelvises but in *Typhlopidae* the evidence is entirely internal. In *Uropeltidae*, however, some of the evidence is visible as vestiges of legs in the form of spur-like appendages on either side of the male's anal opening. Furthermore, *Uropeltidae* has well developed ventral and subcaudal scales, which are absent in *Typhlopidae*.

Genus Cylindrophis

This genus includes snakes whose bodies are cylindrical and have smooth, overlapping scales. They are burrowers, but have narrow ventral scales. "Spurs" flank the anal opening of males and, internally, the left lung is absent.

Their care is similar to that for *Typhlops*, but the aquarium is to be larger. Fill the aquarium, to a depth of several centimeters, with sandy soil and litter with flat stones and pieces of wood or bark. A water dish is required with the rim slightly above the soil level. The soil is kept moist by occasional spraying. These snakes feed upon worms, insect larvae, and other snake, especially *Typhlops*.

Only one species lives Thailand.

Cylindrophis rufus rufus — Figure 10
Red-tailed Pipe Snake — Plates 6 and 7
(Two-headed Snake) — งูก้นขบ. งูสองหัว

This is a medium-sized snake, reaching a length of approximately one meter. The snout is rounded and the front of the head is flattened. The head is not distinct from the stout, cylindrical body. The tail is very short and stout. The dorsal side is deep purple or black. Some specimens have a reddish cross-bar on the neck. Irregular white or cream bars extend up from the ventral surface along the sides of the body, but usually do not encircle the body. The ventral surface is black with a series of rough-edged white or cream bars which cross it and, as mentioned, extend partly up the body sides. The ventral surface of the tail is bright red in juveniles but tends to fade as the animal matures. The eyes are small and the pupil round.

Figure 10. *Cylindrophis rufus rufus.*

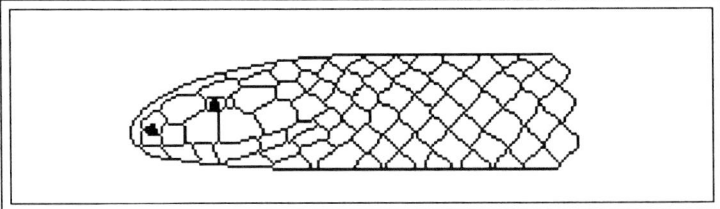

The rostral is high and intrudes between the nasals. The nasals are in contact behind the rostral. Loreals and pre-oculars are not present. The supraoculars are close to the front of the eye. There is one postocular and six supralabials, the third and fourth bordering the eye, and six infralabials.

The body scales are smooth and very iridescent. They are arranged in rows of 19 or 21 at mid-body. Both the anal plate and subcaudals are single. The ventrals range from 186 to 216, the subcaudals from five to seven. A pair of spurs flank the anal plates of males. This species is viviparous, producing 3 to 13 young at a time.

This snake has an interesting defense. When threatened, it raises its tail, making its posterior end look like a cobra in a defensive position. The motion of raising its tail, the flash of red color exposed from the bottom of the raised tail, and the cobra-like position in which the tail is held all combine to make it appear very threatening to a predator. During this time, the real head is fast disappearing under a rock or piece of wood and the tail, fighting a rear guard action, slowly follows. For this reason, *Cylindrophis rufus rufus* is called the "Two-headed Snake" in parts of its range.

This snake is common in all the lowlands of Thailand and is also found on the remainder of mainland Southeast Asia and in Indonesia.

FAMILY XENOPELTIDAE

This family contains one genus, *Xenopeltis*. This genus, in turn, contains a single species. It bears some superficial similarity to *Cylindrophis*, but features of the anatomy, the structure of the skull, and the absence of a vestigial pelvis make this species different. Furthermore, the left lung is large and well developed in *Xenopeltis* but absent in *Cylindrophis*.

Genus Xenopeltis

The body is long and cylindrical and the head is barely distinct from the neck. The eyes are small and the pupils round; the scales are smooth.

The sole species in this genus may reach 1.2 m in length, so a rather large cage is required for its housing. This is a species of burrowers; thus, the floor of the cage must be covered with a reasonable amount of loose, sandy soil. Pieces of bark and wood and some flat stones should be placed in the cage. A large water dish should be placed in the soil with the rim just above the soil level. The soil is to be kept moist but not muddy. This species usually does well in captivity, although its burrowing habits do not make it a good display animal. It has a gentle disposition and eats a wide variety of animals, including mice, other snakes, lizards, frogs, and even birds.

Xenopeltis unicolor
Iridescent Earth Snake
(Sunbeam Snake)

Figure 11
Plates 8 and 9
งูแสงอาทิตย์

This species is well equipped for burrowing. Its head is flat and chisel-shaped and almost indistinguishable from its cylindrical body with a short and blunt tail. Scales are smooth and highly polished. Dorsal color is uniformly dark brown or purplish-black and the ventral surface white. The body is very iridescent and displays various colors when struck by the sun. Newly hatched have a collar of white around the neck that may extend to their head. This disappears as they mature (see Plate 9, page 125). The species is oviparous and females are the larger. Clutches contain from 6 to 17 eggs.

Figure 11. *Xenopeltis unicolor*.

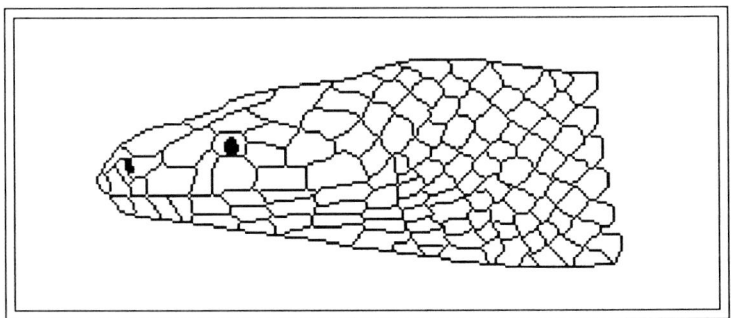

Scale rows number 15 throughout the thickest part of the body. Ventrals range from 173 to 196, and the paired subcaudals from 24 to 31. The anal is divided. The parietals are separated; there are two postoculars, one supraocular, and no suboculars. One scale, either a loreal or preocular is located on the anterior side of the eye. Eight supralabials exist with an anterior nasal wedged between the first and second. The fourth and fifth border the eye. There are eight infralabials.

The Sunbeam Snake is nocturnal, widely distributed, and quite common in Thailand. It is also reported in southern China and the remainder of Southeast Asia.

Superfamily Booidea

The completeness of the bony roofing of the nasal capsule distinguishes *Booidea* from all other forms. The prefrontal bone is broadly expanded over the muzzle. Also, the maxillary and premaxillary bones are not in contact; the maxillaries are toothed and the premaxillary may, or, may not be toothed. Pelvic spurs are present on males, supported internally by vestiges of a hind limb. A large, functional left lung exists in all *Booidea*. A tracheal lung is absent.

This superfamily is divided into Family *Boidae* and Family *Pythonidae*. Only *Pythonidae* is found in Thailand.

FAMILY PYTHONIDAE

Members of this family have a supraorbital bone as a part of the skull. The palatine tooth row is continuous with that of the pterygoid. All *Pythonidae* are oviparous.

This family of constrictors includes snakes of a length of one to ten meters. The females of some genera incubate their eggs and increase body temperatures when doing so. Most possess heat-sensitive sensory pits on their rostrals and/or labials which assist in the finding of warm-blooded prey. Although most are primarily terrestrial, nearly all are good climbers. Females attain a greater length than males.

Two of the three pythons native to Thailand are large and require spacious cages. They would be uncomfortable in a cage of a size less than four by two meters. Both *Python reticulatus* and *Python molurus bivittatus* like to climb; therefore, their cages should have suitable branches for climbing. All three of the native Thai species like to soak in water and cages should contain a water receptacle of sufficient size for soaking their entire body.

Seriously consider the matter of safety, before adding members of this subfamily to your collection. Although not poisonous, the larger members are powerful constrictors and are quite capable of killing a male adult. Ill-tempered captives are an obvious source of danger, but even tame ones can be dangerous. Good feeders, when they have the scent of food, will strike and seize anything that moves. Instances have occured when keepers, working alone, have been killed by tame pythons.

Most members of this subfamily eat warm-blooded prey. Large specimens are capable of swallowing quite large animals but normally they can be sustained on a diet of rats. Of course, it requires many rats to satisfy the needs of a five or seven meter snake. Before deciding to keep such a specimen, a reliable and constant source of food must be easily available. It is also noted that these snakes have the the potential to grow to a great length.

It is generally unwise for a private collector to keep *Pythonidae* that can attain considerable size for the above reasons. Such animals inevitably outgrow the resources of most private collectors and must be disposed of. Most zoos in Europe, Japan, and the United States possess an over-abundance as they often receive unwanted specimens. Due to this, many zoos no longer accept them.

Genus Python

Three members of the genus exist in Thailand. All of them have enlarged scales and sensory pits on their heads. In each case, the upper and lower jaws have teeth, the anterior teeth being long and curved towards the throat.

Python curtus brongersmai   Plate 10
Blood Python
(Short-tailed Python)   งูหลามปากเป็ด

The head of the Blood Python is distinct from its neck. The eyes are somewhat small and the pupil is elliptic. The striking thing is that the body is very thick when compared to the head. The tail is quite short and pointed. The color is variable, ranging from grayish-brown, orangish-brown, pinkish-brown to even bright red. The bright red examples are especially beautiful with a series of spots, sometimes merged, occuring on the body. The spot color varies and may be tan, cream, or yellow. On the neck, head and tail the spots are replaced by tapered lines. The head and tail of red specimens tend to be black with a pale ventral surface and possibly scattered dark blotches. Some may reach three meters in length with the average about two meters. Once established and eating, this species does well in captivity. One captive specimen lived for more than 27 years.

    The scales are smooth, and arranged in 53, 55, or 57 rows at mid-body. The ventrals range from 160 to 175. The subcaudals vary from 26 to 32 and are paired, and the anal is single. There are two preoculars, ten small loreals, and two postoculars. The upper preocular is the larger of the two. The rostral has two transverse pits. Of the ten supralabials, the first and second have deep pits and the fifth and sixth contact the eye. There are 19 infralabials with the second to the fourth and the tenth to fourteenth having shallow pits.

The Blood Python is the most aquatic of the three Thai pythons. It is normally found close to river banks in forested areas. Often it enters the water and spends many hours soaking.

It is nocturnal when hunting its prey of primarily rodents and aquatic birds. Due to its relatively smaller size it does not require a cage as large as is necessary for the other two Thai members of the genus. A cage of 3 x 2 meters is adequate which should include a large receptacle for soaking. Blood Pythons generally do well in captivity, but do have a nasty disposition and rarely become tame.

This species is oviparous with the female producing 10 to 15 eggs in each clutch. The female incubates and the young emerge after approximately 70 days. The first shedding occurs around ten days after hatching when the baby is ready for its first meal.

This species is the least common of the *Pythonidae* of Thailand and found only in the South. It also exists in Burma, Kampuchea, Vietnam, and Malaysia with other subspecies found in Borneo and Sumatra.

Python molurus bivittatus Plates 11 to 15
Burmese Python งูหลาม

This snake, with *Python molurus molurus* of India, is the most gentle and easily tamed of the giant pythons. Consequently, it is the python most often seen in movies and at snake exhibits where tourists are often photographed holding them. Some adult specimens may reach seven meters, but the average is closer to 4.5 meters. The snake is large, heavy bodied and the head distinct from the neck. The eyes are moderate in size with elliptic pupils. The basic ground color of yellowish to light brown with black-edged, dark brown blotches superimposed, creates an attractive snake. On top of the head is an arrowhead mark with the point towards the snout. To each side of the head is a dark brown, black-edged line that passes the eye and the base of the jaw before entering the neck. The ventral surface of white or cream extends to the sides.

The scales are smooth and arranged in rows from 61 to 75 at mid-body. The ventrals number 242 to 265 and the paired subcaudals 58 to 83. The single anal plate is flanked by a pair of spurs. There are two preoculars, two postoculars, and four loreals. The rostral has two deep diagonal pits. The 12 supralabials do not contact the eye. The first two have distinct pits. There are 20 infralabials; the fourteenth to the eighteenth have shallow sensory pits.

This is a forest snake and is most often found in open forest, wooded grassland, and low forest near bases of mountains. In contrast to *Python reticulatus*, it is not seen in, or, near metropolitan areas. Food consists of warm-blooded animals, for example, birds and rodents. Females incubate after laying 30 to 58 eggs per clutch and are capable of maintaining a body temperature of 7.3° C above the ambient temperature for long periods. Incubation lasts approximately 50 to 60 days. Hatchlings measure 0.5 to 0.55 m and are ready for food after the first shedding. This snake is sexually mature after two to three years and approximately two meters in length. Females are the larger.

In 1979, five albino Burmese Pythons were found in the Northeast. Ages were estimated to be between nine months and one year, eyes were pink, the ground color white with a body pattern of beautiful golden yellow. Four, three males and one female, became the property of Siam Farm, Bangkok. The fifth was acquired by Bangkok Zoo. In February 1983, two males and a female, were stolen from Siam Farm. The one at Bangkok Zoo also disappeared. The remaining male was moved from Siam Farm to Pata Zoo in Bangkok, where it mated with a normal female which produced 42 babies in June 1983. None were albinos, but each carries the recessive gene. Later in 1983, the stolen female was recovered and mated with the male. In May 1985, 35 hatchlings were born. The first hatched after 59 days of incubation by the mother; the last on the 64th day. The babies averaged 0.55 m and each was a true albino, quite similar in appearance to the parents. Only a minor difference could be observed. Markings were orangish-yellow, rather than the parent's golden yellow, but on maturity the markings changed to golden yellow. Ten months after birth they averaged 1.45 m. These beautiful albinos are now known as the Golden Thai Python.

Python molurus bivittatus is found in each province except those south of Chumphon. Its range also includes southern China, Java and the remainder of mainland Southeast Asia, except peninsular Malaysia.

Python reticulatus Plates 16 to 18
Reticulated Python งูเหลือม

This is the second largest snake in the world, exceeded in length only by the Anaconda, *Eunectes murinas*, of South America. The Reticulated Python has been reported as exceeding ten meters and the author has seen one dead at a length of a few centimeters less than ten meters. Specimens measuring 7.5 m are not uncommon.

The Reticulated Python is an attractive snake with a head distinct from the neck and relatively stout body, though less so than that of *Python molurus bivittatus*. The tail is long and tapered. The body is tan or yellowish-tan with an intricate series of reticulating black lines extending along the top of the body. These lines extend down the sides where they are wider and contain white spots. The head is tan or yellowish-tan with a black line extending from the snout over the top of the head to the neck. Black lines also extend from the eye to the base of the jaw. The eyes are yellow and the pupil is elliptic. The white belly has black marks along the flanks that intrude slightly upon the body sides. Colors and pattern lose some of their brilliance with age, and older specimens are generally darker than the younger ones. This snake has an iridescent quality which is very noticeable after shedding.

The scales are smooth, and the scale rows number from 69 to 79. Ventrals vary from 297 to 332 and the subcaudals from 75 to 102. Most subcaudals are paired. The anal plate is single and flanked by spurs which are considerably larger on males than on females. There are two preoculars, the upper being the larger, three postoculars, and three to four loreals. The rostral has two large diagonal sensory pits.

There are 13 or 14 supralabials, the first to fourth each having a sensory pit. The sixth or seventh borders the eye. Of the 23 infralabials, the second to fourth or fifth have small sensory pits.

Primarily a nocturnal animal, *Python reticulatus* is often found near human habitation. Large specimens are still caught within the city limits of Bangkok. Prey consists of rats, rabbits, small deer, small jungle cats, pigs, domestic cats, and dogs with most ingestable, warm-blooded animals being acceptable. This species usually has an unpleasant disposition and is quick to bite, which with their great size and strength, make them dangerous display animals.

The Reticulated Python is oviparous and quite prolific, females producing up to 100 eggs in one clutch. Females coil around their eggs until they hatch but, unlike *Python molurus bivittatus*, they do not have the ability to increase body temperature. Hatchlings average 0.6 to 0.75 meters in length and begin to emerge after approximately 90 days of incubation. The *Python reticulatus* becomes sexually mature later than previously discussed species, apparently due to its size. It seems that specimens must be five years of age and approximately four meters in length before being able to reproduce. When mature, females are larger than males.

The Reticulated Python is found throughout Thailand, and Southeast Asia with the exception of eastern Indonesia.

Superfamily Colubroidea

This extremely large superfamily includes many species and subspecies that remain to be thoroughly studied. Hence, the status of many of its members is in dispute. Some speculate that any single characteristic may not distinguish all *Colubroidea*. Nevertheless, the following generalizations can be made. Both pelvic vestiges and cornoid bones are absent. Additionally, the left lung is either a mere vestige or totally absent. Except for a few vipers, only the left carotid artery is present. When a right carotid artery is present it is much smaller. The skull is very flexible and tooth-bearing bones of the upper jaws and palate move freely.

FAMILY COLUBRIDAE

Some, but by no means all, produce a toxic fluid which aids in prey capture. In all there is a mucus-secreting gland, the superior labial gland. This is situated in the upper jaw and extends from the snout to behind the mouth angle with a similar gland also adjacent in the lower jaw. The primary function of these is to produce mucus to lubricate prey and ease passage into the digestive tract. In some genera, *Xenodermus* for example, serous cells are present in the superior labial gland. These produce a toxic fluid that allows the snake to subdue prey. Furthermore, in some genera the posterior portion of the superior labial gland is distinctly enlarged and encapsulated, forming a distinct gland - Duvernoy's gland. The secretions from Duvernoy's gland are toxic and more concentrated than those of serous cells and are a more effective aid in subduing prey. Duvernoy's gland is present in many genera thought of as "mildly venomous," but is also present in some unlikely genera, such as *Oligodon*. Some genera, *Sibynophis* for example, have both serous cells and Duvernoy's gland. If fangs are present, all of the maxillary teeth are anterior to them (Opistoglypha, Figure 7). Fangs will be smooth or grooved, but not hollow.

Primitive features such as premaxillary teeth, coronoid bones, pelvic elements, and a left lung are absent. The skeletal structure of the skull is flexible in most members.

Subfamily Calamariinae

An unusual arrangement of the frontal and sphenoid bones near the orbit distinguishes this subfamily. In addition, the hemipenis lacks spines and is either shallowly bilobate or simple. The nasal and prefrontal bones are in contact, causing suspension of the snout. Genera *Calamaria* and *Pseudorabdion* are the only examples of this subfamily found in Thailand.

Genus Calamaria

Although they are not classified as primitive, the burrowing members of this genus do retain some "primitive" characteristics, for example, a shortening of the supratemporal and quadrate bones in addition to the consolidation of other skull bones. These are obvious advantages to animals whose lives are spent burrowing through soil.

Subfamily Calamariinae

These are small, 0.3 to 0.4 m burrowing snakes and generally found in wooded hill country under stones or among decaying vegetation. Their diet consists primarily of worms and insects. Duvernoy's gland is present, but these animals are not a threat to mankind. Members of this genus are oviparous and adult females are significantly larger than adult males.

In this genus, the number of head scales is greatly reduced. The loreals and internasals are absent, the nasal is single, and the last supralabial is in contact with the parietal. The eyes have round pupils, and the head is indistinct from the neck. Scales are smooth and arranged in rows of 13 at mid-body, subcaudals are paired, and the anal plate is single.

Snakes of this genus are quite easy to maintain in captivity. Procedures for keeping are the same as for other burrowing forms. That is, the floor of the aquarium or cage should be covered with a few centimeters of loose soil and bark pieces, with other forms of vegetation, or stones strewn on the surface. A water dish is required with the rim just above soil level. Spray with tepid water often to keep the soil moist but not muddy. Before attempting to keep members of this genus a reliable food supply is required. Invertebrates are difficult to obtain in areas with severe winters. Food animals should always be available in their cages.

Calamaria lumbricoidea Figure 12
Variable Reed Snake งูพงอ้อหลากลาย

This species is dark brown to black with a ventral surface of yellowish-white crossed by irregular dark bands. Its head is mostly or entirely yellowish-white, and the tail is pointed.

Five supralabials are present, with the third and fourth touching the eye. This species has one preocular and one postocular. Ventrals number from 147 to 210 and the subcaudals from 15 to 26. The Variable Reed Snake may reach 0.4 m in length.

This species is found in the South with a range extending into Malaysia, western Indonesia, and the Philippines.

Figure 12. Heads of Reed Snakes.

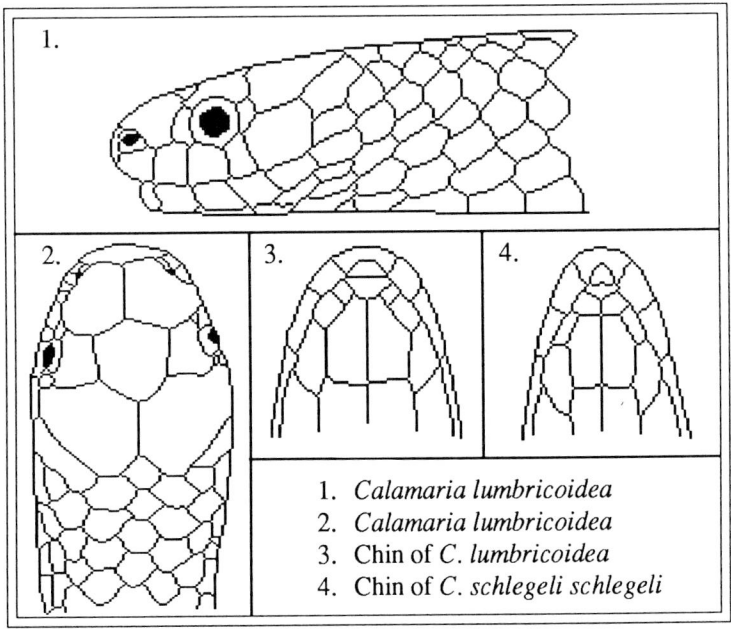

1. *Calamaria lumbricoidea*
2. *Calamaria lumbricoidea*
3. Chin of *C. lumbricoidea*
4. Chin of *C. schlegeli schlegeli*

Calamaria pavimentata
Collared Reed Snake

งูพงอ้อท้องเหลือง

This species is reddish-brown with a number of small black spots on the sides. A black band encircles the neck with a yellow ring immediately behind the black. The head is dark while the chin, throat, and belly are white. The tail is pointed; adults of the species average 0.35 m in length.

There is one preocular and a single postocular. Five infralabials are present and of the four supralabials, the second and third touch the eye. Ventrals number from 133 to 168 and the subcaudals from 13 to 20.

This species is found in hilly country throughout Thailand and is also reported on the remainder of mainland Southeast Asia, Taiwan and the Ryukyu Islands of Japan. It may also occur in western Indonesia and India.

Calamaria schlegeli schlegeli
White-headed Reed Snake

Figure 12

งูพงอ้อหัวขาว

The head is black with a white temporal area. The body is iridescent black with a thin white collar encircling the neck. The chin, neck, and belly are white and the subcaudal area traversed by dark bars. The tail is pointed. It is quite a large member of the genus, with some reaching 0.45 meters.

The ventrals vary from 134 to 149, the subcaudals from 24 to 37. There is one preocular and one postocular which touches the fourth and fifth supralabials. The third and fourth supralabials contact the eye. There are five infralabials; with the first in contact behind the mental scale.

This snake has only been found in the southern province of Pattani. Its range includes peninsular Malaysia, Singapore, and western Indonesia.

Genus Pseudorabdion [2]

This genus consists of small burrowing snakes that are usually found under stones or fallen logs. Bodies are almost cylindrical, and heads are not distinct. They prey upon worms, insect larvae, frogs, and lizards. Only one member of this oviparous genus, *Pseudorabdion longiceps,* is native to Thailand.

Pseudorabdion longiceps are best kept in a small aquarium with several centimeters of loose soil on the floor. Items under which specimens can hide should be placed on the surface and a few plants added. A water dish set in the soil to the rim is required. Spray occasionally throughout with tepid water.

Pseudorabdion longiceps　　　　　　　　Figure 13
Dwarf Reed Snake　　　　　　　　　　งูพงอ้อเล็ก
　　　　　　　　　　　　　　　　　　งูพงอ้อหัวยาว

This small snake has a wide range of colors. It is iridescent and may be uniformly gray, brownish-red, brown, or black. Normally a white or yellowish collar is around the neck, and two yellow spots on the back of the neck. The Dwarf Reed Snake is well named as it rarely exceeds 0.23 m in length.

---

2. This genus was originally described as *Pseudorabdion* and, according to the rules of the International Code of Zoological Nomenclature, the original spelling must be retained. Nonetheless, this genus is often presented as *Pseudorhabdion*.

Figure 13. *Pseudorabdion longiceps.*

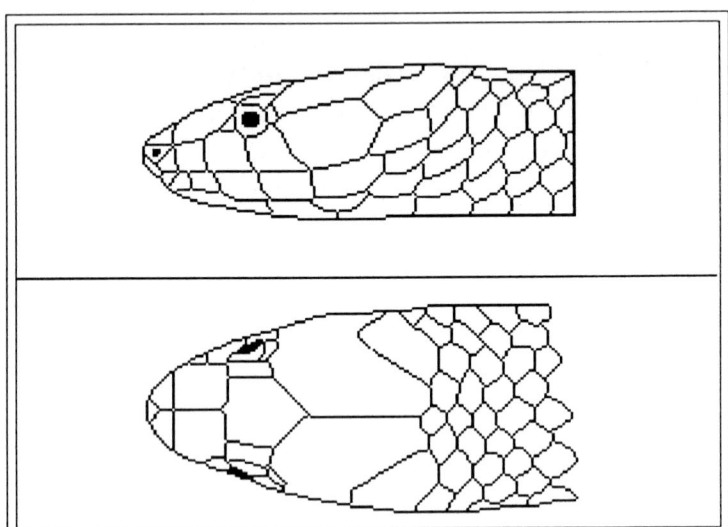

The single preocular is small; the lone postocular is in contact with the fourth and fifth supralabials. The loreal is not present. There are six supralabials, the fifth being the largest, the fourth contacting the eye. There are also six infralabials. The scales are smooth and in rows of 15 at midbody. The ventrals number from 129 to 148; the subcaudals are paired and number from 16 to 31. The anal is single.

The Dwarf Reed Snake is found in the two southernmost provinces of Narathiwat and Yala. Its range continues throughout peninsular Malaysia and Singapore into the Philippines, Borneo, and western Indonesia.

Subfamily Colubrinae

This subfamily has lost the primitive features of premaxillary teeth, the coronoid bone, pelvic vestiges, and the left lung. The bones of the skull are extremely flexible and most move freely on the braincase. Some have developed grooved fangs on the posterior end of the maxillary bone as well as Duvernoy's gland in the supralabial gland.

Eleven genera are found in Thailand.

Genus Ahaetulla

This genus includes long, very slender rear-fanged snakes (Opistoglyphs. Figure 7). All are arboreal, and serious consequences from their bites have not been reported. Heads are elongated and distinct from the neck, and their snouts sharply pointed. The distinctive eyes are horizontally oval, and the pupils laterally elliptic.

Members of this genus generally do not do well in captivity; however, success in their husbandry can be quite rewarding. They are usually not aggressive and make an attractive display. Although not extremely large they should be provided with ample room. If confined within cramped quarters there is considerable risk of injury to their pointed snouts as they search for more space. As they are very arboreal, numerous branches for climbing are necessary. A water dish is also required, as well as frequent spraying of the branches and leaves. Most arboreal snakes obtain their water intake from drinking rain or dew directly from branches and leaves. Frequent spraying and the presence of a water dish will also help to maintain a desirable level of humidity. Prey consists of small animals which share the same arboreal environment, such as birds, tree frogs, and lizards. Members of this genus are viviparous, females being significantly larger than males. *Ahaetulla* are active during the day and make an attractive and interesting display.

Thailand is home to three species of this genus.

Ahaetulla mycterizans   Plate 19
Malayan Green Whip Snake   งูเขียวหัวจิ้งจกมลายู

This is a grayish-green snake with a white line extending along the outer edge of the ventral scales. It is a snake of medium length, averaging one meter.

There is one preocular which contacts the frontal, two postoculars, and three or four small loreals. There are eight or nine supralabials, with the fourth and fifth or the fourth, fifth, and sixth touching the eye. Scales are smooth and in 15 rows at mid-body. Ventrals number 186 to 195, the paired subcaudals 132 to 156, and the anal plate single. This species is very similar to *Ahaetulla prasina* but has an entire anal plate and fewer ventrals and subcaudals.

The Malayan Green Whip Snake occurs only in the South, its range extending through Malaysia into Java.

Ahaetulla nasuta Plate 20
Long-nosed Whip Snake งูเขียวปากแหนบ

This particular snake is unique among native members of the genus as it has a relatively long, pointed appendage protruding from the rostral. The body is usually a bright green and the tail is reddish. The sides of the head are yellow.

The scales are smooth and in 15 rows at mid-body. Ventrals number from 135 to 207, the paired subcaudals from 156 to 180. The anal is divided. The internasals and prefrontals contact the supralabials. The eye is distinct with a horizontal pupil and yellow iris. The supraocular shield forms a ridge above the eye. There is one preocular and two postoculars, the upper being the larger. There are no loreals. Supralabials number eight, with the fourth and fifth contacting the eye. There are nine or ten infralabials.

Males average 1.3 m in length, females 1.9 meters. *Ahaetulla nasuta* is normally found in bushes or on trees. When seeking prey, it slowly moves its body back and forth, like a twig being moved by the wind. Females bear up to 23 young at a time. The babies are light yellow, light gray, or pale green at birth and approximately 0.3 m in length.

This species is found in the North, Northeast, the Central Region, and in the South as far as Phuket Province. It is also native to India, Sri Lanka, and Burma.

Ahaetulla prasina Figure 14
Oriental Whip Snake Plates 21 and 22
งูเขียวหัวจิ้งจก. งูง่วงกลางดง

The head is long and the snout pointed, but it lacks the appendage present in *Ahaetulla nasuta*. The head is quite distinct from the neck, and the body is long and slender. The color is pale to dark green, and the ventral surface pale green. The eyes are quite distinct as they are large and the pupils horizontal. The supraoculars form a ridge above each eye. Although usually green, the color apparently varies depending upon the geography of the native habitat. For instance, those in the Central and Northeast regions are usually gray. Some found in Kanchanaburi, in the West, are orange or yellow. A female of this species gives birth to seven to ten babies each time. The new born snakes are the same color as the mother and average 0.24 m at birth.

The scale rows number 15 at mid-body and the scales are smooth. One large preocular is present, along with two postoculars and two small loreals. There are eight or nine supralabials with the fourth, fifth, and sixth bordering the eye. Infralabials number eight. Ventrals vary from 194 to 235, the subcaudals are paired and range from 151 to 187. The anal is divided. The skin to which the scales are attached is both black and white. When *Ahaetulla prasina* is provoked into a defensive position the neck inflates exposing this black and white skin. The result is an apparently larger, more fearsome animal which might give a potential predator second thoughts. This snake averages 1.2 m in length.

This species is widely distributed throughout Thailand. It is also found in other parts of Southeast Asia, and India.

Figure 14. *Ahaetulla prasina*.

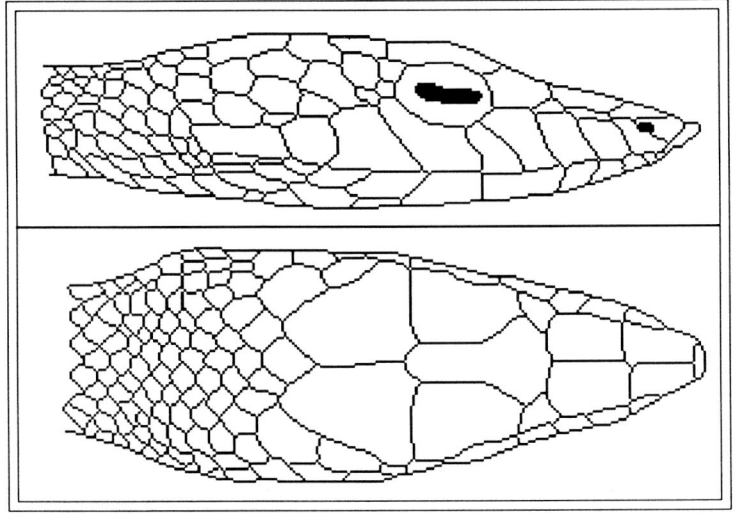

Genus Boiga

This genus contains medium-sized to large arboreal snakes. Members have a large, rounded head that is distinct from the slender neck. Their eyes are large and vertically elliptical, but dilate at night when hunting. The internasals, prefrontals, frontals, supraoculars, and parietals are large. There are two postoculars and seven to ten supralabials. The body scales are smooth and arranged in rows of 17 to 31 at mid-body. The subcaudals are paired. Bodies are generally long and slender. Each genus is oviparous with females larger than males. Thailand is the home to nine members of this genus.

Since these snakes are primarily tree dwellers, cages are to be equipped with something on which to climb. Branches will suffice, but live plants make for a more attractive display and create a more natural environment.

As many species, especially the young, drink rain water from leaves, plants should be periodically sprayed with water for the benefit of both the snake and the plants. Cages of large horizontal and vertical dimensions are necessary for the larger members of this genus. Do not house *Boiga* with other snakes as they have been known to eat other snakes, as well as birds, eggs, lizards, rats and mice. They envenomate by biting and chewing. Large, grooved fangs located at the rear of the upper jaw conduct venom into the wounds. These are rear-fanged snakes (Opistoglyphs), but their dispositions are generally mild.

Boiga cyanea Plate 23
Green Cat-eye Snake
(Green Cat Snake) งูเขียวดง

This species is uniformly green above and greenish-white on the ventral surface. The chin is bluish-white. The long, thin body is vertically compressed. This species averages 1.5 m in length at maturity, but some may reach two meters. Juveniles are brownish-yellow with a brown pattern. Clutches usually contain from 9 to 13 eggs.

The scales are smooth and arranged in rows of 21 at mid-body. Ventrals vary from 237 to 257, the paired subcaudals from 144 to 158, and the anal single. A single, large preocular reaches the top of the head with one loreal and two postoculars. There are 11 or 12 infralabials and 8 supralabials, with the fourth to the sixth bordering the eye.

This species frequents forests near mountains with numerous waterways and has been found at elevations of 1875 meters. It inhabits every province in Thailand and is also found in India, China, and mainland Southeast Asia.

Plate 1.
Ramphotyphlops braminus
Common Blind Snake
Flower Pot Snake
งูดินธรรมดา

This small species is quite common and widely distributed but rarely seen as it is a devout burrower. The shielded eyes, cylindrical body, smooth scales, and spine at the end of the tail are excellent adaptations for a subterranean existence.

Plate 2. *Typhlops lineatus*. Striped Blind Snake. งูดินลายขีด. (Rarely photographed). One of the larger members of the genus. Range southern Thailand.

Photo: Dr. Lim Boo Liat

Plate 3. *Acrochordus granulatus*. File Snake. Granular Snake. งูผ้าขี้ริ้ว. Less common in Thailand than the Elephant Trunk Snake.

Photo: Dr. Lim Boo Liat

Photo: J. P. Kleinman

Photo: J. P. Kleinman

Plates 4 and 5.
Acrochordus javanicus
Elephant Trunk Snake
Wart Snake. งูงวงช้าง

Plate 4. (top)
Heavy-bodied, rough skin. A marine animal, nearly helpless on land as lacking ventral scales.

Plate 5. (center)
The head is covered with many small scales rather than plates. Both eyes and nostrils are prominent from above and nostrils are located on snout front - useful adaptations for marine existence.

Plate 6. *Cylindrophis rufus rufus*. Red-tailed Pipe Snake.
Two-headed Snake. งูก้นขบ. งูสองหัว.
Secretive, burrowing snake. When disturbed, tail is raised to display red subcaudals to distract attention from the head.

Photo: Suthigit Patramangorn

Plates 125

Plate 7. *Cylindrophis rufus rufus*. Red-tailed Pipe Snake.
Two-headed Snake. งูก้นขบ. งูสองหัว. Brown variety.
Usually black, brown sometimes found. Compare with Plate 6.

Plates 8 and 9. *Xenopeltis unicolor*. Iridescent Earth Snake. Sunbeam Snake.
งูแสงอาทิตย์. A burrowing species common and widespread in
Thailand. Skin colorful; docile and thrives in captivity.

Plate 8.

Specimen freshly killed.

Plate 9. (below) White collar, prominent on juveniles, disappears on maturity.

Photo: Jarujin Nabhitabhata

Photo: Jarujin Nabhitabhata

Plate 10. *Python curtus brongersmai*. Short-tailed Python.
Blood Python. งูหลามปากเป็ด. (Red Phase).
Smallest of genus *Python* in Thailand. Body quite thick. Several colors. Nasty disposition but thrives in captivity.

Plates 11 and 12. *Python molurus bivittatus*. Burmese Python. งูหลาม.
Plate 11. Albino male (smaller) and normal color female breeding.

Photo: Preecha Varavichit

Plate 12. An albino. First found in Thailand. Now being bred in captivity. Pairs produce only albino offspring. Juveniles have orangish-yellow pattern which becomes golden yellow when mature.

Plates   127

Plates 13, 14 and 15.
Python molurus bivittatus. Burmese Python. งูหลาม

Plate 13. (top)
Color mutation; some species occasionally found in Thailand. This beautiful snake was 3 m when caught in north Thailand. Now in English zoo.

Photo: Suthigit Patramangorn

Plate 14.
A color mutation - most of the darker pigment missing. This yearling was found in the northeast of Thailand.

Plate 15. (below)
A mutation, lighter, with normal specimen.
Left: Normal
Right: Mutation

128   *Plates*

Plates 16 to 18. *Python reticulatus*. Reticulated Python. งูเหลือม.

Photo: Suthigit Patramangorn

Plate 16. (top)
1 year old. Attractive. Thrives in captivity if kept well.

Plate 17. (center)
Female laying eggs.
Females of this genus fast, and coil around eggs throughout incubation of about 90 days.

Plate 18. (bottom).
Hatchling. 1 of 26 hatched June 1987. Some heads out after 88 days, finally left shell after 4 - 5 days.

Av. length 0. 65 m.
1st shed and food, adult mouse, 11 - 12 days later.
Those refusing food released.

Plate 19.

Ahaetulla mycterizans
Malayan Green Whip Snake
งูเขียวหัวจิ้งจกมลายู

Least common of native Thai genus members. Differs from *Ahaetulla nasuta* - lacks rostral appendage; differs from *Ahaetulla prasina* - has fewer ventrals.

Photo: Dr. Lim Boo Liat

Plate 20.

Ahaetulla nasuta
Long-nosed Whip Snake
งูเขียวปากแหนบ

Attractive, unusual, arboreal snake. Rostral appendage and horizontal elliptical pupils. Fascinating head features. Thin body, ventral surface attractive green.

Photo: Suthigit Patramangorn

Plate 21.

Ahaetulla prasina
(Gray-brown)
Oriental Whip Snake
งูเขียวหัวจิ้งจก. งูง่วงกลางดง.
Usually green, but some gray-brown. Plate shows defensive position - neck inflated.
Note: black interstitial skin and horizontal elliptical pupil.

130   *Plates*

Photo: Jarujin Nabhitabhata

Plate 22.

Ahaetulla prasina
Oriental Whip Snake
งูเขียวหัวจิ้งจก
งูง่วงกลางดง

Yellow phase.

Plate 23.

Boiga cyanea
Green Cat-eye Snake
Green Cat Snake
งูเขียวดง

Arboreal snake found throughout Thailand.

Photo: Suthigit Patramangorn

Plate 24.

Boiga cynodon
Dog-toothed Cat Snake
Horse-tail Whip Snake
งูแส้หางม้า. งูกะปิ. งูกินไข่

Largest of Thai *Boiga*. Probably the most gentle. Note: typical *Boiga* head - the vertebral ridge is extremely prominent.

Boiga cynodon Plates 24 to 26
Dog-toothed Cat Snake งูแส้หางม้า
(Horse-tail Whip Snake) งูกะปี. งูกินไข่

This is a large snake. A length of 2.765 m has been recorded with the average probably around 1.8 meters. The head is large and round; its large eyes protrude. The body is long, slender, and vertically compressed. The tail is especially long and slender with a texture similar to a horse's tail. Hence, one of the Thai common names for this species, Horse-tail Whip Snake. The front maxillary teeth are enlarged and similar to the canine teeth of a dog, resulting in another common name, Dog-toothed Cat Snake. *Boiga cynodon* also has large, grooved rear fangs.

The Dog-toothed Cat Snake has variable patterns. The majority have a brown head and a yellow or creamish-brown jaw. A black stripe extends from behind the eye to the base of the jaw. The body is yellowish in color with a variable pattern of dark brown or black cross-bars that become closer posteriorly until the tail is virtually a solid color. The ventrals are dull gray-brown and the subcaudals black. The color and patterns of infants are similar to those of adults.

Scales are smooth and arranged in rows of 23 or 25 at mid-body; the vertebral scales are enlarged. There are eight or nine supralabials, with the fourth, fifth, and sixth in contact with the eye, and 13 or 14 infralabials. The ventral scales number 262 to 289, the paired subcaudals 125 to 163. The anal is single. There is one large preocular, one loreal, and two postoculars. The upper is the larger and touches the parietal.

This is a large, gentle snake which seems reluctant to bite, although it will if handled roughly. Persons who have been bitten have experienced slight pain and swelling in the area of the bite. It is a nocturnal hunter and preys upon a wide variety of animals including birds and small mammals. Although an animal which spends a lot of time in trees, *Boiga cynodon* will descend to hunt for food. It seems to be fond of eggs and has been found frequently near chicken coops, where it eats chicks, ducklings, and eggs.

Boiga cynodon should be provided with a large cage equipped with branches as well as a hiding box.

This species is found in the South from the province of Prachuap Khiri Khan to the Malaysian border. It is also found in Malaysia, Indonesia and the Philippines.

A melanistic form with the additional common name of Black Dog-toothed Cat Snake frequently occurs in the South. This differs from the more common form described earlier by being much darker. Its head has the same shape and pattern, but is darker. The body is dark brown with a pattern of irregular black cross-bars. The tail is solid black. Irregular yellow blotches occur randomly on the dorsal surface and they intrude, but less frequently, onto the gray ventral surface. This form is also found in Malaysia and Singapore.

Boiga dendrophila melanota Figure 15
Mangrove Snake Plate 27
งูปล้องทอง

This attractive snake is often seen in collections in Europe, Japan, and the United States of America. It is slightly more heavily bodied than other members of the genus and quite long, often attaining a length in excess of 1.7 meters. Its head is large, round, and very distinct from the neck. The body is slightly compressed vertically. Scales are smooth and glossy, making this snake's black and yellow colors very attractive. The eyes are large and prominent and the pupils vertically elliptic. The head is black; the supralabials are yellow, but separated by intrusions of black between each labial scale. The body is also black but crossed by a series of yellow bands which extend into the yellow ventral surface. The tongue is red.

Mangrove Snakes have eight supralabials; the third, fourth, and fifth border the eye. The sixth and seventh are the largest. There are 11 infralabials. The internasals are much smaller than the prefrontals, and there are two rather small postoculars, as well as a preocular and a loreal. The scale rows number 21 or 23 at mid-body, the ventrals number from 209 to 239, and the paired subcaudals from 89 to 110. The anal is single.

Mangrove Snakes are fond of humid forests and are often found in branches near or overhanging water. They are nocturnal with a diet consisting of a variety of animals, including small rodents, birds, lizards, fish, and other snakes. When first approached, they display a fearsome defense by flattening their necks vertically and repeatedly striking.

Figure 15. *Boiga dendrophila melonata*.

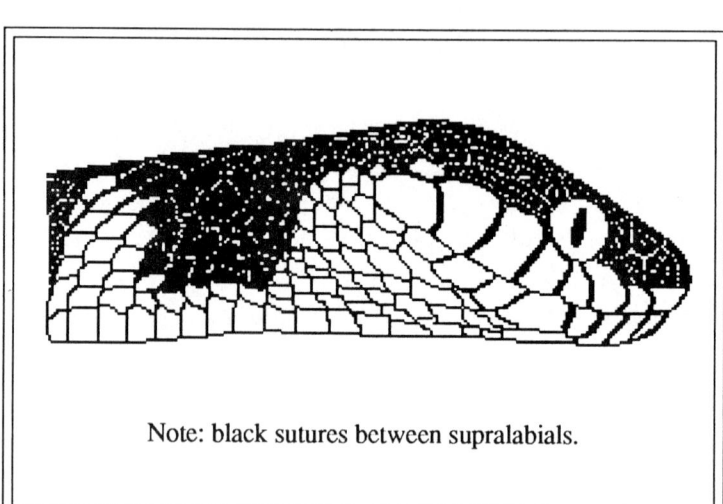

Note: black sutures between supralabials.

To date, serious consequences to man from the bite of Mangrove Snake have not been reported. Their venom is effective against their natural prey, however. In addition to venom, constriction is also used to subdue prey. They are oviparous, laying 4 to 15 eggs at a time. The cylindrical eggs are quite large, measuring approximately 30 x 50 mm.

Such snakes are comfortably housed in a large cage with numerous branches. A hiding box should also be made available, at least initially. A high-level of humidity should be maintained as this species seems to have special difficulty in shedding when the humidity is low. Do not house this snake with others as the Mangrove Snake will readily eat them.

Boiga dendrophila melanota is abundant in the area south of Chumphon Province. Its range continues through Malaysia into Indonesia.

Boiga drapiezii Plates 28 and 29
White-spotted Cat Snake งูดงคาทอง

Boiga drapiezii is reported to reach a length of 2.1 m but the average is less. The head is very distinct from the neck, and the body is vertically compressed. There are two color phases and both are unusually attractive. The first is predominantly green. The head is dark green with areas of pink and white on the supralabials. The infralabials, chin, and throat are dirty-white. The body has very irregular bands of dark green, light green, tan, and pink. The pink bands become wider and more dominant lower on the body, close to the ventral surface with moderately large, irregular pink spots in the same area. The ventral surface is cream and mottled with numerous small green and pink specks. A thin, broken dark green line extends along each flank of the ventral surface. The posterior edge of each ventral scale is black. This pattern continues into the subcaudals when the green specks become blotches and green becomes the dominant color.

The second color phase is predominantly brown. The head is dark brown, but the supralabials, infralabials, chin, and throat are white. The body has alternating dark and light brown bands, the light brown bands being widest near the vertebral ridge and narrowest low on the sides of the body. The ventral and subcaudal surfaces are light brown. Thin, broken dark brown lines extend along the flanks of the ventral surface. The edges of the ventral scales are light in color.

The preocular of this species is particularly large. There are two postoculars, the uppermost being the larger, and one small loreal. Of the eight supralabials, the third, fourth, and fifth touch the eye. There are 11 infralabials. Scales are arranged in rows of 19 at mid-body and the vertebral row is enlarged. Ventrals number from 250 to 279 and the paired subcaudals from 114 to 168. The anal is single.

This is a species which is difficult to maintain in captivity for several reasons. They often refuse to eat, have difficulty shedding, and often suffer from dehydration. Although a fairly long snake, they are also quite slender, so it seems unlikely they consume large prey. Lizards are their natural prey, but they might be induced to eat other prey, such as tree frogs, very small birds, bird's eggs or other snakes. The problems of shedding and dehydration may be solved by the presence of a large water receptacle and by frequent spraying of the plants. The combined effect will maintain a high-level of humidity and provide droplets of water on the plant leaves and the snake's body which it may drink. Refer to: Appendix 7. Observe the captive closely and if you do not see it drink, lead it to water. Gently place its head in the water receptacle and it will probably drink. This species is oviparous, the females laying up to ten cylindrical eggs per clutch. The eggs hatch in about 90 days; hatchlings are approximately 0.3 m in length and closely resemble their parents.

In Thailand, *Boiga drapiezii* is found in the South, from Pattani Province, south to the Malaysian border. Its range continues throughout Malaysia and Singapore into western Indonesia.

Boiga jaspidea   Plate 30
Jasper Cat Snake   งูดง

This snake is not often seen outside Southeast Asia. An adult attains a length of approximately 0.9 meters. The eyes are quite large; the head is distinct from the neck. It is a rather thin snake compared to the other members of the genus, the body being vertically compressed. The body is reddish-brown with a series of irregular black blotches throughout and a series of thin pinkish-white lines on the vertebral ridge. On the body sides, near to the ventral margin, are white or yellow dots with black bases. These bases often intrude onto the ventral edges. The head is brown with a pattern of white-edged black spots. The labials are white with black spots. The chin and the underside of the head are white. The throat and one third of the ventral surface are yellow. There is a gradual transition and then the remainder of the ventral surface is light brown.

Of the eight supralabials, the third, fourth, and fifth border the eye. There are 12 infralabials. The ventrals number from 243 to 267 and the divided subcaudals from 140 to 166. The anal is also divided. Mid-body scales are in 21 rows, the vertebral row is slightly enlarged, and body scales smooth. One or two preoculars exist and of the two postoculars, the upper is the larger. There is one small loreal.

This snake preys primarily upon lizards but will accept mice occasionally. It is arboreal and found in humid, well-watered areas and should be housed in a humid cage.

Boiga jaspidea is found at elevations up to 800 m in the South from Phuket province to the Malaysian border. The range continues through Malaysia and Singapore into western Indonesia. It is rare everywhere within its range.

Boiga multomaculata [3]
Marble Cat Snake
(Many-spotted Cat Snake)
[Spotted Cat Snake]

Plate 31

งูแม่ตะงาวรังนก

This is one of the smallest members of the genus to be found in Thailand. Although it is reported to have attained the record length of 1.87 m the average length, on maturity, is one meter. The head is more triangular than most other members of the genus, but is distinct from the neck. The body is vertically compressed and slender. The gray-brown color of the body extends to the top of the head. A black line extends back from the large eye, with its elliptical pupil, to the base of the jaw. Two elongated, black-edged brown markings extend from the snout to the base of the head. A series of irregularly shaped, black-edged, brown markings cover the body. Those on the flanks are slightly smaller than those on the back. The ventral surface is gray-brown and peppered with small brown spots.

The scales are smooth and arranged in 19 scale rows at mid-body. There are eight supralabials with the third, fourth, and fifth touching the eye. Infralabials number 11. There is one preocular, one loreal, and two postoculars. Vertebral scales are enlarged. Ventrals number from 199 to 245, subcaudals from 78 to 109. The anal scale is single.

---

3. This species is often presented as *Boiga multimaculata*. However, it was first reported as *Boiga multomaculata*. The rulings of the International Code of Zoological Nomenclature state that an original name may not be changed for any reason, even for misspelling. For that reason, I have retained the original name for this species.

Females are reported to produce from five to seven eggs per clutch which require an incubation period of approximately 60 days.

This snake is rather docile and makes an interesting pet. It has the advantage of being small, therefore, it does not require a large cage. Conversely, it feeds primarily on lizards, which may present a problem during cold months. Reports state that this snake frequently enters houses in search of house lizards. It is arboreal and is to be provided with branches.

Boiga multomaculata is especially common in the forests of the Southeast but has been reported as far north as Chiang Mai. It has also been seen at sea level and at elevations up to 1000 meters. The species has been observed in India, Burma, Kampuchea, Laos, Vietnam, Malaysia, southern China, and western Indonesia.

Boiga nigriceps   Plate 32
Red Cat-eye Snake   งูต้องไฟ

This is one of the more beautiful members of the genus. Their ground color ranges from reddish-brown to intense red. The head is usually a dark brownish-red. The ventral surface is pink or dark yellow. Normally there is no pattern, although some specimens do have small black dots on the body sides. The labials are creamish-white. Juveniles are the same color as adults. Adults average 1.75 m in length.

The Red Cat-eye Snake possesses all of the general characteristics of the genus. The head is quite distinct from its neck, eyes are large with vertically elliptical pupils, and the body and tail long and slender. It is arboreal and oviparous.

The scales are smooth with those of the vertebral ridge enlarged. Scale rows number 21 at mid-body. the ventrals number 240 to 263, the subcaudals 140 to 154. The anal is single. There are eight supralabials, with the third, fourth, and fifth contacting the eye. There is one preocular, one loreal, and two postoculars.

In June 1988, a healthy, 21-year old man working at Siam Farm was bitten on the right thumb by a medium-sized *Boiga nigriceps*. He had been bitten several times by *Boiga dendrophila melanota* without any ill effects and, therefore, ignored this, his first bite from *Boiga nigriceps*. Within an hour he complained of pain and swelling in his thumb. He was taken to a hospital for treatment when the pain and swelling began to enter his lower arm. His upper arm became swollen before the symptoms of envenomation began to disappear. Red blotches, indicating subcutaneous hemorrhaging, remained on his arm long after the bite. Such a severe reaction to a bite from this species is rare, but this case should serve to remind us that any "mildly venomous" snake is potentially dangerous. Perhaps a "mildly venomous" snake does not exist.

This snake does well in captivity and makes for an attractive display animal. Due to its length, it requires a large cage, and its arboreal nature is to be accommodated. Its wide range of food preferences lessens the problem of obtaining food, as it eats lizards, birds, mice, and probably frogs and other snakes as well.

This is a snake of the South and has been found at elevations 800 m above sea level. It is found in the province of Surat Thani south to the Malaysian border with a range continuing through Malaysia into western Indonesia.

Boiga ocellata งูแส้หางม้าเทา
Gray Cat Snake งูแส้หางม้าเล็ก

This snake was once considered a color form of *Boiga cynodon*, but is now recognized as a separate species. The characteristics listed in Table 6 differentiate it from *Boiga cynodon*.

Table 6. Characteristics of *Boiga ocellata* and *Boiga cynodon*.

Boiga ocellata	*Boiga cynodon*
1. Parietal ridges do not meet	parietal ridges meet and form a medial crest approximately two-thirds down parietal
2. medial arm of the ectopterygoid extends forward of palatine-pterygoid junction	medial arm of the ectopterygoid lies posterior to the palatine-pterygoid junction
3. rostral square	rostral higher than wider
4. ventrals average 256.5	ventrals average 278.7
5. subcaudals average 122.1	subcaudals average 151.1
6. grayish-brown with light and dark cross-bars and numerous flank markings	yellowish with dark cross-bars or dark with light cross-bars
7. eight supralabials	nine supralabials
8. supralabials entering eye: 3, 4, & 5	supralabials entering eye: 4, 5, & 6
9. infralabials: 12-13	infralabials: 13-14

The Gray Cat Snake has the typical *Boiga* body form. The head is distinct from the neck, the eye is large and contains a vertical pupil. Mature length averages 1.3 m, but specimens of 1.67 m have been observed. The dorsum is light brown or gray and crossed by a series of 87 to 98 black or dark brown, narrow cross-bars that are most distinct anteriorly. Alternating dark and light spots extend along the sides, a dark-edged white spot often extending from below each dark spot onto the edges of the ventrals. The ventral surface is light brown or gray, although lighter anteriorly and darker posteriorly. The chin and throat are white, the head darker than the dorsal surface of the body. A dark streak extends from behind the eye and terminates behind the last supralabial. Two dark lines extend from the base of the head and merge with the first dark cross-bar.

The body scales are smooth and in 23 rows at midbody; those on the vertebral ridge are enlarged. The single loreal and preocular are large, the preocular extending to the upper surface of the head. Of the two postoculars, the upper is the larger. Eight supralabials are present and either 12 or 13 infralabials. The third, fourth, and fifth supralabials contact the eye, the seventh is the largest. Ventrals range from 247 to 270, the paired subcaudals from 116 to 129, and the anal plate is single. There are 13 maxillary teeth. The first ten are followed by a short space and then three enlarged grooved fangs, the first being the largest.

Boiga ocellata is found in every province except those of the South. It has been recorded in the provinces of Kanchanaburi, Ratchaburi, Chiang Mai, Loei, Nakhon Ratchasima, Chumphon, Tak, Chantaburi, and Trat. Also, it has been found in Vietnam, Kampuchea, Burma, and the Assam and Bengal regions of India.

Boiga saengsomi Plate 33
Banded Green Cat Snake งูเขียวดงงลาย

This species was originally described as *Boiga saengsomi* [4] and was then changed to *Boiga mahasomi* in later publications. However, under Article 23 (m) of the International Code of Zoological Nomenclature, the name first used in the description of a species is not to be changed arbitrarily. Therefore, *Boiga saengsomi* is the correct taxon for this species. *Boiga saengsomi* is a new species proposed by Wirot Nutphand. The Banded Green Cat Snake has the distinct characteristics of genus *Boiga*. The head is large, rounded and quite distinct from the neck with large eyes and an elliptic pupil. The body is long and moderately slender with predominant yellowish-green color, but the interstitial skin is black. There are small alternating yellow bands around the body approximately two centimeters apart. The scales on the tail are black with a central yellow dot. The head is olive, the supralabials yellow, and the black interstitial skin is visible between each scale. The chin and throat are white but the ventral surface of the body is yellowish-white.

The dorsal scale rows number 21 at mid-body. They are smooth and generally uniform in size, but the vertebral scales are enlarged posteriorly. There are eight supralabials, with the third, fourth, and fifth touching the eye; 12 infralabials are present. The scales on the tail are slightly larger than those on the body. There are 245 ventrals

---

4. Wirot Nutphand, Subfamily *Boiginae*.

and 117 subcaudals. The divided subcaudals are arranged in a zigzag pattern, the anal is single. There is a large preocular, a small loreal which contacts both the prefrontal and the second supralabial, and two postoculars.

To date, only two specimens have been observed. The first arrived at Pata Zoo in Bangkok on December 5, 1985, and is still alive and on display. It is a male 1.56 m long and was found at noon, coiled in a tree four meters above the ground. The site of capture was a thick forest 170 m above sea level in the southern province of Trang. The second specimen to arrive at the Pata Zoo died 60 days after arrival as it was badly injured during capture. This was a female of 1.62 m and was also found in the South in Krabi Province. This specimen was captured when coiled in a bush adjacent to a creek at an elevation of 150 m above sea level. Specimen number one, the male, is doing well on a diet of lizards, small birds, and mice.

During August 1989, the Pata specimen was kept at the home of the author for observation. It readily ate a duckling, mice, and chicken eggs, but could not kill and eat a guinea pig. It seized the guinea pig, but made no attempt to constrict. *Boiga saengsomi* also attempted to eat a *Boiga cyanea*, but was prevented by the author. In August 1989, the specimen exceeded two meters in length.

Certain reports state that *Boiga saengsomi* closely resembles *Boiga cyanea*, but the resemblance is remote (compare Plates 23 and 33, pages 130 and 173). The shape of their heads, coloration and pattern are all different. Furthermore, *Boiga saengsomi* has proportionally larger vertebral scales and its temporals are arranged in a 3+3+3 formulation rather than the 2+3+3 formulation of *Boiga cyanea*.

Genus Chrysopelea

This genus is noteworthy as its members possess a remarkable ability to climb. In fact, they also have a limited "flight capability," and are sometimes known as "flying snakes". "Gliding snakes" would be more descriptive as they are not really capable of flight, but can glide. They do so by drawing in the ventrals to create a concave belly. By keeping the body rigid as they glide from a limb, they are able to glide several meters from one limb to another, or, from a tree to the ground. The ventrals and subcaudals are keeled and notched to facilitate climbing. These snakes have been seen upside down clinging to, and crawling along the wire of the roofs of cages, openly defying gravity. Their beauty and agility make them extremely attractive display animals. They have grooved fangs on the posterior end of the maxillary bone.

Chrysopelea usually thrives in captivity. They are not extremely large, but cages should be sufficiently large to enable freedom of movement and provide them with an opportunity to display their climbing and gliding abilities. Obviously, they need objects to climb upon, and cages should be frequently sprayed with tepid water and contain a water bowl. They appear to hunt their food during the morning and are often found on the ground at this time. Prey is varied and consists of birds, mice, lizards, and frogs. They are oviparous with females of *Chrysopelea ornata* and *Chrysopelea paradisi* being slightly larger than the males. This is probably also true of *Chrysopelea pelias*.

Chrysopelea ornata   Plate 34
Golden Tree Snake   งูเขียวพระอินทร์
                                           งูเขียวลายดอกหมาก

The head is slightly flattened and quite distinct from the neck. The Golden Tree Snake has a long and tapered body. The head is green with black markings, the eyes large and with rounded pupils. The labials are light green or greenish-yellow, the body dark green and with each scale having a black edge and a central black dash. The ventral surface is light green to greenish-yellow. This species is oviparous, females producing from 6 to 12 elongated eggs in one clutch. The hatchlings are from 0.115 to 0.15 m in length.

    Chrysopelea ornata reaches a maximum length of 1.2 meters. Its scale rows number 17 at mid-body with smooth or weakly keeled scales. There is one large preocular, one loreal, and two postoculars. The supralabials number nine with the fifth and sixth touching the eye. Infralabials number ten. Ventrals range from 213 to 234, the subcaudals are divided, have black edges, and number from 110 to 138. The anal plate is divided.

    This diurnal species is quite common and found throughout Thailand with regular sightings in metropolitan Bangkok. The range extends to India, Bangladesh, Sri Lanka, and the rest of Southeast Asia.

Chrysopelea paradisi Plate 35
Paradise Tree Snake งูเขียวร่อน

This is a very beautiful snake with a black head crossed by five narrow, greenish-yellow bands. The band that crosses the head behind the eyes bows towards the snout. The bands become large, greenish-yellow blotches on the neck; occasionally the color is reddish-orange. On the body the blotches are replaced by red or reddish-orange spots. These distinct, bright, four-pointed spots are in one row that extends along the vertebral ridge. The colorful spots terminate above the anus where they are replaced on the tail by smaller, light green dots. The remaining body scales are black, with a central green spot creating a speckled appearance on the body sides. The labials, throat, and ventral surface are yellowish-green. Few specimens exceed 0.6 m in length. Clutches of five eggs have been observed.

The scale rows number 17 at mid-body and may be gently keeled. Nine or ten supralabials are present with the fourth, fifth, and the sixth, or the fifth and sixth bordering the eye. There is one preocular, one loreal, and two postoculars. The ventrals number 198 to 238, and the paired subcaudals range from 106 to 139. The anal is divided.

This species is confined to the South and is increasingly difficult to find. The range continues into Burma, and through peninsular Malaysia, into western Indonesia and the Philippines.

Chrysopelea pelias   Plates Cover and 36
Barred Tree Snake   งูดอกหมากแดง

This snake is quite rare in Thailand and is only occasionally found in the provinces adjacent to the Malaysian border. It is rather small compared to the other members of the genus, averaging a length of only 0.55 meters. It is also quite beautiful. The top of the head is light brown and crossed by three black-edged, orange bars. The narrowest passes in front of the eyes, the second passes behind the eyes, and the widest crosses the base of the head. A narrow black line separates the colorful top of the head from the white labials and chin. Much of the remainder of the dorsal surface is reddish-orange. The vertebral area is crossed by a series of black-edged white bars that do not extend down the body sides. The bars become less distinct posteriorly; the black edges become thicker and more prominent and the white interiors gradually become orangish-white. The flanks are light brown but heavily speckled with small, bluish-white blotches. The ventral surface is yellowish-white. A thin black line extends the length of the body on the edges of the ventral scales.

    This species has from 181 to 199 ventrals and 98 to 120 paired subcaudals. The anal is divided. The body scales are smooth and arranged in 17 rows at mid-body. Of the nine supralabials, the fifth and sixth or the fourth to the sixth border the eye.

    The Barred Tree Snake is also found in Burma and peninsular Malaysia but, as in Thailand, it is the rarest of the members of *Chrysopelea*.

Genus Dendrelaphis

Members of this genus are well suited to an arboreal life. Their bodies are extremely thin and they are very agile. Their ventral scales are keeled, and they can move over tree branches and shrubs with startling speed. Their heads are distinct, and eyes large and round; the vertebral scales are enlarged. Although they are timid snakes who would rather flee than fight, they will assume a defensive position if denied an escape route. When doing so, they inflate their slender necks, thus increasing their apparent size and revealing the bright colors of the edges of their scales, which are not normally seen. The flash of color and apparent size of the snake often combine to intimidate aggressors. These snakes do not have enlarged posterior fangs.

Members of this genus are active during the day and do quite well in captivity. A fairly large cage should be provided as they are moderately active. The cage should contain a number of plants and branches upon which they can move. Daily spraying of tepid water will maintain a high-level of humidity in addition to providing drinking water. A bowl of water should also be placed in the cage. Members of this genus require a diet of lizards and frogs. They are also reputed to eat the eggs of birds. Six members of the genus live in Thailand and each species is oviparous. Females are larger than males.

Dendrelaphis caudolineatus   Plate 37
Striped Bronzeback   งูสายม่านแดงหลังลาย

This is a very colorful snake with a bronze head and yellowish-green supralabials. The body is gray-green. A white stripe, bordered by a narrow black stripe above, and a broad one below, extends along the body sides. The ventral surface is greenish-yellow with a black line extending through the center of the subcaudals until the tip of the tail. Mature specimens attain a length of approximately 1.35 meters. Clutches of eight eggs have been observed.

The scales are smooth and arranged in rows of 13 at mid-body. Ventrals range from 171 to 188, the paired subcaudals from 100 to 170, and the anal divided. There is one preocular, one elongated loreal, and two postoculars. Of the nine supralabials, the fifth and sixth touch the eye. There are 11 infralabials.

This species is confined to the southern provinces of Thailand. It is also found in India, Burma, peninsular Malaysia, and Indonesia.

Dendrelaphis cyanochloris
Wall's Bronzeback   งูสายม่านคอขีด

The body is olive, and the smooth scales have distinct black edges. A black stripe extends from the head through the neck onto the body, where it scatters and finally disappears. The belly and the throat are yellowish. This snake may reach a length of 1.3 meters.

The ventrals vary from 186 to 211, the paired subcaudals from 135 to 159, and the anal divided. Supralabials number nine, with the fourth, fifth, and sixth bordering the eye. Mid-body scale rows number 15, the vertebral scales being enlarged.

This species is found from the northern provinces of Thailand as far south as the province of Phuket. It is also found in India, Burma, Laos, and possibly Vietnam.

Dendrelaphis formosus   Plate 38
Elegant Bronzeback   งูสายม่านหลังทอง

This attractive snake has a variety of body hues. The overall body color is bronze to light brown. Small blue or green spots may be present on the sides, and black lines may occur on the posterior sides of the body. The head is reddish-brown with a neck of a deeper shade. A black stripe that continues along the neck, runs from the rostral through the very large eyes. The tongue is red. The belly and neck are light green, the posterior section of the belly is dull green or brownish. Adults average one meter in length.

The vertebral scales are greatly enlarged and the body scales smooth. Mid-body scale rows number 15. The ventrals range from 174 to 205, the paired subcaudals from 132 to 158, and the anal is divided. There is one preocular, an elongated loreal, and usually three postoculars. Supralabials number eight or nine, with the fourth and fifth, or, the fifth and sixth touching the eye. The eye is exceptionally large.

The Elegant Bronzeback is a snake of the South with a range extending as far north as the province of Phuket. Its range also extends south through Malaysia to the islands of western Indonesia.

Dendrelaphis pictus pictus Plate 39
Common Bronzeback งูสายม่านธรรมดา
งูสายม่านพระอินทร์

As the name implies, the body of this snake is bronze in color. The top of the head is of a lighter shade, and the sides of the head and chin are white. The tongue is red. A black line extends from the side of the head to the eye, along the neck, finally disappearing on the body. The ventral surface is yellow to pale green. This species reaches an adult length of one meter. Females produce from five to eight eggs per clutch.

Ventrals number from 167 to 200, the paired subcaudals from 127 to 164, and the anal is divided. The scales are smooth and arranged in 15 rows at mid-body. There is one preocular, one loreal, and two postoculars. Infralabials number from 10 to 15 and there are either eight or nine supralabials, with the third, fourth, and fifth or the fourth, fifth, and sixth touching the eye.

This species is widely distributed in Thailand and has also been reported in the other nations of Southeast Asia.

Dendrelaphis striatus   Plate 40
Cohn's Bronzeback   งูสายม่านลายเฉียง

The dorsal surface is bronze with a series of narrow black vertical bars to each side of the anterior portion of the body. Black and white lateral stripes are not present on this species. The head is distinct from the neck, the snout rather short and broad, and a rather wide, black line extends from the snout back through the eye to the neck. The labials and throat are yellow and the ventral surface gray or very light brown. Adults average one meter in length.

There are nine supralabials, although occasionally individual specimens may possess eight or ten. The fifth and sixth, and occasionally the fourth, contact the eye. The vertebral scales are slightly enlarged; mid-body scale rows number 15. The keeled ventrals range from 152 to 163, the paired subcaudals from 103 to 142. The anal is divided.

This species was only recently discovered in Thailand; several specimens have been collected in the South. It also has been recorded in Malaysia and western Indonesia.

Dendrelaphis subocularis   Plate 41
Mountain Bronzeback   งูสายม่านเกล็ดใต้ตาใหญ่

This is a rather small snake, averaging a mature length of only 0.7 meters. The body color is bronze but the scales have black edges, giving the appearance of thin, discontinuous lines along the body. A cream colored line runs the length of the body through the scales bordering the ventrals. The head and neck are olive-green and the ventral surface is white.

There are 15 scale rows at mid-body with the vertebral scales only slightly enlarged. Ventrals range from 153 to 175, the paired subcaudals from 85 to 105, and the anal divided.

Supralabials vary from seven to nine; the fifth supralabial is elongated and in contact with the entire lower edge of the eye. There are 10 or 11 infralabials in addition to one preocular, one loreal, and two preoculars.

Dendrelaphis subocularis is found north of 11° N latitude. It is also found in Burma, Kampuchea, Vietnam, and probably in Laos.

Genus Dryophiops

This genus contains only two species, one of which, *Dryophiops rubescens*, has only a limited range in southern Thailand. *Dryophiops rubescens* is a small arboreal snake; its reproductive habits are unknown.

A small cage will be adequate but it should be provided with ample facilities for climbing. Occasional spraying is recommended and a water dish should be available at all times. This snake is known to eat lizards.

Dryophiops rubescens
Red Whip Snake

Plate 42
งูสายน้ำผึ้ง
งูเถา

The elongated head is distinct from the slender body but the snout is not pointed. The pupil is horizontally elongated. The dorsal color is gray with a series of irregular, small, black dots. The head is olive, and a faint brown stripe extends from the tip of the snout through each eye and down the back. The supralabials are white, and the ventral surface is light yellow. *Dryophiops rubescens* is a rear-fanged snake, but the venom is not considered to be a threat to man. Adults grow to a length of approximately one meter.

Scale rows number 15 at mid-body and the scales are smooth. One large preocular, one elongated loreal, and three postoculars are present. Of the nine supralabials, the fourth to the sixth touch the eye with the seventh being the largest. Infralabials number ten. The ventrals are keeled and notched and range from 190 to 200; the divided subcaudals vary from 111 to 136 and are also keeled and notched. The anal is divided. The vertebral scales are not enlarged.

The Red Whip Snake has a limited range in Thailand and is extremely rare throughout. The species is found in the province of Phuket and south to the Malaysian border with a range extending through Malaysia into western Indonesia and the southern Philippines.

Genus Elaphe

This is a genus of fairly large constrictors with some members reaching a length of two meters. All are oviparous and prey upon warm-blooded animals. The pupils are round and the head distinct from the neck. Scales may be smooth or keeled and are arranged in mid-body rows of 19 to 27. The subcaudals are paired. Posterior maxillary teeth are not enlarged.

Members of genus *Elaphe* are commonly found in forested country. Although primarily terrestrial, they are good climbers and often climb trees in search of food. Most are believed to be nocturnal. These are large, attractive snakes and generally do well in captivity. At first, they are quite nervous and tend to inflate their necks before striking and then strike at any movement. They should be provided with a hiding box and, if this is done, they usually become calm. Because they are large and like to climb, their quarters should be spacious. Branches for climbing should be included in addition to a large water receptacle to permit soaking. A diet of rats, occasionally supplemented with a meal of chicks, is most suitable. Each member of the genus is oviparous, and, although statistics for *Elaphe taeniura ridleyi* are not available, it appears that females are slightly larger than males.

Six members are native to Thailand.

Elaphe flavolineata Figure 16
Common Malayan Racer Plate 43

งูทางมะพร้าวดำ. งูหลุนชุน

The anterior of the body has a pair of black stripes flanking the bright yellowish-orange vertebral column. The stripes do not begin on the neck but begin in the area above the twentieth ventral and continue to the area above the one hundredth ventral. There is a series of black spots along the edges of the ventrals on the anterior portion of the body. The color of the areas in between the black body markings is yellowish-brown. Beyond the one hundredth ventral, the yellowish-brown color disappears, and the body becomes black. The head is olive-brown, but the supralabials are the color of old ivory. A black line extends down from the eye to the fifth and sixth supralabials with another line extending back from the eye through the postoculars to the eighth supralabial. A black line also extends from the base of each jaw along the neck to the area of the tenth ventral. The ventral surface is pale white. This large constrictor may reach a length of two meters. Females usually produce from five to ten eggs per clutch.

The scales are keeled and arranged in rows of 19 at mid-body. The ventrals range from 193 to 234, the subcaudals are divided and number 89 to 115, and the anal single. There is one large preocular, one loreal, two postoculars, and nine supralabials. The fourth, fifth, and sixth supralabials touch the eye; the sixth also contacts the anterior temporal scale. There are either 10 or 11 infralabials.

In Thailand, this species is found in the South from the province of Phuket to the Malaysian border at sea level up to 490 meters. It is also found in Burma, peninsular Malaysia, western Indonesia, and the Andaman Islands of India.

Figure 16. Heads of Racers:
1. *Elaphe flavolineata*. 2. *Elaphe radiata*.
3. *Elaphe porphyracea porphyracea*.

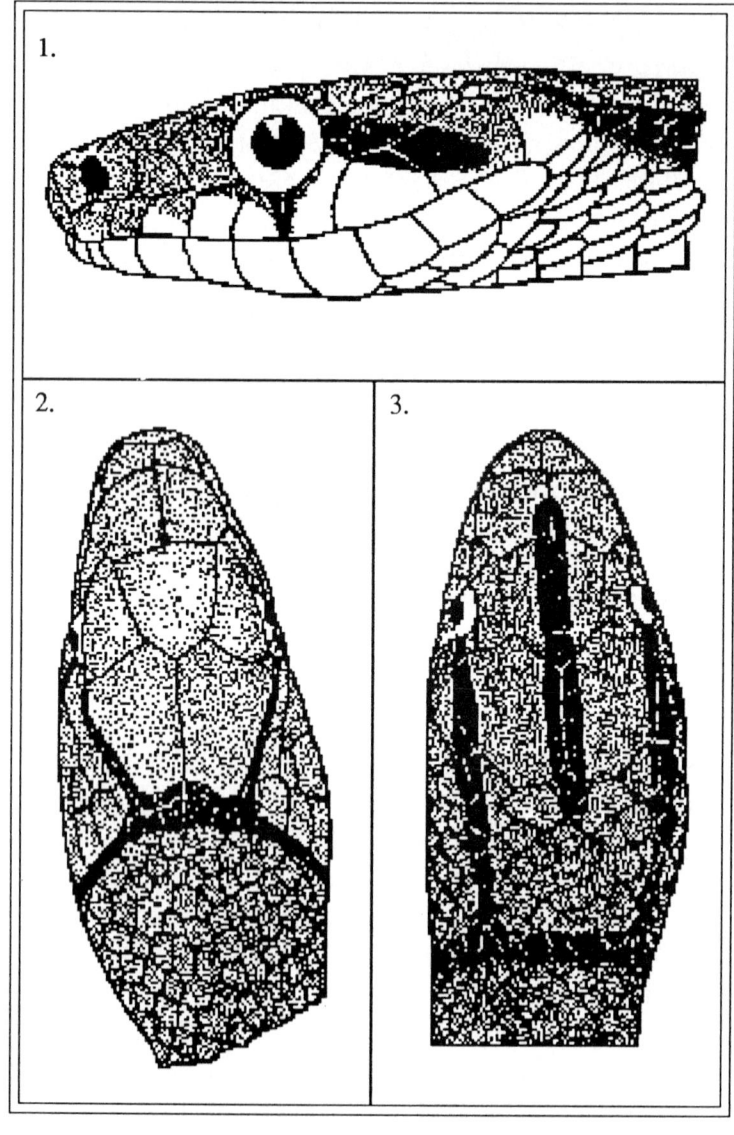

Elaphe porphyracea nigrofasciata
Black-striped Mountain Racer   งูทางมะพร้าวแดงแถบดำ

This attractive snake may reach an adult length slightly in excess of one meter, females being slightly larger than males. This snake is slender with the head being only slightly distinct from the neck. It is reddish-brown in color with a series of 9 to 12 dark cross-bars extending down the back. Two distinct black dorsolateral lines extend from the eye to the tip of the tail. The ventral surface is light in color. This snake is oviparous, clutches of two to three eggs having been reported. The incubation period is approximately two months. *Elaphe porphyracea nigrofasciata* differs from *Elaphe porphyracea porphyracea* in having fewer cross-bars and two dorsolateral lines.

There are eight supralabials, the fourth and fifth contacting the eye, as well as a single loreal. The dorsal body scales are smooth and in 19 rows at mid-body. The ventrals range from 190 to 218, the anal is divided, and the subcaudals range from 52 to 76.

Elaphe porphyracea nigrofasciata has only recently been found in the North near the Laotian border. It is a rare Thai snake, normally found in forested areas at high elevations. Therefore, it should be kept at temperatures slightly lower than those indicated within Appendix 2.

A single specimen was recently found in the Phu Luang Wildlife Sanctuary, Loei Province at an elevation of 1300 meters. This snake is also native to Burma, northern Vietnam, southern China, including Hainan, as well as Hong Kong.

Elaphe porphyracea porphyracea Figure 16
Red Mountain Racer Plate 44

งูทางมะพร้าวแดง. งูบ้องไฟ

This is a very attractive snake. Adults are dark red and have a series of 14 to 16 white-edged, black cross-bars extending around the body. On the posterior portion of the body, these cross-bars are connected by a thin black line extending along the body to the tail. The head and supralabials are red. A black stripe extends from the prefrontals to the parietals with abrupt termination at the first body scales. A black line extends back from each eye to merge with the first cross-bar on the body. The ventral surface is pale reddish. This is a small member of the genus, with adult length averaging only one meter. It is a mountain form, rarely being found below 1200 meters. Females produce two to five eggs per clutch which require approximately 60 days to hatch.

The scales are smooth and arranged in rows of 19 at mid-body. Ventrals number from 190 to 218, the subcaudals from 52 to 76, and the anal single. There is one large preocular, an elongated loreal, two postoculars, and eight supralabials. The fourth and fifth supralabials border the eye. There are ten infralabials.

In Thailand, this species is only found in the northern provinces of Mae Hong Son and Chiang Mai. The range extends across Thailand's border into Burma, the Assam region of India, Laos, and southern China. The range is discontinuous, as it reappears in peninsular Malaysia and on the Indonesian island of Sumatra. Perhaps this is due to its preference for higher elevations and, consequently, relatively cool temperatures. This being so, specimens should be kept at temperatures a few degrees cooler than those detailed in Appendix 2.

Elaphe radiata
Copperhead Racer

Figure 16
Plate 45
งูทางมะพร้าวธรรมดา

This large snake reaches a length of two meters or more. Its body color is gray-brown or yellow-brown with four black stripes along the anterior portion of the body. The upper two are quite wide and flank a light vertebral stripe. The lower two are narrower and may be broken and discontinuous. The head is copper colored. Three black lines radiate from the eyes. One reaches down through the supralabials and the infralabials, a second is set diagonally to the base of the jaw. The third runs back to the rear of the head where it merges with the narrow black band that encircles the base of the head.

The scales are keeled and arranged in rows of 19 at mid-body. The ventrals number from 222 to 250 and the paired subcaudals from 82 to 108. The anal is single. There is one preocular, a loreal, two postoculars, and eight or nine supralabials. The third through the fifth, or the fourth through the sixth supralabials touch the eye. The sixth supralabial also contacts the anterior temporal. There are ten infralabials. Females produce 5 to 12 eggs at a time.

The Copperhead Racer is found in the lowlands throughout Thailand and in India, southern China, and the rest of Southeast Asia.

Elaphe taeniura ridleyi   Plate 46
Cave Dwelling Snake   งูกาบหมากหางนิล
(Malayan Stripe-tailed Racer)   งูใบ้

This attractive snake grows to a length of 1.5 meters. The anterior of the body is light gray to gray-brown, but this gradually changes posteriorly. The posterior portion of the body is striped. There is a white vertebral stripe flanked by two wide black stripes, beneath each of which there is another white stripe. The ventral surface is white. The head is dull gray, and there is a wide black line extending from the eye to the base of the jaw. The neck is light reddish-brown.

    There are two preoculars, the upper being the larger, one loreal, and two postoculars, as well as nine supralabials. The fifth and sixth supralabials border the eye. There are usually 11 or 12 infralabials. The body scales are smooth, but may be gently keeled anteriorly. They are arranged in rows of 23 or 25 at mid-body. The ventrals range from 276 to 305 and the subcaudals from 89 to 112. The anal is divided; the tongue is red.

    Thai specimens are only found in Yala Province, which borders Malaysia. They are said to be found frequently in, or, near to caves, where they prey upon bats. For this reason they are sometimes referred to as the Cave Dwelling Snake. They are also native to Malaysia and Indonesia.

Elaphe taeniura taeniura Plate 47
Stripe-tailed Racer งูกาบหมากดำ

This is a large snake which may reach the length of nearly two meters. The top of the head is black and the labials are white with no dark sutures. There is a pale yellow vertebral stripe, broken by narrow black bands, on the anterior portion of the body. The yellow stripe intensifies in color and the narrow black bands gradually disappear posteriorly until only a bright yellow vertebral stripe remains on the posterior portion of the body. The sides of the body are gray with small, irregularly shaped black markings anteriorly. The black markings gradually expand until the sides of the body are solid black posteriorly. Similarly, a pale yellow anterior dorsoventral line becomes bright yellow on the posterior portion of the body. The ventrals have black edges which increase in size posteriorly; the subcaudals are solid black.

The body scales are smooth or slightly keeled and arranged in rows of 25 at mid-body. There are 230 to 305 ventrals and 89 to 114 paired subcaudals. The anal is divided. There are two preoculars, the upper being the larger, and one loreal. Of the two postoculars, the upper is the larger. Supralabials number eight, with the fourth and fifth bordering the eye. There are 10 to 12 infralabials.

This is a snake of the North, Northeast, and Central Regions with a range extending into the northern half of the South. It is also found in Burma, Laos, northern Vietnam, and southern China.

Genus Gonyosoma

Only one member of this genus lives in Thailand. It is a large arboreal snake and captives should be provided with large cages equipped with branches for climbing. Prey, which consists of rodents and birds, is constricted. Females are slightly larger than males. The posterior maxillary teeth are not enlarged.

Gonyosoma oxycephalum   Plate 48
Red-tailed Racer   งูเขียวกาบหมาก

This is a large attractive snake exceeding 1.75 m in length. The head is long and pointed. A thin dark stripe passes through the eye and separates the dark green of the top of the head from the lighter green of the remainder. The body is also green, but the scales have black edges which may be noticeable. The tail may be reddish-orange or red. The ventral surface is yellowish-green and the throat yellowish-white. The tongue, which is bluish-black, is flicked and moved very slowly. From five to twelve eggs are produced at a time; they require an incubation period of approximately 45 days. The young average a length of 0.24 m at birth.

    The body scales are smooth or faintly keeled and arranged in rows of 23, 25, or 27 at mid-body. The ventrals are faintly notched and keeled. In males they number from 230 to 240 and in females from 243 to 253. The subcaudals range from 131 to 145 in males and from 134 to 141 in females. In both cases, they are paired and the anal is divided. There is one large preocular, one elongated loreal, and two postoculars. Supralabials number from seven to eleven. If there are seven or eight supralabials, the fifth and sixth contact the eye. If there are nine or ten, the sixth and seventh touch the eye, and if there are eleven supralabials, the seventh, eighth, and ninth are in contact with the eye. If there are only seven supralabials, the seventh is very large. Twelve infralabials are present.

This species is found throughout Thailand and is also native to Burma, Kampuchea, Laos, Malaysia, Singapore, the Philippines, and western Indonesia.

Genus Psammodynastes

Only two species are recognized, *Psammodynastes pulverulentus* from Thailand, and *Psammodynastes pictus* from western Indonesia. *Psammodynastes pulverulentus* is a small snake which thrives in a medium-sized aquarium. It is primarily terrestrial, but climbs occasionally. Therefore, a few plants should be added to the cage. Its disposition is mild, although freshly caught specimens often bite. The last two maxillary teeth are enlarged and grooved, but there is no record of their bites being harmful to humans. Its diet consists primarily of small frogs and lizards.

Psammodynastes pulverulentus   Figure 17
Common Mock Viper   Plates 49 and 50

งูหมอก

The Common Mock Viper is usually less than 0.45 m in length. Its eyes are large with vertical pupils. The head is quite distinct from the cylindrical body. The basic color of this species is brown or reddish-brown. The head has the hint of a dark pattern, and the supralabials are cream colored. Small white dots extend from the supralabials through the neck into the body. The ventral surface is light brown or pink. This basic pattern, however, is quite variable.

Figure 17. *Psammodynastes pulverulentus.*

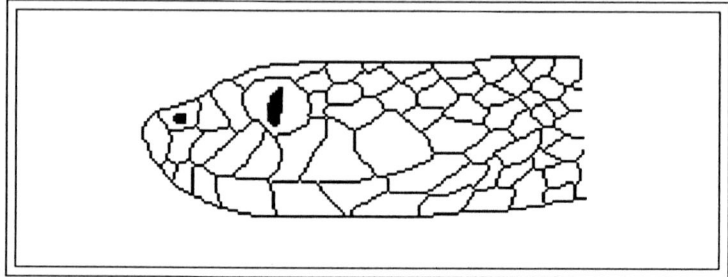

The scales are smooth and arranged in rows of 17 at mid-body. There are eight supralabials with the third, fourth, and fifth bordering the eye. There is one large preocular, one loreal, and two postoculars. Infralabials number seven or eight. Ventrals number 146 to 175 and the paired subcaudals 44 to 71. The anal is single.

This snake gives birth to living young which are miniature replicas of the parents. Females produce from five to ten young at a time, and females tend to be slightly larger than the males.

Psammodynastes pulverulentus is usually seen at elevations above 700 meters. It prefers wooded, hilly country. It is widely distributed in Thailand, and is also native to Nepal, India, all of mainland Southeast Asia, southern China, Taiwan, the Philippines, and Indonesia.

Genus Psammophis

Only one subspecies of this genus occurs in Thailand, *Psammophis condanarus indochinensis*. It is arboreal and does well in a moderately large aquarium. This genus is oviparous and in most species males are larger than females. The last two maxillary teeth are quite enlarged, grooved, and sharply directed posteriorly.

Psammophis condanarus indochinensis — Plate 51
Indochinese Sand Snake — งูม่านทอง

This attractive snake averages over a meter in length. The head is tapered and distinct from the neck. The eyes are distinct and the pupils are round. The body is cream colored with three brown stripes, each extending from the head to the tail. The widest of the three is on the vertebral ridge, and the remaining two are on the sides. The lateral stripes continue through the eye to the snout. The supralabials are cream, the belly white.

The scales are smooth, shiny, and arranged in rows of 17 at mid-body. There are eight supralabials, the fourth and fifth touching the eye, and ten infralabials. In addition, there is one large preocular, an elongated loreal, and two postoculars. The ventrals number from 156 to 173 and the paired subcaudals vary with sex; males have from 75 to 85 and females from 66 to 75. The anal is divided.

This is a mild mannered snake, and it does well in captivity. It feeds primarily on lizards and frogs, but captives also have eaten small mice.

The Indochinese Sand Snake is not found in the South, but it is found in every other part of the Kingdom at elevations up to 2000 meters. It is also found in Burma.

Genus Ptyas

These are large to very large snakes, some exceeding three meters in length. *Ptyas mucosus* has achieved a record length of 3.7 meters. Their heads are distinct from their bodies, and the large eyes have round pupils. A small presubocular scale, easily confused with a preocular, is often present. The tail is extremely long, and the subcaudals are paired. This oviparous genus preys upon a wide variety of animals including rodents, birds, frogs, lizards, and perhaps smaller snakes. They are not constrictors. If a prey offers resistance, it is seized and pinned down and held with the body and swallowed alive. Smaller prey is simply seized and swallowed. Members of this genus are active during daylight hours. The posterior maxillary teeth are only slightly enlarged and are not considered fangs. Obviously these snakes need large cages. They are terrestrial, therefore, it is not necessary to provide branches. They are, however, very nervous and quick to strike when first caught. Hiding places are necessary, and precautions should be taken to prevent snakes from striking against the glass of the cage and injuring themselves.

Three species are native to Thailand.

Ptyas carinatus งูสิงหางดำ
Keeled Rat Snake งูบองหมายควาย

This is the largest of the Asian colubrids, reaching a length of 3 meters. The anterior portion of the body is black to olive-brown and has indistinct yellow cross-bands. This gradually changes posteriorly until the ground color becomes light brown with a black pattern. The tail is black with yellow spots. A gradual change is also present on the ventral side. The color is dull yellow from the chin to mid-body, but the posterior portion gradually becomes black and yellow.

The fourth to the sixth scale rows on the sides of the body are heavily keeled, the adjoining scale rows slightly keeled. Scales are arranged in 16 or 18 rows at mid-body. An even number of scale rows is unusual but applicable to this species causing some taxonomists to classify it as *Zaocys carinatus*. Ventrals number from 208 to 215, and the paired subcaudals range from 109 to 118. The anal is divided. There are two to four loreals, one large preocular, and two postoculars. Supralabials number from eight to ten with the fourth and fifth or the fifth and sixth touching the eye. There are 11 infralabials.

Ptyas carinatus occurs in many parts of Thailand and in most of mainland Southeast Asia.

Ptyas korros
Indochinese Rat Snake

Figure 18
Plate 52

งูสิงธรรมดา

The head and anterior of the body are gray to olive-brown. The posterior of the body gradually becomes dark brown or black. The edges of the scales are white and become more distinct posteriorly until they appear as white lines. The chin, labials, and ventral surface are brownish-white. The subcaudal area is white. The eyes are proportionally larger than those of other members of the genus. Males average just over one meter in length; females average 0.94 meters. Clutches contain an average of six to nine eggs.

Figure 18. Head of the Indochinese Rat Snake, *Ptyas korros*.

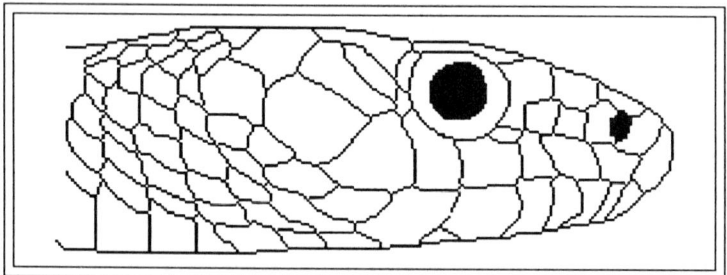

There are two or three loreals, one large preocular, and two postoculars, the lower being the smaller of the two. Of the eight supralabials, the fourth and fifth border the eye. There are ten infralabials. The scales are smooth and arranged in rows of 15 at mid-body. The subcaudals are paired and number from 120 to 147. The ventrals number from 160 to 187 and the anal is divided.

This snake lives in lowland areas throughout Thailand. *Ptyas korros* is also found in southern China, on the rest of mainland Southeast Asia, and in western Indonesia.

Plates 25 and 26. *Boiga cynodon*. Dog-toothed Cat Snake.
Horse-tail Whip Snake. งูแส้หางม้า. งูกะปิ. งูกินไข่.

Plate 25. (top)
Melanistic, or black, variety. Fairly common. Habits and characteristics similar to normal coloration.

Plate 26. (below)
4 month juvenile. Juvenile coloration lighter than adult - darkens with maturity.

Plate 27.

Boiga dendrophila melanota
Mangrove Snake
งูปล้องทอง

An attractive Thai snake.
Note: glossy scales, fairly thick body.
Eats wide variety of food.
Thrives in captivity.

Photo: Suthigit Patramangorn

172   *Plates*

Plates 28 and 29. *Boiga drapiezii*. White-spotted Cat Snake. งูงวดทอง

Photo: Suthigit Patramangorn

Photo: Suthigit Patramangorn

Plate 28. (top) Green phase. Predominant pink and green attractive. Common in south Thailand, but rare in foreign collections.

Plate 29. (center) Brown phase. Less attractive than green phase. Habits, characteristics similar to green phase.

Plate 30. *Boiga jaspidea*. Jasper Cat Snake. งูกระ.
Beautiful snake from south Thailand. Small and rare.

Photo: Suthigit Patramangorn

Plate 31. *Boiga multomaculata*. Spotted Cat Snake. Marble Cat Snake. Many Spotted Cat Snake. งูแม่ตะงาวรังนก. Small genus member. Note: *Boiga* eye, ventral color. Pattern and color similar to *V. r. siamensis*. Docile, although rear-fanged. Interesting pet.

Photo: Suthigit Patramangorn

Plate 32. *Boiga nigriceps*. Red Cat-eye Snake. งูต้องไฟ.
Attractive. Thrives in captivity. Note dark color - top of head and pink ventral surface. Diet mice, birds, lizards.

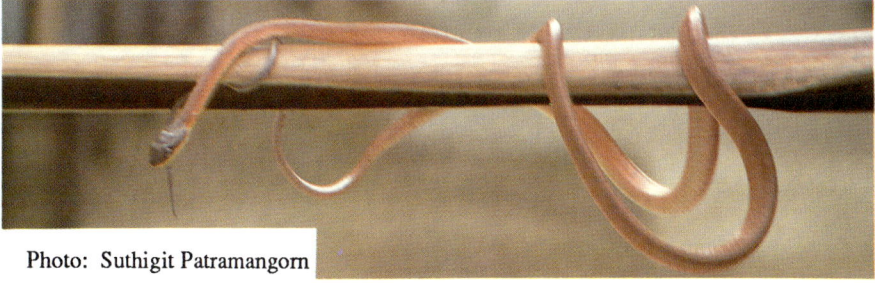

Photo: Suthigit Patramangorn

Plate 33. *Boiga saengsomi*. Banded Green Cat Snake. งูเขียวดงลาย.
New species found S. Thailand. Large colorful. Ventral yellow-green.

Photo: Suthigit Patramangorn

Photo: Raynoo Cox

Plate 34. (above)
Chrysopelea ornata
Golden Tree Snake
งูเขียวพระอินทร์, งูเขียวดอกหมาก
Common Thai species. A "flying snake." Arboreal - sometimes hunts for food on ground. Prey - lizard, tree frog, mice.

Plate 36. (above)
Chrysopelea pelias
Barred Tree Snake
งูดอกหมากแดง
Beautiful snake.
Extremely rare within range.

Plate 35. (below). *Chrysopelea paradisi*. Paradise Tree Snake. งูเขียวร่อน
Beautiful. Rare throughout range. Nervous but thrives in captivity.
A "flying snake", arboreal - climbs with speed and agility.

Plates 175

Plate 37. *Dendrelaphis caudolineatus*. Striped Bronzeback. งูสายม่านแดงหลังลาย.
Attractive. Found S.Thailand. Agile climber - needs items to climb.

Plate 38. *Dendrelaphis formosus*. Elegant Bronzeback. งูสายม่านหลังทอง.
Attractive "flying snake". Thrives in captivity. When disturbed, inflates neck to threaten. Note: enlarged vertebral scales.

Plate 39. *Dendrelaphis pictus pictus*. Common Bronzeback.
งูสายม่านพระอินทร์. งูสายม่านธรรมดา.
Most common Thai member of genus. Preys upon lizards.

Photo: Suthigit Patramangorn

Plate 40. *Dendrelaphis striatus*. Cohn's Bronzeback. งูสายม่านลายเฌียง.
Recent addition to Thai herpetofauna. Note: absence - lateral stripes.

Plate 41. *Dendrelaphis subocularis*. Mountain Bronzeback. งูสายม่านเกล็ดใต้ตาใหญ่.
Northern species. Distinguished by single large subocular scale.

Plate 42. *Dryophiops rubescens*. Red Whip Snake. งูสายน้ำผึ้ง. งูเถา.
Rare in Thailand - found only southern provinces.
Note: long, slender arboreal body - yellow markings extending
from lower head to throat. Dark eyes prominent.

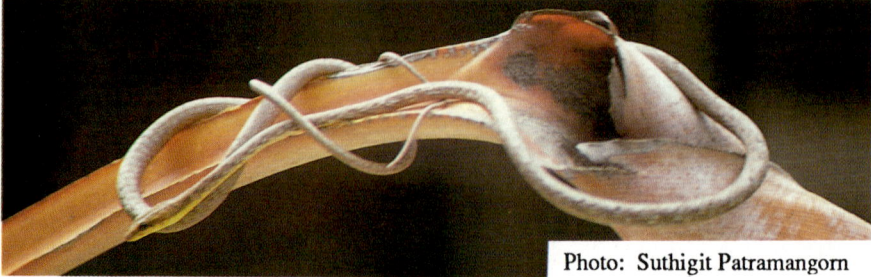

Photo: Suthigit Patramangorn

Plate 43.
Elaphe flavolineata
Common Malayan Racer
งูทางมะพร้าวดำ. งูหลุนซุน
No black line on the back of head distinguishes it from *Elaphe radiata*. A constrictor. Prey - rats and mice.

Photo: Jarujin Nabhitabhata

Plate 44. (above)
Elaphe porphyracea porphyracea. Red Mountain Racer. งูทางมะพร้าวแดง. งูบ้องไฟ
Beautiful species.
Found only in the North.

Plate 45. (right)
Elaphe radiata. Copperhead Racer. งูทางมะพร้าวธรรมดา
Typical defensive posture of Thai *Elaphe*. Menacing in appearance but harmless.

Photo: Suthigit Patramangorn

Plate 46.
Elaphe taeniura ridleyi
Cave Dwelling Snake
Malayan Stripe-tailed Snake
งูกาบหมากหางนิล. งูใบ้.
Uncommon cave dweller from S. Thailand. Posterior black and white stripes- easy to identify.

Plate 47. *Elaphe taeniura taeniura*. Stripe-tailed Racer. งูกาบหมากดำ. Attractive constrictor. Found in North, West, and South.

Photo: Jarujin Nabhitabhata

Plate 48. *Gonyosoma oxycephalum*. Red-tailed Racer. งูเขียวกาบหมาก

Large constrictor. Climbs well. Prey - rodents, birds. Thrives in captivity with proper care.

Photo: Suthigit Patramangorn

Ptyas mucosus Plate 53
Oriental Rat Snake
(Common Rat Snake) งูสิงหางลาย

This large snake has a brown head with a yellowish snout. The labials are yellow with black sutures. The body color is light brown to brown. The posterior is marked with transverse black bands that cross the back and sides to, but not crossing, the ventrals. The ventrals are light yellowish-brown but become more yellow posteriorly; the subcaudals are yellowish-white. The posterior ventrals and subcaudals have small brown spots. Males average 1.3 m in length, females 1.22 meters. Therefore, on average, males tend to be slightly larger than females. Females produce ten or more eggs which hatch in approximately 60 days. The hatchlings are from 0.38 to 0.4 m in length and reach maturity in about 20 months.

The scales are smooth or faintly keeled and arranged in rows of 16 or 17 at mid-body. The vertebral ridge is noticeable. Ventrals number from 109 to 213, and the paired subcaudals from 121 to 146. The anal is divided. There are three small loreals and two preoculars, the upper being the larger. Two postoculars are present and also eight supralabials, with the fourth and fifth touching the eye. Infralabials number ten.

The Oriental Rat Snake is common and has been recorded in all regions excepting the South. This species is also found in Pakistan, India, Bangladesh, Sri Lanka, southern China, mainland Southeast Asia, and western Indonesia. It is quite common throughout its range.

Genus Xenelaphis

The head is distinct from the neck and contains a rather large eye with a round pupil. The body is cylindrical, the body scales smooth with apical pits and arranged in 17 rows at mid-body. The vertebral scales are somewhat enlarged and hexagonal. The tail is quite long and the subcaudals divided.

Only one of the three known species of this genus is known to be native to Thailand. It is terrestrial and does not need items for climbing but adequate hiding facilities are required in addition to a large water receptacle. Prey consists primarily of rats and mice, which are killed by constriction. Birds are also eaten.

Xenelaphis hexagonotus
Malayan Brown Snake   งูควนขนุน

This large snake may reach an adult length of two meters. The head is distinct, the pupil round, the body cylindrical and slightly stout, and the tail quite long. Juveniles are uniformly light brown with a series of black bands crossing the body. When maturity is reached, however, the body color becomes dark brown, although somewhat lighter posteriorly. Additionally, the black bands become indistinct and only remain as dark blotches on the neck and anterior portion of the body near the ventral scales. Posterior body scales have black spots. The labials are yellow with black edges. The ventral surface is yellow, each ventral having a small black spot on its margin. The subcaudals have black edges.

The smooth body scales are in 17 rows at mid-body. Ventrals range from 185 to 198, the paired subcaudals from 140 to 179. The anal is divided. Four or five of the ten infralabials contact the anterior chin shields, and of the eight supralabials, only the fourth contacts the eye. The nasal is divided, there is one nearly rectangular loreal, and two postoculars. A few specimens have three postoculars. There is a subocular below the preocular and a large one beneath the postoculars; occasionally a third subocular is present, separating the fourth supralabial from the eye.

This species is relatively rare in Thailand. It has been reported from Surat Thani Province south to the Malaysian border. The range extends through Malaysia into Java and Borneo. It is also found in lower Burma and southern Vietnam, and possibly also Kampuchea. This species is terrestrial and commonly found in mangrove forests.

Subfamily Homalopsinae

Posterior hypapophyses, a series of internal ventral projections from the vertebral column, is a characteristic of this subfamily. In addition, members have a large tracheal lung. All have adapted, in varying degrees, to life in close association with water. For instance, the nostrils may be situated near to the top of the snout for easier breathing when near the surface of water. In some species, the nostrils have valves which close to prevent a water inflow. The eyes are often high on the head, and some species have extremely narrow ventral scales. Each member is rear fanged and has a venom producing capability, but none have proven to be fatal to humans. In some species traditional head scales have been greatly reduced or entirely replaced by smaller scales. The anal is divided and the subcaudals paired. The eyes are small and the pupils usually elliptical. In this subfamily females are viviparous and larger than males.

Small to medium aquariums will suffice for housing most members of this subfamily. The aquarium should be equally divided between dry surface and clean water at all times. The water should be maintained at a temperature between 27° C to 29° C. Most species eat fish, frogs, tadpoles, and shrimp.

Twelve species occur in Thailand.

Genus Bitia

This genus contains a single species. The head is indistinguishable from the neck, and the parietals are fragmented into a series of smaller scales. The anterior third of the body is narrow, but the remainder of the body is thicker. The narrow ventral scales have two strong lateral keels, similar to those found in *Chrysopelea* and *Dendrelaphis*. Specimens have a pair of enlarged, grooved teeth.

The sole member of this genus is more aquatic than most other members of the subfamily, and inhabits coastal water areas. Its quarters, therefore, should contain brackish water. The diet consists of fish, shrimp, and crabs.

Bitia hydroides งูปากกว้างท้องสัน
Keel-bellied Water Snake งูเบี้ยว

This species averages 0.5 m in length. The head is small and the eye is located dorsally. The pupil is elliptical. The cylindrical body is at its thickest posteriorly, and the tail slightly compressed. The head is gray or dark brown, each scale and the suboculars containing one or two small, dark spots. The chin, throat, ventrals, and subcaudals are pale yellow. The dorsal surface of the body is pale brown or gray and is marked with a series of dark gray or black cross-bars which extend a little beyond halfway down the body sides. The cross-bars and their interspaces are nearly equal in width and the coloration bright. The pattern extends to the tip of the tail. This species is viviparous.

The body scales are smooth and roughly triangular in shape with a noticeable gap between the base and apex of the two adjacent scales. Each scale is entirely attached to the interstitial skin with scale rows ranging from 37 to 43 at mid-body. Ventrals number from 157 to 165, the paired subcaudals from 24 to 34 with females having fewer subcaudals. The anal is divided and longer than any of the ventral scales. The nostrils are narrow slits separated by a single internasal scale. There is a single loreal which is in contact with the first three supralabials, but occasionally contacts only the second and third. There is one preocular, extending halfway under the eye, and two postoculars, the lower the larger. Of the seven supralabials, none contact the eye; the fourth is directly below the eye and the fifth and sixth are the largest. Infralabials number from 10 to 12, the sixth being the largest.

The Keel-bellied Water Snake inhabits sea coasts and the mouths of rivers, but it is not a common Thai snake. Its slightly compressed tail and banded pattern sometimes cause difficulty in distinguishing it from some of the very poisonous sea snakes that share the same habitat. This species is, however, harmless. It is found in the coastal areas of the South with a range extending into Burma and Malaysia.

Genus Cantoria

In this genus the head shields are large and distinct. The head is slightly flattened and indistinct from the neck. The body is long and compressed. The dorsal scales are smooth and the ventrals fairly narrow.

Of the two members of genus *Cantoria,* only one is found in Thailand. It is a rather long, slender snake and requires a cage of moderate size with a surface equally divided between land and water. The water should be slightly brackish, and the dry surface equipped with hiding places. Members of this genus produce living young. Both juveniles and adults do well in captivity on a diet of fish occasionally varied with shrimp or frogs.

Cantoria violacea
Cantor's Water Snake   งูปลาหลังม่วง

This is a long slender snake whose head is indistinct from its neck. It averages one meter in length. The body is blackish-brown and crossed by a series of narrow, pale yellow bands which broaden and merge near the ventral surface. These bands are more distinct on the anterior part of the body and tail than on the posterior portion. The head is dark brown with yellow cross-bars in front of the eyes. The head of some specimens may have irregular yellow spots. The labials, throat, and ventral surface are yellow. Faint black bands cross the ventral surface, and the subcaudals have black edges. A second color phase is predominantly brown with white bands and a white ventral surface.

The head scales are prominent. The mouth is below and behind the rostral. There is a single internasal, and the nostrils are placed dorsally. The prefrontals are in broad contact. The single loreal is square; there is one preocular and a small postocular. The preocular and subocular prevent any contact between the eye and the supralabials. There are five supralabials, the third and fourth situated directly below the eye. Eight infralabials are present. The scales are smooth and in rows of 19 at mid-body. Ventrals range from 260 to 291, and the divided subcaudals number from 56 to 69. Males usually have more subcaudals but fewer ventrals than females. The anal and subcaudals are divided.

This snake lives in coastal areas and river mouths, but is not common in its range. It has been found in Phuket Province, on the west coast of Thailand and probably also occurs in provinces to the south. It is also found in Burma, Malaysia, western Indonesia, and on the Andaman Islands.

Genus Cerberus

This genus is represented by only one species in Thailand. The parietals are divided into smaller scales, and the body scales are keeled. It frequents mangrove areas adjacent to coastlines and, if disturbed, is quick to bite. Brackish water and a diet of fish should be provided.

Cerberus rynchops   Figure 19
Dog-faced Water Snake   Plate 54
งูปากกว้างน้ำเค็ม

The head is distinct from the neck, and the body cylindrical and rather stout. The color is greenish-gray or gray. A black line extends from the snout through the eyes to the neck. Faint black bands are present across the back. The ventral surface is dull gray with traces of yellow at the edges. The subcaudals are gray-black. This species reaches a length of one meter. Between 8 to 26 young are born in each litter, the offspring measuring from 0.175 to 0.2 m in length.

Body scales are strongly keeled and in rows of 23 or 25 at mid-body. The ventrals range from 136 to 152, the subcaudals from 50 to 66, and are divided, as is the anal.

Parietals do not exist and are replaced by small, irregularly placed scales. There are 11 infralabials, one preocular, two loreals, one postocular, and two suboculars. There are 11 supralabials and none touch the eye. The fifth and sixth, however, are directly below the eye.

This is a common snake in the mangrove areas of the coastal provinces of Thailand. It is found in similar environments in India, Bangladesh, Sri Lanka, and both mainland and insular Southeast Asia.

Figure 19. Heads of Homalopsine Snakes.
Progressive degeneration of the shields.
1. *Cereberus rynchops*.
2. *Enhydris plumbea*.
3. *Homalopsis buccata*.

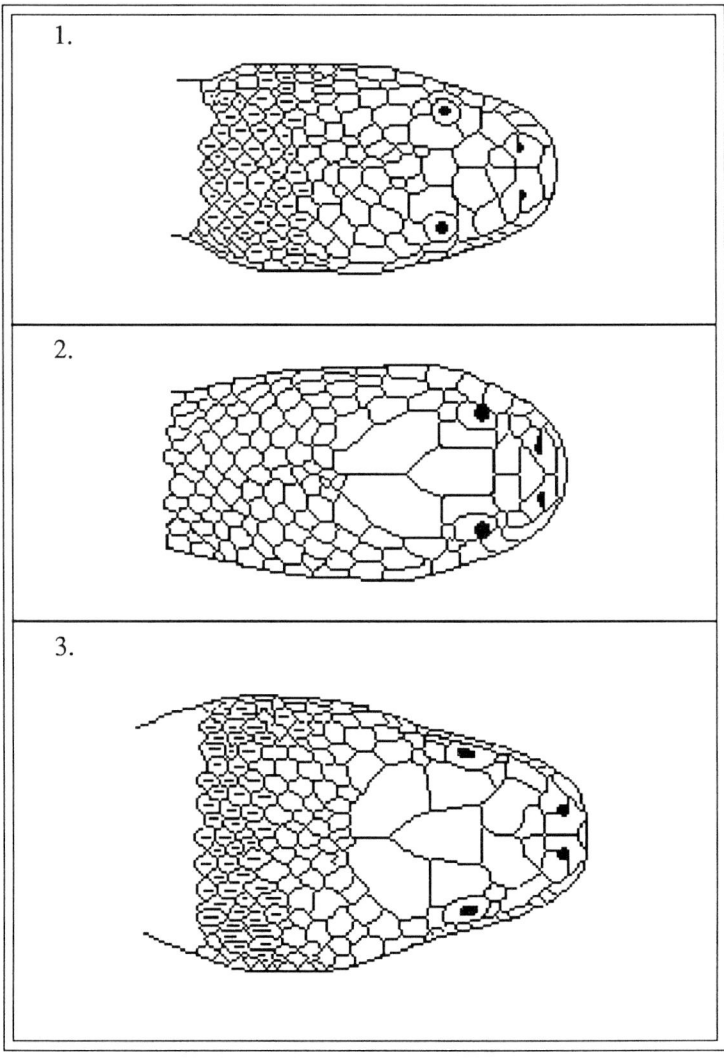

Genus Enhydris

Characteristically, the members of this genus have cylindrical bodies. Their heads are flattened and only slightly distinct from their necks or, in some cases, indistinct. Native Thai members have smooth scales in rows of 19 to 29 at mid-body. The subcaudals and anal plate are divided. The eyes are small and the pupils vertical. They usually have only one preocular, one loreal, and two postoculars.

Enhydris bocourti   Plate 55
Bocourt's Water Snake   งูใน

This is the largest member of genus *Enhydris* native to Thailand, sometimes reaching 1.1 meters. Its head is large, blunt, and barely distinct from its thick neck. The body is stout and round; its tail is short.

The color is reddish-brown with transverse, black-edged, light bars which narrow towards the edges of the ventral surface where they terminate. The head is a uniform greenish-brown and supralabials cream with black edges.

The scales are smooth and in 27 rows at mid-body. There is one preocular, one loreal which is greater in height than width, and two postoculars. The supralabials number nine, with the fourth bordering the eye. There are 14 to 15 infralabials. Ventrals range from 123 to 132, the subcaudals from 32 to 47, and males have longer tails than females. Females give birth to 15 to 20 young at a time, each of a length averaging 0.22 meters.

This snake is found from Nakhon Sawan in the Central Region to Pattani Province in the South. It is also native to Kampuchea, central Vietnam, and northern peninsular Malaysia.

Enhydris enhydris Plate 56
Rainbow Water Snake งูสายรุ้งธรรมดา

This is one of the more attractive members of the genus. The pattern on the back consists of alternating stripes of dark and light brown extending from head to tail. The stripes on the sides are alternating pink and yellow; the head is small, narrow, and distinct from the neck. Supralabials are light brown; the belly is white.

The average length is around 0.5 m, but some reach a length of 0.8 meters. The scales are smooth and scale rows usually number 21 at mid-body, but occasionally there are 23. There are eight supralabials; the fourth touches the eye and the sixth and seventh are the largest. The single, large loreal is in contact with the first three supralabials. Nine infralabials, one preocular, and two postoculars are present. Occasionally only one postocular is present. Ventrals range from 134 to 172, subcaudals from 50 to 78, and the anal is divided. Males have more subcaudals than females.

The Rainbow Water Snake is gentle and rarely bites. Litters contain 6 to 18 young. The newborn babies are pink with a reddish pattern.

This species is found in humid, well-watered places throughout Thailand. It is also found in India, Bangladesh, southern China, and most of Southeast Asia.

Enhydris jagori   Plate 57
Striped Water Snake   งูสายรุ้งลาย

This is a rather small snake with a generally nasty disposition. It averages 0.4 m at maturity. The body is gray with lines of black dots running from the sides of the neck to the tail. In some specimens, the body color becomes white or pink below the line of black dots.

There are 21 scale rows at mid-body; the body scales are smooth. The ventrals number from 119 to 146 and the subcaudals from 50 to 65. There are eight supralabials; the fourth touches the eye with the sixth being the largest. There are ten infralabials. The loreal is in contact with the first three supralabials. There is one preocular which is greater in height than length, and two postoculars. Males have much longer tails and more subcaudals than females.

This species is widely distributed in Thailand. Its range extends into Laos, Kampuchea, and Vietnam.

Enhydris plumbea   Figure 19
Plumbeous Water Snake   Plate 58
                                               งูปลิง

Although occasionally exceeding 0.5 m, this is one of the smaller members of the genus, averaging only 0.35 meters. The body is short but stout, the blunt head distinct from the neck and with relatively small eyes. Body color is olive but this becomes yellow low on the sides. The ventral surface is dull white. The labials are yellow. Females produce an average of nine young per litter.

The scales are smooth, polished, and arranged in rows of 19 at mid-body. There is one preocular, one loreal, which is in contact with the second and third supralabials, and two postoculars. The parietals are well developed. There are eight supralabials, with the fourth and fifth bordering the eye, and ten infralabials. Ventrals number from 112 to 132 and the subcaudals from 29 to 43. The anal is divided.

This nocturnal snake likes fresh water and is found where an abundance exists, sometimes at high elevations. Its diet consists primarily of fish and frogs. It has a nasty disposition and strikes vigorously, often thrusting its body forward a short distance.

Enhydris plumbea is found throughout Thailand, as well as in southern China, the remainder of mainland Southeast Asia, and in Indonesia as far east as Celebes.

Enhydris smithi
Smith's Water Snake   งูสายรุ้งดำ

This snake averages 0.6 m in length. Its body is short and stout and its head barely distinct from its neck. The dorsal surface and sides are black; the ventral surface is gray. A series of 38 or 39 narrow bands, pink on the back and yellow on the sides and ventral surface, extend from the ventral surface around the body sides to unite near the vertebral area forming festoons. The head is black with pink or light gray markings. The tail is encircled by alternating wide black and narrow yellow rings.

Smith's Water Snake has eight supralabials. The fourth touches the eye with the fifth and sixth in contact with the subocular. There is a single loreal which contacts the first three supralabials. Usually, there is only one preocular present, but occasionally specimens have two. Similarly, two postoculars are normal, but occasionally three are present. There are 10 or 11 infralabials. The scales are smooth and arranged in rows of 21 at mid-body. The ventrals number from 118 to 127 and the subcaudals from 49 to 56. The anal is divided.

This is generally a mild mannered snake and rarely bites. It has a limited distribution within Thailand, having only been found in Bangkok, nearby Samut Prakan Province in the Central Region, and Prachuap Khiri Khan Province in the West. It is not found outside of Thailand.

Genus Erpeton

There is only one member of this genus, *Erpeton tentaculatum*. The body is narrow and partially compressed anteriorly but becomes thicker posteriorly. The head is distinct from the slender neck; the eyes are small with rounded pupils. Scales are heavily keeled and quite rough to the touch. The ventral scales are very narrow. The unique feature of this snake is the presence of two soft, tentacle-like protuberances from the rostral. Their function is unknown but they may serve to lure prey within striking distance. See Plate 60, page 222.

This snake usually does not thrive in captivity as it is vulnerable to skin diseases; however, some keepers consider that this problem is alleviated by maintaining *Erpeton* in water of pH 6.0. Others consider they do well in aquariums with duckweed. It may be suitable to keep them in medium-sized aquariums filled with water of pH 6.0 and plenty of duckweed. A dry surface is not necessary as this species is nearly helpless on dry surfaces. They spend their time in water. Attach branches or similar objects to the bottom of the aquarium for the snake to attach its tail and patiently wait for prey, small fish or shrimp, to pass within striking range. Aerate and filter the aquarium in order to maintain maximum cleanliness.

Erpeton has grooved teeth and a mild venom which is not harmful to man. In fact, these snakes are mild mannered and rarely bite. They become rigid when handled but seldom make any threatening moves. This genus bears living young, giving birth to up to 12 babies per litter, each a miniature replica of the parents.

This species has been successfully bred at the New York Zoological Park. A group of six females and one male was maintained in a large, filtered aquarium. The water temperature was maintained at a relatively constant 26° C to 27° C throughout the year. Seven litters were born over a period extending from August 20, 1980 to October 1, 1986. As few as 5 and as many as 12 were born in any single litter, the average being eight. Snout to vent lengths of the new-born ranged from 0.143 to 0.168 m, total lengths from 0.197 to 0.244 meters. Body weights ranged from 2.4 to 4.2 grams. Births occurred between early July and early October. Interestingly, the first shedding occurred one to three days after birth and the young began eating small goldfish only one day after birth. Both events usually occur 10 to 14 days after birth in most other species.

Erpeton tentaculatum Plates 59 and 60
Tentacled Snake งูกระด้าง

The Tentacled Snake averages less than 0.5 m, although a few specimens may measure 0.7 m in length. The head is small and distinct from the neck. The eyes are small and round. There are two "tentacles" on the rostral of a length of 0.13 to 0.19 m in adults which do not seem to serve a specific function. The body color is light brown. Two dark brown or black stripes extend down the neck to the tail, where they divide into spots. Two lines extend from the snout through the eyes and along the length of the body. The ventral surface is brownish-yellow and covered with brown, black, or white spots. There is also a black, or melanistic, form of *Erpeton tentaculatum,* being entirely black with faint traces of a pattern. A series of fairly large grayish blotches is present along the vertebral ridge in addition to a series of small grayish blotches on the flanks. Anteriorly, there are some white streaks but these become spots posteriorly.

There are 20 infralabials and 14 supralabials, none of which border the eye. Small scales are often present between the typical head scales. The loreal area has four to eight small scales without a "true" loreal. There is one preocular and two postoculars; the nostrils are located on top of the snout near the base of the tentacles. Body scales are in rows of 35 or 37 at mid-body. The narrow ventrals number from 108 to 133. They are little more than twice the width of the body scales. Females have more ventrals than males. The paired subcaudals are not very distinct but vary from 87 to 126, males having more than females. The anal is single.

This is a snake which prefers living in slow moving water. It can remain submerged for more than six minutes before having to surface to breathe. During the dry season it will, if necessary, bury itself and rest quietly under mud until the rainy season brings water back to its environment.

The Tentacled Snake is found in the Central Region and the South. It is also native to Kampuchea and Vietnam.

Genus Fordonia

The head is not distinct from the stout body. The head shields are large and distinct; the body scales are smooth and the ventrals broad and unkeeled.

There is but one species in this genus. It is a fairly large snake and requires a cage of moderate size. The water should be brackish. *Fordonia* thrives on a diet of fish and crabs. Females of the species produce 10 to 15 young per litter, the newborn averaging 0.18 m in length. It is said to have a mild disposition.

Fordonia leucobalia
Crab-eating Water Snake   งูปลาหลังเทา

This species is generally less than one meter in length. The head is short, wide, and indistinct from the cylindrical body. Dorsal colors range from gray through olive-brown to purplish-black. The ventral surface, throat, and chin are ivory-white. The labials are yellowish-white.

The nostrils, located dorsally, are slits with the nasals separated by a single internasal. There is one preocular and two postoculars, the lower extending slightly under the eye. No loreals are present. Of the five supralabials, the fifth is the largest. The third supralabial, and in some specimens the second, borders the eye. The first three infralabials contact the first pair of chin shields. The scales are smooth and are in rows of 25 or 27 at mid-body. The ventrals are wide and number from 138 to 156; the paired subcaudals range from 27 to 43. The anal is divided.

The Crab-eating Water Snake is found in coastal areas where it preys on small fish and crabs. It frequently takes refuge in the burrows of crabs. It is found along the west coast of Thailand and is also native to Burma, Malaysia, Kampuchea, the Philippines, Indonesia, New Guinea, and the northern coast of Australia.

Genus Gerarda

This genus also has only one member. It has a cylindrical body, and the head is only slightly distinct from the neck. The snout is rounded, the eyes small, and the pupils vertical. The scales are smooth; the tail is short and stout. Two grooved fangs are on the roof of the mouth at the back.

Gerarda prevostiana
Cat-eyed Fishing Snake งูปลาตาแมว
(Gerard's Water Snake) งูปากกว้างตาเล็ก

This species averages a length of 0.45 meters. The color is grayish-green, and the skin between the scales dark gray. Its chin is yellow and both the ventral scales and subcaudals are gray. The supralabials are ivory.

Mid-body scale rows number 17. The number of the ventrals and subcaudals is approximately the same for both sexes. The ventrals number from 144 to 151 and the subcaudals from 29 to 36. The ventrals are fairly wide, the subcaudals paired, and the anal divided. One large preocular extends to the top of the head. One loreal, and two postoculars, the lower one extending under the eye, are present. The parietals are well developed. There are eight supralabials, the fourth touching the eye. The seventh is the largest, the eighth the smallest. There are seven infralabials.

The Cat-eyed Fishing Snake prefers coastal and tidal areas, including mangrove forests. It is quite at home in brackish water. Fish, shrimp, and other aquatic animals comprise the bulk of the diet.

This species is found in the coastal areas of Thailand and in similar environments of India, Sri Lanka, Burma, Bangladesh, and peninsular Malaysia.

Genus Homalopsis

This genus also has only one member. The head is large and broad posteriorly and, thus, quite distinct from the neck. The eyes are small, the pupils elliptical.

Homalopsis attains a length of one meter and requires a fairly large cage, half of which should be dry surface. This snake is most often caught in crevices or at tree bases near to water. It is quick to bite when first caught but the venom is mild without noticeable adverse effect on humans. It adjusts to captivity and thrives on a diet of fish and frogs.

Homalopsis buccata — Figure 19
Masked Water Snake — Plates 61 to 63
(Puff-faced Water Snake) — งูหัวกระโหลก. งูเหลือมอ้ำ

The first common name is the result of the white pattern on the top of the head, which gives the appearance of a mask. The body is reddish-brown with a pattern of creamish-white bands. The belly is white with small black dots. Juveniles have a black body with a pattern of white bands.

There are one or two preoculars, one elongated loreal, and two postoculars. Three or four suboculars separate the 13 supralabials from the eye. The parietals are present. The heavily keeled dorsal scales are in rows of 43 to 47, usually 45, at mid-body. The ventrals vary from 160 to 176; males have from 78 to 103 subcaudals, on females they range between 70 and 91. Both the subcaudals and anal plate are divided. Females produce an average of 10 living young at a time, in groups ranging from 2 to 20.

The Masked Water Snake is found in almost every well-watered part of Thailand, the rest of mainland Southeast Asia and also in parts of China and India.

Subfamily Lycodontinae

This is a rather diverse subfamily and with its validity under challenge. Indeed, not all of its members share numerous common characteristics. Nevertheless, Subfamily *Lycodontinae* provides a convenient grouping for certain Thai forms and will be retained in this study.

Genus Dinodon

In this genus, the pupil is vertical. The maxillary bones extend farther anteriorly than the palatine and the posterior maxillary teeth are fang-like.

Members of this genus prefer well-watered areas and prey upon mice and probably frogs and lizards as well. They have no venom producing glands; prey is swallowed alive. Each species is oviparous, and it seems that males are larger than females.

Captives should be provided with a large cage as *Dinodon* grows to a moderately large size. A large water receptacle suitable for soaking should be provided. As they are nocturnal animals their activity during the day will be limited.

Only one species is found in Thailand.

Dinodon septentrionalis
Hill Wolf Snake   งูปล้องฉนวนภูเขา

The ground color of this snake is grayish-black with a number of transverse white bars. Most of the ventral surface is white and the tail bottom is black. The head is broad, somewhat flattened, and distinct from the neck. This species has large, bulging eyes and the pupil is vertical. Adults average 1.15 m in length. Prey consists primarily of cold-blooded animals.

The scales are smooth and arranged in rows of 17 at mid-body. Ventrals range from 207 to 212, the paired subcaudals from 81 to 92, and the anal is single. There are two fairly small postoculars and a small, narrow loreal. Of the eight supralabials, the third, fourth, and fifth border the eye. There are seven or eight infralabials.

The Hill Wolf Snake has only been found in the northern province of Chiang Mai. Outside of Thailand, it is found in northeastern India, Burma, Laos, and southern China. It is probably native to northern Vietnam as well.

Genus Dryocalamus

This genus contains nocturnal, oviparous snakes which prey upon cold-blooded prey. They are excellent climbers and have a very gentle disposition. Adults attain a length of approximately 0.7 m, the body being quite slender. Hypapophyses is absent and the maxillary teeth increase in size posteriorly.

Housing can be of moderate size but should contain ample facilities for climbing as well as a large water dish. An adequate supply of lizards or tree frogs should be assured before members of this genus are included in any collection.

Thailand is home to two members of genus *Dryocalamus*. *Dryocalamus davisonii* is found to the north of 11°N latitude. *Dryocalamus subannulatus* has a more limited range in southern Thailand.

Dryocalamus davisonii
Common Bridle Snake   งูปล้องฉนวนธรรมดา

This small snake averages 0.6 m in length. The head is depressed and distinct from the neck; the eyes are large, the pupils vertical. The body has numerous black cross-bars astride the vertebral ridge which extend approximately halfway down the body sides. They are quite distinct anteriorly but become less distinct posteriorly. The remainder of the body is white, as is the ventral surface. The snout and parietal areas are black but the remainder of the top of the head, especially the area posterior to the eyes, has a heavy dusting of light pigment. The labials, chin, and throat are white. Females produce three or four eggs per clutch which require an incubation period of roughly 70 days. Hatchlings are approximately 0.25 m in length.

The ventrals are notched on the sides, strongly keeled, and number from 233 to 255. The paired subcaudals range from 90 to 108 and the anal plate is single. The body scales are smooth and arranged in rows of 13 at mid-body. Of the seven or, rarely, eight supralabials, the third and fourth touch the eye. There are usually eight infralabials, no preocular, an elongated loreal which is in contact with the eye, and either one or two postoculars.

This species is found in the lowland regions of every part of the Kingdom north of 11°N latitude. It is also found in Burma, Laos, Kampuchea, and southern Vietnam.

Dryocalamus subannulatus [5]　　　　　　Plates 64 and 65
Malayan Bridle Snake　　　　　　　　　งูปล้องฉนวนมลายู

This is an attractive arboreal snake that has only recently been found in Thailand. The head is somewhat flattened, ovate, and distinct; the eye is large, dark, and prominent. The body is slender and the tail long and tapered. Adults attain an average length of 0.6 meters. Chocolate brown and yellow are the primary colors, but two varieties of pattern have been observed. In both, the top of the head is brown, however, a dusting of yellow pigment predominates above and behind the eye extending back through the base of the jaw, creating a masked appearance. In each, the labials, chin, throat, and ventrals are white, although the ventrals may have very faint, light brown blotches. Dorsal

---

　　5. Wirot Nutphand, **Wolf Snakes**, page 7.
　　　Wirot accounts here for a proposed new subspecies, *Dryocalamus davisonii tungsongensis*. This proposed new subspecies, however, appears to be identical to this species.

body patterns, however, differ. In one form, the dorsal surface is chocolate brown with a series of yellow, rough-edged, parallel stripes extending along the body. One stripe commences a few millimeters from the head, and extends along the vertebral ridge to the tip of the tail. The second, usually two scale rows wide, is found at roughly mid-point on the body sides. This stripe commences at the base of the jaw, and completes at the tail tip. The third, sometimes peppered with small brown spots, extends along the first and second scale rows.

The second variety appears to be banded when viewed from above. The chocolate brown bands cross the vertebral ridge, extend halfway down the body sides, leaving the remainder of the body yellow. The cross-bands are widely spaced with distinct edges anteriorly but posteriorly they become more closely spaced, smaller, and the edges are less distinct. Small brown blotches appear on the lower body side at the first third of its length. They are usually located between and below the cross-bands and increase in size posteriorly. The posterior pattern is very irregular, creating a lacey appearance.

The single loreal is elongated and contacts the eye beneath the lone preocular. There are two small postoculars, seven infralabials, and also seven supralabials, the third and fourth being in contact with the eye. The scales are smooth and in 15 rows at mid-body. The keeled ventrals range from 225 to 244, the paired subcaudals from 88 to 107, and the anal is single.

In Thailand, this attractive snake has only been found south of 11°N latitude, most being found in the province of Nakhon Si Thammarat. It is also native to peninsular Malaysia, western Indonesia, and Singapore.

Genus Lepturophis

This genus contains a single species. Certain characteristics of dentition and scalation are similar to those of genus *Lycodon*. The sole member of this genus, however, has ventral scales which are keeled, and notched laterally, a characteristic shared with members of arboreal genera, such as *Chrysopelea* and *Dryophiops*, but absent in the terrestrial genus *Lycodon*. Data from a limited number of specimens indicate that males may be larger than females.

Specimens can grow to moderate lengths and are probably arboreal. Thus, captives require a moderately large cage as well as facilities for climbing. As the single member of this genus is found in the south of Thailand, the climatic conditions outlined within Appendix 7 should be simulated and supplemented with occasional spraying. Frogs and lizards are known to be at least part of the diet of this snake. It may eat mice and birds as well. Specimens have been seen foraging on the ground at night.

Lepturophis borneensis   Plate 66
Slender Wolf Snake   งูปล้องฉนวนบอร์เนียว

This species has been equated with *Lycodon albofuscus* by at least one respected taxonomist. I have been unable to determine the basis for this apparent reclassification and, therefore, retain the above taxon.

The head is flattened and quite distinct from the neck; the rostral is barely visible from above. The large nostrils are located above the first supralabial. The eye is vertically elliptic. The circular body is long and slender, as is the tail;

adults may achieve a length in excess of 1.6 meters. Specimens found outside of Thailand have been described as brown or dark gray dorsally and white or yellowish-white ventrally. Each specimen found in Thailand, however, had a black dorsal and a white ventral surface. In each case, the labials, chin, and throat were white. It is not known whether this species is oviparous or viviparous.

The body scales have finely serrated keels and are arranged in 17 rows at mid-body. In the limited number of specimens examined to date (12), the ventrals have ranged from 243 to 247 in males and from 225 to 238 in females. The ventrals are keeled and notched in both sexes. The subcaudals are paired and range from 189 to 206 in males and from 113 to 182 in females. The anal is divided. There is a single loreal, one preocular, and two postoculars. Of the eight supralabials, the third, fourth, and fifth border the eye. The first five of the nine lower labials are in contact with the anterior chin shields. The first lower labials are in contact along the mental groove, separating the mental from the anterior chin shields.

This snake is most frequently found in forests near the banks of steams. *Lepturophis borneensis* is native to Sabah and Sarawak, on the island of Borneo. In 1967 a specimen was found on peninsular Malaysia in the state of Pahang. Subsequently, other specimens have been found on peninsular Malaysia. This species has now extended its range into Thailand. Specimens have been found as far north as Songkhla Province.

Genus Liopeltis

These snakes are small, usually under 0.8 m in length. The body is cylindrical, the eye large, and the pupil round. The dorsal scales are smooth, the tail long, and the subcaudals paired. Maxillary teeth are roughly equal in size.

Members are terrestrial and most often found under debris on forest floors, sometimes on high elevations. They can be kept in small aquariums, but pieces of bark and similar litter should be provided for them. The aquarium should be sprayed with tepid water frequently to help maintain the humidity at a high level. The food preferences of *Liopeltis* are not well known, but the remains of a snake were found in the stomach of one. Perhaps they will also accept lizards and frogs.

Females are oviparous and larger than males.

Liopeltis baliodeira cochranae
Striped Ringneck งูสายทองลายแถบ

The head, scarcely distinct from the neck, is nearly uniformally light brown. A series of eight or nine black bars, separated by cream or light brown areas, extends from the neck onto the anterior portion of the body, where they become indistinct and blend into the violet-brown body. Some of the black bars are broken by a light mid-dorsal line, others are not. One violet-brown line extends along the outer edge of the ventrals, separating the white ventrals from the dorsal coloration. The ventral side of the head, throat, and subcaudals are also white.

The nasals are fused above the nostrils. There are two preoculars and two postoculars, but no loreal is present. Of the eight supralabials, the fourth and fifth are in contact with the eye. There are seven infralabials. The smooth dorsal scales are in 13 rows at mid-body. The type specimen has 118 ventrals, 72 subcaudals, and the anal is divided. Its snout to vent length is 0.256 m and the tail 0.118 m, giving a total length of 0.374 meters.

The type specimen, the only one recorded to date, is now cataloged as specimen No. 94826 in the United States National Museum in Washington, D. C. It was found on Khao Soi Dao (12^0 55' N, 102^0 14' E) in the southeastern province of Chanthaburi. Khao Soi Dao or, Soi Dao Mountain, reaches a height of 1525 meters. The surrounding terrain is heavily forested and rugged, with many peaks reaching elevations of 1250 meters.

Liopeltis scriptus
Common Ringneck   งูสายทองคอแหวน

This small snake, with a length of approximately 0.35 m, has a cylindrical body and a head which is only slightly distinct from the body. The dorsal color is gray. The head is gray to brownish-gray. A narrow black ring encircles the eye, but widens on the supralabials below the eye. A cream line extends from the fifth and sixth supralabials to behind the eye, and a black-edged cream line extends from the eighth supralabial nearly to the parietal. The side of the snout is very light brown. A narrow dark band, with a white posterior edge, encircles the neck. The ventral surface is cream.

There is a loreal and a preocular, as well as two postoculars. The supraoculars are approximately half the width of the frontal. Of the eight supralabials, the third, fourth, and fifth contact the eye, and the seventh is clearly the largest. Eight infralabials are present. There are 13 scale rows at mid-body, the scales being smooth. Ventrals vary from 122 to 145, the anal is divided, and the subcaudals range from 87 to 103.

Thai specimens have been found from Kanchanaburi in the West to Nakhon Si Thammarat in the South. Apart from Thailand, this species has only been found in southern Burma.

Liopeltis tricolor
Malayan Ringneck   งูสายทองมลายู

This is the largest member of the genus that resides in Thailand with adults reaching a length of 0.5 meters. The color is greenish or olive dorsally. A distinct black line passes through the eye and extends two to five centimeters behind the head prior to tapering and finally disappearing. The ventral surface is yellow.

The ventrals range from 140 to 187, the anal is divided, and subcaudals vary from 103 to 137. Unlike the other two members of the genus, this species has 15 scale rows at mid-body. Of the eight supralabials, the fourth and fifth touch the eye.

Thai specimens have been found in the South. The Malayan Ringneck is also native to peninsular Malaysia, Singapore, western Indonesia, and the island of Borneo.

Genus Lycodon

This is a genus of relatively small snakes. Although some of the anterior teeth on the upper jaw are enlarged and fang-like they are not real fangs. Snakes in this genus do not possess an ability to produce venom and are completely harmless. Their heads are flattened and slightly distinct from their bodies, they have an elongated loreal, and their eyes have vertically elliptical pupils. All are nocturnal and feed primarily upon lizards, which are grasped and swallowed alive. Of the four species found in Thailand, all are oviparous with females slightly larger than males.

These snakes thrive in relatively small cages. They are terrestrial and do not need objects for climbing, but the addition of one or two small plants will make the cage attractive and provide hiding areas.

Lycodon capucinus Plate 67
Common Wolf Snake
(House Snake) งูสร้อยเหลือง

The head and body are brown. The supralabials are white with brown sutures, and normally there is a residual light band around the neck. Some body scales have white edges which form a pattern of irregular, wavy white lines on the body. The ventral surface is dull white. This species averages 0.47 m in length; females produce 3 to 11 eggs at a time. Skinks and geckos form the bulk of its diet.

The scales are smooth and arranged in rows of 17 at mid-body. The ventrals are keeled and number 189 to 205; the paired subcaudals range from 61 to 71. The anal is divided. There is one large preocular, two postoculars, an elongated loreal, ten infralabials, and nine supralabials. The third, fourth, and fifth supralabials are in contact with the eye.

The House Snake is found throughout Thailand and has also been reported on the remainder of mainland Southeast Asia in addition to the Philippines and western Indonesia. It frequently enters houses in search of house geckos, a staple of its diet.

Lycodon fasciatus
Banded Wolf Snake   งูปล้องฉนวนเมืองเหนือ

This is an attractive snake. The head is brown with each supralabial having a light spot. The shiny black body has a series of yellow bands. On the anterior of the body these bands narrow and have an abrupt edge, but posteriorly they become slightly wider with fuzzy edges. These bands continue across the ventral surface. The throat and posterior infralabials are cream colored. Specimens may achieve an adult length of 0.8 meters. Females produce from 4 to 14 eggs at a time and other snakes, as well as lizards are reported to be a part of the diet.

The scales are slightly keeled and arranged in rows of 17 at mid-body. Ventrals vary from 197 to 220 while the paired subcaudals range from 69 to 94. The anal is single. The loreal contacts the eye and there is one preocular above. There are two postoculars and eight supralabials, of which the third to fifth border the eye. Nine infralabials are present.

This snake is found at elevations between 1000 to 2300 m in the North. It has also been reported in northern India, Burma, Laos, northern Vietnam, and southern China.

Lycodon laoensis   Plate 68
Laotian Wolf Snake   งูปล้องฉนวนลาว
(Indochinese Wolf Snake)   งูปล้องฉนวนลายเหลือง

This is a very beautiful snake. The top of the head and the anterior supralabials are deep blue. The posterior labials and the chin are cream colored. The body is a glistening black and crossed by a series of yellow bands with white edges. These bands occasionally widen and enclose areas of black. Anterior bands are wider than posterior ones, which often enclose black areas. The ventral surface is entirely white. Adults average 0.4 m in length with females producing approximately five eggs per clutch.

There are nine supralabials, with the third, fourth, and fifth touching the eye. There are also ten infralabials, one large preocular, two postoculars, and one elongated loreal. The scales are smooth and the scale rows number 17 at midbody. Ventrals range from 163 to 187, the paired subcaudals from 60 to 76, and the anal is divided.

The Indochinese Wolf Snake is common in the South, Southeast, and Northeast. It is also found in Laos, southern Vietnam, Kampuchea, and peninsular Malaysia.

Lycodon subcinctus   Figure 20
Malayan Banded Wolf Snake   Plates 69 and 70
                                             งูปล้องฉนวนบ้าน

This species is the largest of the genus *Lycodon* found in Thailand, attaining an adult length of approximately one meter. Adults are dark brown or black with a broken white patch on the sides of the neck. There are a few widely spaced white bands on the anterior portion of the body. Proceeding posteriorly, the white bands become progressively indistinct to create a posterior that is uniformly black. The chin and throat are gray-white, as are the ventrals. The edges of most ventral scales are black. Young specimens sometimes have up to 20 white bands which can allow them to be mistaken for *Bungarus candidus*. As they mature, however, the bands gradually disappear, commencing with the posterior ones. Refer to: Plate 70, page 226.

Figure 20. *Lycodon subcinctus*. The loreal is hatched.

The preocular is absent in this species; the loreal touches the eye. There are two postoculars, nine infralabials, and eight supralabials. The third, fourth and fifth supralabials border the eye. Body scales are faintly keeled and arranged in rows of 17 at mid-body. The ventrals number from 197 to 230, the subcaudals are paired and range from 71 to 90, and the anal divided. Females produce 5 to 11 eggs per clutch which hatch in about 70 days.

Lycodon subcinctus is found throughout the Kingdom. The range includes Laos, Vietnam, Kampuchea, southern China, Malaysia, western Indonesia, and the Philippines.

Genus Oligodon

The snakes of this genus are short to medium in length. They are diurnal, terrestrial, and generally found under debris. Their rostrals are quite large and intrude slightly between the nasals. The upper rear teeth are enlarged but are not real fangs. Members of this genus have a venom producing gland, but are not yet proven to be dangerous to man. The subcaudal scales are paired. The genus is oviparous, with female members producing three to six eggs at a time. Size differences between the sexes varies from species to species.

This genus generally does not do well in captivity. Specimens seem vulnerable to stress and are often reluctant to eat and are usually quick to bite. They are also nocturnal and spend most of the day hiding. Small cages will suffice, but should have numerous places to hide in order to minimize stress. Plants and pieces of bark will serve this purpose well. A water dish is recommended in addition to the occasional spraying with tepid water. Their prey consists of frogs and lizards.

This is one of the more abundant genera in Thailand. Twelve members of the genus have been found in the Kingdom to date.

Oligodon barroni   Plate 71
Barron's Kukri Snake   งูปี่แก้วหัวลายหัวใจ

This small snake averages 0.36 m in length. Its head is light brown, and a dark brown stripe passes from the area of the frontal through the eye to the fifth supralabial. A diagonal stripe extends from the parietal to the third ventral, and a heart-shaped spot is present on the neck. A series of dark brown lines extends the entire length of the body, often parallel to a series of small spots. The chin is white and the ventral surface coral red with two rows of brown marks extending onto the subcaudals. Females are slightly larger than males.

There are seven or eight supralabials, of which the third and fourth or the fourth and fifth touch the eye. Infralabials number nine, and there is one preocular and two postoculars, plus a loreal. The scales are smooth and arranged in rows of 17 at mid-body. Ventrals number from 135 to 160 and the subcaudals from 32 to 44, the anal is single.

This is not a common snake, but this may be due more to its size and secretive nature rather than to a lack of numbers. Specimens have been found in the Southeast in Chon Buri and Rayong Provinces, and in Saraburi Province of the Central Region. It has also been found in southern Kampuchea.

Oligodon cinereus multifasciatus
Ashy Kukri Snake   งูปี่แก้วลายกระธรรมดา

This snake reaches the adult length of 0.68 meters. The color is gray-brown. Many scales have black edges and tend to form a series of broken lines along the body. On the neck is a narrow chevron with the arms extending a short distance onto the body. The head is a darker shade of brown, and there is a dark line extending from the supralabials through the eyes to the frontal. There are dark spots on the parietals and frontal and a diagonal dark line above the mouth. The female of the species is slightly larger than the male.

The scales are smooth and in rows of 17 at mid-body. Ventrals number from 190 to 197 and the paired subcaudals from 32 to 42. The anal is single. There are two preoculars, the upper considerably larger than the lower, and one elongated loreal. There are also two postoculars, eight infralabials, and seven supralabials. The fourth and fifth supralabials are in contact with the eye.

This subspecies is found in the North in Chiang Mai Province and in the Northeast in Loei Province. It has also been recorded in Burma and Kampuchea. Most probably this snake will also be present in Laos, although sightings have not yet been reported.

Oligodon cinereus swinhonis
South Chinese Kukri Snake   งูปี่แก้วลายกระเหมืองจีน

This subspecies reaches a mature length of approximately 0.63 meters. The dorsal color is light brown to gray. There is a dark chevron on the nape, the apex extending on to the parietal and arms reaching the sides of the neck. The only other marking on the head is a faint dark mark just below the eye. A series of approximately 21 irregular bands extend across the body to the first dorsal scale rows. The ventral surface is yellowish-white.

There is one loreal which is longer rather than higher, a preocular, and two postoculars. Of the seven supralabials, only one touches the eye; there are nine infralabials. The smooth dorsal scales are in 17 rows at mid-body. The ventrals range from 174 to 196, the anal is single, and the subcaudals number from 34 to 42.

To date, only one specimen has been recorded in Thailand. The location where the specimen was sighted is situated in the Chiang Mai Province of the North. This subspecies has been reported in southern China and most probably the snake will also occur in Laos and northern Vietnam.

Oligodon cyclurus smithi Plate 72
Common Kukri Snake งูปี่แก้วธรรมดาลายจาง

This subspecies may reach the adult length of one meter. The overall color is yellowish-brown with a series of dark-edged brown or black blotches that are centered on the vertebral column. Each of these blotches is separated by three rows of narrow black bands. There is a narrow chevron on the back of the neck with the arms extending down the body sides to the tenth ventral level. The apex ends on the frontal. A dark stripe extends from above the eye to the base of the jaw. In addition, a dark line passes through the eye to the labials. The ventral surface is white and unmarked. Males are significantly larger than females; clutches contain from 3 to 16 eggs.

Scale rows number 21 or 23 at mid-body and the scales are smooth. The ventrals number from 161 to 185 on males and from 170 to 195 on females; males have from 42 to 58 subcaudals, females from 36 to 46. The anal is single. There is one loreal, two preoculars, a pre-subocular, two postoculars, and nine infralabials, the fourth and fifth border the eye.

This is a common snake in Thailand north of 11° 15' N latitude. Its considerable range also includes the areas of northeastern India, Bangladesh, Burma, Laos, Kampuchea, and Vietnam.

Oligodon cyclurus superfluens
Taylor's Kukri Snake   งูปี่แก้วธรรมดาลายเข้ม

This subspecies reaches an adult length of just over 0.5 meters. The body color is a light brownish-gray with numerous transverse lines formed by the black edges of the dorsal body scales. A chevron mark on the nape of the neck has its apex on the frontal scale with the arms reaching the level of the ninth ventral scale. Two bars cross the snout, the foremost bar passes through the eye, the second passes behind the eye.

There is a single quadrangular loreal, two preoculars, and two postoculars. The lower preocular intrudes between the third and fourth supralabial. Of the eight supralabials, the fourth and fifth contact the eye; nine infralabials are present. There are 21 scale rows at mid-body, 179 ventrals, 47 subcaudals, and the dorsal scales are smooth.

To date, just one specimen has been found in Thailand. This specimen was captured and subsequently described by Edward H. Taylor in **The Serpents of Thailand and Adjacent Waters** (page 768). Taylor's specimen, a male, was captured at an elevation of 150 m in the northern province of Chiang Mai.

Plates 49 & 50. *Psammodynastes pulverulentus*. Common Mock Viper. งูหมอก.

Photo: Jarujin Nabhitabhata

Plate 49. (above)
Small, slender and despite viper-like head and rear fangs, not yet proven harmful to man. Pattern variable.
Prey - lizards.
Thrives in captivity.

Plate 50. (right)
Reddish color phase.
Color and pattern variable.
Compare to Plate 49.

Plate 51. *Psammophis condanarus indochinensis*
Indochinese Sand Snake. งูก่านทอง
Attractive, quick, agile, common and harmless. Thrives in captivity.

Photo: Jarujin Nabhitabhata

Plate 52.

Ptyas korros
Indochinese Rat Snake
งูสิงธรรมดา

Captured in Pattani Province. Very fast and difficult to catch. Thrives in captivity when well housed.

Photo: Jarujin Nabhitabhata

Plate 53.
Ptyas mucosus
Oriental Rat Snake
Common Rat Snake
งูสิงหางลาย
Large, very fast, harmless and very beneficial. Typical large eyes, round pupils. Specimen recently killed.

Photo: Jarujin Nabhitabhata

Plate 54.

Cerberus rynchops
Dog-faced Water Snake
งูปากกว้างน้ำเค็ม

Common in areas with brackish water.

Plate 55. *Enhydris bocourti*. Bocourt's Water Snake. งูโซ.
Short, heavy-bodied, harmless and quite common. Prey - fish, frogs.

Photo: Suthigit Patramangorn

Plate 56.

Enhydris enhydris
Rainbow Water
Snake
งูสายรุ้งธรรมดา

Note: Head barely distinct from thick, striped body. Thrives on a diet of fish, and frogs.

Photo: Suthigit Patramangorn

Plate 57. *Enhydris jagori*. Striped Water Snake. งูสายรุ้งลาย.
Small, less common water snake. Captives need ample dry surface.

Photo: Suthigit Patramangorn

222   *Plates*

Photo: Suthigit Patramangorn

Plate 58.

Enhydris plumbea
Plumbeous Water Snake
งูปลิง

Thick-bodied with fairly large head. Common. Thrives in captivity.

Photo: Suthigit Patramangorn

Photo: Jarujin Nabhitabhata

Plates 59 and 60.
Erpeton tentaculatum
Tentacled Snake
งูกระด้าง

Plate 59. (above) Two Thai color phases. Small rostral "tentacles" - easy to identify.

Plate 60. (left) Head showing unique rostral appendages.

Plates 61 to 63.
Homalopsis buccata
Masked Water Snake
Puff-faced Water Snake
งูหัวกระโหลก. งูเหลือมอ้อ

Plate 61. Gentle, attractive. Thrives in captivity. Common in well-watered areas.

Photo: Suthigit Patramangorn

Plate 62. Juvenile marked black and yellow, dissimilar to parents. Young born live - average 10 per litter.

Plate 63. Head of an albino. Lack of pigmentation in this specimen makes the head scales noticeable. Note: small parietals.

Plate 64.

Dryocalamus subannulatus
Malayan Bridle Snake
งูปล้องฉนวนมลายู

Striped form. Recent addition to Thai herpetofauna. Small, arboreal. Found in South.

Plate 65. (below)

Dryocalamus subannulatus
Malayan Bridle Snake
งูปล้องฉนวนมลายู

Banded form. Shares range with striped variety.

Plate 66. (below)

Lepturophis borneensis
Slender Wolf Snake
งูปล้องฉนวนบอร์เนียว

Commencement of shedding.

Photo: Raynoo Cox

Plate 67.

Lycodon capucinus
Common Wolf Snake
House Snake
งูสรอยเหลือง

Widespread throughout the Kingdom.

Photo: Jarujin Nabhitabhata

Plate 68.
Lycodon laoensis
Laotian Wolf Snake
Indochinese Wolf Snake
งูปล้องฉนวนลายเหลือง
งูปล้องฉนวนลาว
Beautiful, small, terrestrial. House in small aquarium.
Diet small lizards.

Photo: Suthigit Patramangorn

Plate 69.

Lycodon subcinctus
Malayan Banded Wolf Snake
งูปล้องฉนวนบ้าน

Largest of genus found in Thailand.
Common. Does well in captivity.

Photo: Suthigit Patramangorn

226   Plates

Plate 70.
Lycodon subcinctus
Malayan Banded
Wolf Snake
งูปล้องฉนวนบ้าน
Juvenile is banded throughout body. Bands fade with maturity - some remain on the body anterior.

Photo: Jarujin Nabhitabhata

Plate 71.

Oligodon barroni
Barron's Kukri Snake
งูปี่แก้วหัวลายหัวใจ

Defensive position. Small, harmless - can inflict a painful bite.

Plate 72. *Oligodon cyclurus smithi*. Common Kukri Snake. งูปี่แก้วธรรมดาลายจาง. Attractive, attains 1 m. Note: typical head, neck markings. Author found and photographed in Bangkok.

Oligodon dorsalis
Gray's Kukri Snake   งูปี่แก้วภูหลวง

This species is dorsally dark brown to purple. There is a light vertebral stripe which is either black-edged or contains small black spots. A second black stripe extends along the second and third dorsal scale rows. The dorsal surface of the tail usually has two or three large black spots. The head is dark brown and usually has hints of two crossbars, one before, and one behind the eye, as well as a chevron. The ventral surface is mostly black anterior to the vent, and yellowish-orange posterior to it. Males are approximately 0.5 m in length, but the females tend to be larger.

There are 15 scale rows at mid-body, ventrals range from 162 to 188, the subcaudals from 27 to 51, and the anal paired. There is one loreal, one preocular, two postoculars, and seven supralabials. The third and fourth supralabials are in contact with the eye.

This species was only recently found in Thailand. A single specimen was collected in the Phu Luang Wildlife Sanctuary, Loei Province in the Northeast, at an elevation of approximately 1300 meters. It is also native to northeastern India, Bangladesh, and Burma. Its presence in Thailand indicates that its range extends into Laos as well.

Oligodon dorsolateralis
Wall's Kukri Snake   งูปี่แก้วหลังจุดวงแหวน

This species is light brown in color with four stripes extending along the body. The two uppermost stripes extend to the tail, the two on the sides extend to the vent. Ten blotches also occur on the back. The two uppermost stripes merge with a chevron located on the back of the neck. The apex of the chevron extends to the frontal. One dark stripe passes behind the eye to the base of the mouth with another passing from the snout through the eye to the fifth supralabial. The chin and ventral surface are ivory in color. *Oligodon dorsolateralis* achieves a mature length of approximately 0.6 meters.

There are two preoculars, two postoculars, a loreal, and nine infralabials. Of the eight supralabials, the fourth and fifth border the eye. The scales are arranged in rows of 21 at mid-body. Approximately 176 ventrals and 53 subcaudals are present. The anal is single.

This snake has only been found in the northern province of Chiang Mai. It also occurs in Burma, northeast India, and perhaps Bangladesh.

Oligodon inornatus Plate 73
Inornate Kukri Snake งูปี่แก้วสีจาง

This is a small snake, reaching a length at maturity of 0.6 meters. The head and body are brown, but the supralabials are light brown with dark sutures. There are no markings on the head or the body. The ventral surface is yellowish-white.

There are eight supraoculars with the fourth and fifth touching the eye. Preoculars are absent but there is a loreal and two postoculars. The scales are arranged in rows of 15 at mid-body. The ventrals number approximately 170 and subcaudals 40. The anal is single and the tail long and thin.

This species is found in forested highlands in Tak (North) Uthai Thani (Central Region) and Chon Buri (Southeast). It is also found in Kampuchea.

Oligodon joynsoni Plate 74
Grey Kukri Snake งูปี่แก้วใหญ่

Adult length is approximately 0.76 meters. It is purplish-brown with a series of faintly defined bands across the back. Alternate bands are enlarged to become dorsal blotches. A large brown band passes across the prefrontals and through the eyes. A dark stripe exists behind the eyes with a narrow chevron at the back of the neck; the apex extends to the frontal. Ventrals are yellow-white with large dark blotches.

The scales are smooth and in rows of 17 at mid-body. Ventrals range from 187 to 197 and the subcaudals from 40 to 50. The anal is single. There is one preocular, a loreal, a subocular, and two postoculars. The supralabials number eight; the fourth and fifth touch the eye.

The Grey Kukri Snake is only found in Lampang Province. It is not seen outside Thailand.

Oligodon mouhoti
Cambodian Kukri Snake   งูขอดเขมร

This small species probably averages a length of 0.25 m and is gray in color with two light brown stripes extending along the flanks of the vertebral ridge to enclose a series of yellow spots. There are two dark spots on the dorsal surface of the tail, one at the base, the other at the tip. The ventral surface is pink. A dark band extends across the tip of the snout through the eyes into the supralabials. The rostral and the first few supralabials are white. A black nuchal collar reaches the lateral edges of the ventral scales. A black blotch on the back of the neck extends to the parietals.

The scales are smooth and arranged in rows of 17 at mid-body. The ventrals range from 146 to 169 and the subcaudals from 30 to 47. The anal is single. There is one loreal and of the two preoculars, the lower is the smaller. A pre-subocular scale is present. There are also two post-oculars, nine infralabials, and eight supralabials, with the third, fourth, and fifth bordering the eye.

Errors in early taxonomic descriptions have resulted in confusion regarding *Oligodon mouhoti* and *Oligodon taeniatus*. There are clear differences, however. *Oligodon mouhoti* has a nuchal collar which extends to the ventral scales, two dark spots on the dorsal surface of its tail, and 17 rows of scales at mid-body. On *Oligodon taeniatus*, the nuchal collar stops short of the ventral scales, there are no spots on the tail, and there are 19 scale rows at mid-body.

Oligodon mouhoti is found in the Northeast Region of Thailand and also in Kampuchea.

Oligodon purpurascens purpurascens
Brown Kukri Snake งูกด

This snake is purplish-brown with a series of dark-edged bands across the back. Narrower, less clearly defined bands are often present between the broader, more clearly defined ones. There is a narrow chevron on the back of the neck with the apex passing through the parietals to the frontal, where it broadens and covers most of the prefrontals and part of the internasals. One wide, dark band extends down from the eyes into the fourth and fifth supralabials and a second thinner one passes behind the eye to the base of the jaw. The ventral surface is pink or yellow, and there are dark spots on alternate ventrals. As is often the case, juveniles are more brilliant than adults. This is a rather large member of the genus and may reach a length of 0.8 meters.

Mid-body scale rows number either 19 or 21. The ventrals number from 160 to 210 and the subcaudals from 40 to 60. The anal is divided. There is one preocular, one loreal which is greater in height than width, one small subocular, a supraocular, and two postoculars. Of the eight supralabials, the fourth and fifth or only the fifth touches the eye. Infralabials number eight or nine.

This species is found in the South from Surat Thani Province south to Pattani Province. It may eventually be found in the provinces which border Malaysia as it is reported to be a common snake on peninsular Malaysia. It is also found in Singapore and western Indonesia.

Oligodon taeniatus Plate 75
Striped Kukri Snake งูขอด

This species is quite small, usually around 0.25 meters. Specimens are gray with two narrow, light brown stripes extending along the body flanking the vertebral ridge. The vertebral ridge is light in color with small yellow spots. A broad, dark brown band extends from the snout through the eyes on to the supralabials. The rostral is light brown with black edges; the first four supralabials are tan, the remainder white. A dark brown band extends from the parietals down and along the base of the jaw, but does not reach the ventral scales. The chin is white and the ventral surface orangish-red with black markings. The black markings are small and widely spaced anteriorly but they become larger and more frequent posteriorly to eventually become crossbands near to the anal opening. On the subcaudals they become small, widely spaced black dots again. A white line extends along the edges of the ventral scales and continues as a narrow line along the ventral edges. Females are slightly larger than males.

There are two preoculars, the lower being the smaller, but no pre-subocular scale. There are also two postoculars, a loreal, eight or nine infralabials, and eight supralabials. The third, fourth, and fifth supralabials are in contact with the eye. The ventrals range from 146 to 169, and divided subcaudals from 30 to 47. The anal is single. The scales are smooth and arranged in rows of 19 at mid-body.

Oligodon taeniatus is a common snake in the Central Region, and has also been found as far south as the southern province of Pattani. Its range extends into Kampuchea, Laos, and southern Vietnam.

Subfamily Natricinae

This is a large and fairly diverse subfamily which includes members from both temperate and tropical areas. The structure of the hemipenis and strong hypapophyses unite Subfamily *Natricinae*. Dentitional characteristics and the shape of the internasals, resulting in either lateral or dorsolateral location of the nostrils, are also significant factors in classifying members of this subfamily.

Most members native to Thailand have moderate to large eyes, round pupils, distinct heads, and elongated, cylindrical bodies. Most have enlarged, fang-like teeth in the back of the upper jaw and some produce a toxic fluid, or venom, which helps them to subdue their cold-blooded prey. The body scales are usually keeled and both the subcaudals and anal plate are usually divided.

These snakes may be either terrestrial or semi-aquatic and exist at low or high elevations. Thai specimens are oviparous and, in general, females are larger than males.

In Thailand, this subfamily is represented by 6 genera and 15 species and subspecies.

Genus Amphiesma

In this genus, either the maxillary teeth become increasingly larger posteriorly, or the last two teeth are abruptly enlarged. The internasal scales are broad anteriorly and the nostrils are located laterally, on the side of the head. The scales are keeled.

Members of this genus are of moderate size, terrestrial, and oviparous. A cage of medium size with ample hiding places will be adequate for most specimens. These snakes are often nervous and will strike repeatedly if a refuge is not provided. Eventually they settle and do well in captivity, if provided with clean water and a diet of lizards, frogs, and fish.

Amphiesma deschauenseei Plate 76
Northern Keelback งูลายสาบท้องสามชิด

This is a small member of the genus, whose adults average only 0.48 m in length. The head is distinct from the neck. The top of the head is black, and the black color extends through the neck and down the vertebral ridge. The sides of the head are brownish-yellow with thin black lines extending from the eye to the sides of the neck. The supralabials are brownish-yellow with black sutures. The sides of the body are also brownish-yellow but heavily mottled with irregular black blotches. The ventral surface is light with a series of light brown spots that become denser towards the tail.

There is one large, elongated loreal, two preoculars, the uppermost being the larger, and two postoculars. Of the nine supralabials, the fifth and sixth touch the eye. Infralabials number nine or ten. Anteriorly the internasals are not tapered and the nostrils are to the side of the snout. The eyes are round. The scales are keeled and in rows of 19 at midbody. There are approximately 159 ventrals and 140 subcaudals. The subcaudals are paired and the anal divided.

The Northern Keelback is found in the northern province of Chiang Mai and as far south as Uthai Thani Province of the Central Region. Its range may extend into Burma. This species is most often found during the night alongside streams in forested areas.

Amphiesma groundwateri
Groundwater's Keelback   งูลายสาบท่าสาร

This small slender snake has black and yellow markings. The black head is slightly distinct from the neck. The yellow labials are black-edged. A yellow stripe extends from the jaws to the back of the neck to merge into a V-shaped yellow mark. Festooned, yellow lines flank the vertebral ridge and extend from the base of the neck to the tail. The ventral surface is yellow. Each ventral and subcaudal scale has a black spot. Adults average 0.45 meters.

Body scales are smooth on the anterior, however, they are feebly keeled on the posterior portion. Mid-body scale rows number 17. The ventrals number from 147 to 154 and the subcaudals from 120 to 132. The subcaudals are divided and the anal single. The internasals are truncated and the nostrils on the side of the snout. There is one elongated loreal, two preoculars, and two or three postoculars. Nine supralabials, with the fourth, fifth, and sixth touching the eye, and either nine or ten infralabials are present.

This small snake has only been found in Chumphon Province in the South.

Amphiesma inas   Figure 21
Malayan Mountain Keelback   งูลายสาบมลายู

This is a small, slender, olive-brown snake. A series of light spots is located anteriorly on each side of the body which tend to merge and form one line that continues to the posterior. The body has a speckled appearance due to an abundance of black-edged scales. The head is brown with small dark spots. The supralabials have light cream stripes

Figure 21. *Amphiesma inas.*

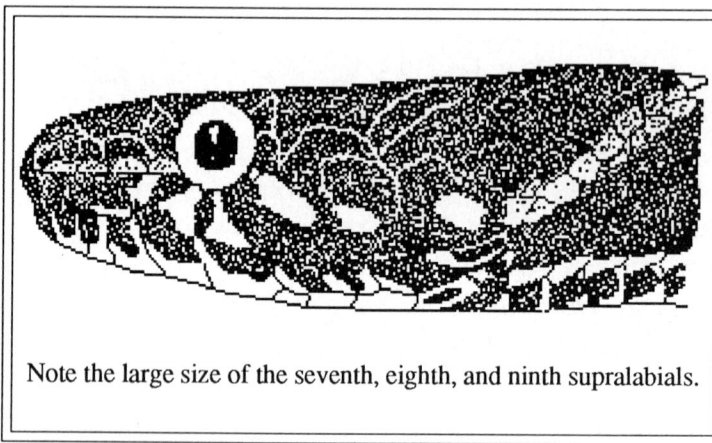

Note the large size of the seventh, eighth, and ninth supralabials.

which extend posteriorly to join the stripes on the body sides. A pair of cream spots exist on the parietals. The ventral surface is light yellow but each ventral scale is edged in black. The eyes are large with the nostrils located to the sides of the snout. The largest specimen recorded was 0.6 m in length.

There are 19 scale rows at mid-body, and the scales are keeled. The ventrals range from 143 to 151, the subcaudals from 93 to 109, and the anal divided. There is one pre-ocular, one loreal, and three postoculars, the uppermost being the largest. Nine supralabials are present, with the fourth to the sixth being in contact with the eye. There are ten infralabials, the tenth being the smallest.

In Thailand, this species is found in the mountainous areas of the South. It is also found in peninsular Malaysia. Throughout its range, it is found at high elevations. Therefore, this species should be maintained at slightly lower temperatures than those indicated within Appendix 7.

Amphiesma stolata chinensis   Plate 77
White-striped Keelback   งูลายสาบคอกหญ้า

The head is greenish-brown and the supralabials are white except for the posterior ones, which are gray. Each supralabial is black-edged. Two wide black lines extend from the eye to the fifth and sixth labials. The body is light brown, with two light dorsolateral stripes extending the full length of the body. Black spots occur between the stripes and along each of the body sides. The chin, throat, and infralabials are white. Adults reach a length of approximately 0.55 meters. Females produce 3-12 eggs in each clutch that normally require 60 days to hatch. Hatchlings are between 0.133 and 0.171 m in length. The prey of this species is primarily frogs.

    The keeled scales are in 19 rows at mid-body but the first row is smooth. Ventrals vary from 118 to 158 and the subcaudals from 50 to 89. The internasals narrow anteriorly, and the nostrils are located toward the top of the snout. The posterior teeth on the upper jaw are abruptly enlarged. This subspecies has one large preocular, one loreal, and also three postoculars. Of the eight supralabials, the third, fourth and fifth are in contact with the eye. There are ten infralabials.

    This subspecies is terrestrial but is found in well-watered lowlands and on elevations up to 2000 meters. In Thailand, it is found in most provinces north of Bangkok. This snake has also been sighted in Laos, southern China, Vietnam, Kampuchea and Burma.

Genus Opisthotropis

In this genus, the head is either distinct or barely distinct from the cylindrical body. The eyes are moderate to small in size, and have either rounded or vertically sub-elliptic pupils. The nostril and nasal are directed upwards and outwards. The scales may be smooth or keeled and are in 15, 17, to 19 rows at mid-body. The anal is divided and the subcaudals paired.

These are terrestrial snakes usually found in or near to mountain streams and should be maintained at temperatures slightly lower than those listed in the appendixes. Medium-sized aquariums provided with numerous hiding places will comfortably accommodate them, provided a water receptacle is included. Their diet probably consists largely of cold-blooded prey, such as frogs, lizards, and perhaps fish. Their breeding habits are unknown.

Only two members of this genus have been found in the Kingdom, each in very small numbers.

Opisthotropis praemaxillaris
Angel's Mountain Keelback   งูลายสอลาวเหนือ

The head is barely distinct from the cylindrical body. The dorsal color is uniformly brown, the ventral surface is yellowish. The supralabials are white with brown edges. The length of each of the three specimens examined to date were 0.21, 0.214, and 0.277 meters. The two smaller specimens each had an egg tooth, indicating that this species might be oviparous.

The internasals are nearly twice as long as broad; the rectangular loreal is not in contact with the internasal. There is one preocular, two postoculars, and nine supralabials. The fourth and fifth supralabials contact the eye, but the sixth is excluded from contact by the lower postocular. The dorsal scales are smooth and in 19 rows at mid-body. There are from 145 to 149 ventrals and 63 to 67 subcaudals.

The only Thai specimen was found in the northern province of Chiang Mai. The specimen was taken from a fast moving stream in a forested area on Saket Mountain (19° N, 99° 15' E) at an elevation of approximately 1000 meters. The two type specimens were found, roughly 450 km to the the east, in a similar environment at Chiang Kuang in Laos (19° 26' N, 103° 10' E). This is a very rare species; only the three specimens having been found.

Opisthotropis spenceri
Smith's Mountain Keelback   งูลายสอสองสี

Smith's Mountain Keelback reaches an adult length of over 0.7 meters. The dorsal surface is olivaceous, the ventral yellowish-white and the subcaudal area gray.

The internasals are more broad than long and are in contact with the rectangular loreal. There is a single preocular, two postoculars, and eight supralabials; the fourth and fifth supralabials contact the eye. The body scales are smooth and in 17 rows at mid-body. There are 183 ventrals and more than 33 subcaudals. This information was taken from the only two specimens found in Thailand.

The two Thai specimens were found in the northern province of Lampang. Any sightings elsewhere have not been reported.

Genus Parahelicops

In this genus, the last two or three maxillary teeth are enlarged to become ungrooved fangs. The body scales are keeled and in rows of 17 or 19 at mid-body. The anal may be double or single with the prefrontals fused into one single scale. The nostrils are located dorsolaterally. An aquarium of medium size is adequate to house members of this genus. A fairly large water receptacle is required and also an adequate number of hiding places, perhaps in the form of plants. Lizards, frogs, and possibly fish provide an adequate diet. The breeding habits of this genus are unknown.

Parahelicops boonsongi
Boonsong's Keelback   งูลายสอหมอบุญส่ง

Boonsong's Keelback reaches an adult length of approximately 0.64 meters. The tail is quite long. The eye is large and the pupil round. The suboculars and supralabials are cream; the ventral surface is generally creamy-white. It is nearly uniform gray-olive dorsally, becoming a little lighter on the flanks.

The single loreal is rectangular; there is a single large preocular that is not in contact with the frontal, two or three postoculars, and two suboculars. The anterior subocular reaches beyond the middle of the eye. Of the nine supralabials, the first three are in contact with the nasal, the fourth and fifth contact the eye, and the seventh the largest. There are ten infralabials. The dorsal scales are in 19 rows at mid-body, all being keeled except the anterior portions of those in the first row. There are 141 ventrals, the anal is single, and more than 33 subcaudals are present.

The single known Thai specimen was found in Loei Province in the Northeast. This species also probably exists in Laos, southern China, and Vietnam.

Genus Rhabdophis

This is a genus of terrestrial forest snakes, some of which are found at high elevations. Although some are found in or near to mountain streams, others inhabit areas quite far removed from streams. In each of the Thai species, the internasals are broad anteriorly and the nostrils located laterally. Furthermore, the eye is large, the pupil round, the scales heavily keeled and in 19 rows at mid-body, and both the subcaudals and anal plate are divided.

The last two maxillary teeth are strongly enlarged and recurved. These snakes also have labial glands that produce a toxic fluid which is guided into wounds created by the "rear-fangs" and is very effective in helping to overpower natural prey. It has long been thought that this toxic fluid was too mild to have any serious effect on man. However, at least two cases have recently been reported where two people nearly died from bites inflicted by members of genus *Rhabdophis*. It would be wise to approach these snakes with caution and make every effort to minimize the risk of being bitten. A very fine line separates a serious and a non-serious bite.

Their possible toxicity aside, these snakes make an excellent display and are easy to maintain in captivity. They are attractive snakes, extremely active, and eat frogs readily. Some also accept fish. Few exceed one meter in length and most will be comfortable in a medium-sized cage. Their quarters will require several plants for hiding and a large water receptacle.

Rhabdophis chrysargus Plate 78
Speckle-bellied Keelback งูลายสาบจุดดำขาว

This snake grows to a length of little less than one meter. The head is dark brown or black and is distinct from the neck. The supralabials are white with black sutures. The chin and throat are also white. The body is greenish-brown with a series of small red spots along the sides which become less prominent posteriorly. The ventral surface is white, and a series of black spots appear on the lateral margins of the ventral scales, hence, the common name Speckle-bellied Keelback. Juveniles differ slightly. In young specimens, the white color of the supralabials continues through the temporals to merge and form a white chevron, its apex pointing posteriorly, on the nape of the neck. Also, black lines extend over the back, connecting the small red spots on the sides. These markings disappear as the juveniles mature.

Scale rows number 19 at mid-body. Ventrals number from 139 to 175, and the subcaudals range from 60 to 101. There is one preocular, one loreal, three postoculars, and nine supralabials. The fourth, fifth, and sixth supralabials contact the eye. There are ten infralabials.

The Speckle-bellied Keelback is reported to have a varied diet that includes mice, small birds frogs, and lizards. It is riparian and most often found along streams.

This nocturnal species is widely distributed throughout the lowlands of Thailand. It is also native to Burma, Kampuchea, Laos, Malaysia, the Philippines, Singapore, and western Indonesia.

Rhabdophis nigrocinctus
Green Keelback   งูลายสาบเขียวขั้นดำ

The head is blackish-brown and distinct from the neck. The supralabials are gray, and there is a black line from the eye to the base of the jaw in addition to a black spot that extends from the parietal to the base of the jaw. There is also a black spot on the back of the neck. The body is dark green but becomes brownish-green posteriorly. On each side of the body is a series of black bars which may join on the vertebral ridge. The chin, throat, and ventrals are white, but the ventrals tend to become grayish posteriorly. The subcaudals are grayish-black with white edges.

There is one preocular that extends to the top of the head without contacting the frontal. There are four postoculars, one loreal, ten infralabials, and nine supralabials. The fourth, fifth, and sixth supralabials border the eye with the seventh and eighth being the largest. The scales are arranged in rows of 19 at mid-body; all but the lower rows are strongly keeled. The ventrals number from 150 to 170 and the subcaudals from 80 to 97. This snake reaches a length of 0.6 meters. It has a venom producing gland as well as rear fangs and should be dealt with cautiously.

Rhabdophis nigrocinctus is nocturnal and widely distributed in Thailand. It is also native to Burma, Laos, Kampuchea, and Vietnam.

Rhabdophis subminiatus helleri
Schmidt's Red-necked Keelback   งูลายสาบสีจาง

The head is olive, but the body is a lighter shade of green. Some specimens have a black line which passes from the eye to between the fifth and sixth supralabials. This line may be indistinct on some or absent on others. The supralabials are gray and the infralabials white. The neck is reddish in color and the body uniformly olive-green. The ventral surface is gray. This subspecies averages 0.9 m in length, but may reach a length of 1.3 meters.

There are eight supralabials, with the third to the fifth touching the eye; the sixth is the largest and the seventh is the highest. There are ten infralabials, one preocular, one loreal, and either two or three postoculars. Scale rows number 19 at mid-body, all but the first strongly keeled. The ventral scales number from 157 to 173, and the subcaudals range from 72 to 95. This subspecies also produces venom and possesses rear fangs; it should be approached with caution.

Rhabdophis subminiatus helleri is found in both lowland and upland areas, but always near streams, rivers, or ponds. In Thailand, it is found in Chiang Mai and surrounding northern provinces. It is also native to Burma and southern China.

Rhabdophis subminiatus subminiatus   Figure 22
Red-necked Keelback   Plates 79 and 80

งูลายสาบคอแดง

The head is brownish-green or olive, and the labials are whitish-yellow. A black line extends down from the eye between the fifth and sixth supralabials. The body is uniformly greenish-brown, but the skin between the scales is black and yellow. These colors sometimes encroach upon the edges of the body scales, giving the appearance of a

Figure 22. *Rhabdophis subminiatus subminiatus.*

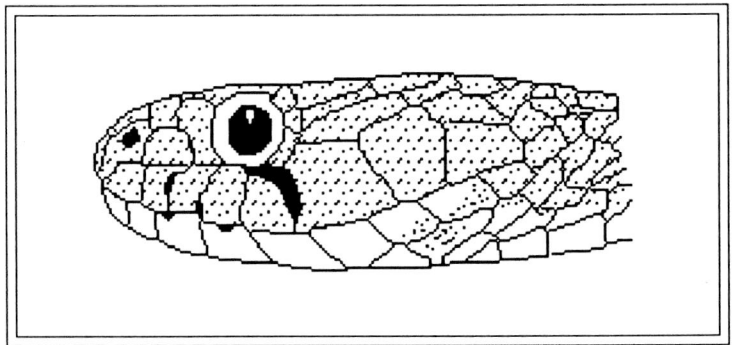

faint pattern. The neck area is red, giving this snake its common name. Juveniles may have a black band between the base of the head and the red area. See Plate 80, page 269. The ventral surface is yellowish. Females produce 5 to 10 eggs at a time; the male hatchlings reach maturity in approximately 13 months and females in 17.5 months.

The scales are in rows of 19 at mid-body; ventrals number from 132 to 175 and the subcaudals from 65 to 89. There is one preocular, a loreal, three postoculars, and eight supralabials. The third, fourth, and fifth supralabials contact the eye. The seventh is the largest. There are nine infralabials. This snake has a venom producing gland and enlarged teeth to the rear of the upper jaw. Its bite has not yet proven lethal, but it has produced serious reactions in humans. Adults reach a length of 0.8 meters. This terrestrial species is active during the day.

The Red-necked Keelback ranges from Chiang Mai Province in the North to the Thai-Malaysian border in the south. It is also native to Burma, Kampuchea, Laos, Vietnam, and peninsular Malaysia.

Genus Sinonatrix

Genus *Natrix* was originally a large, cumbersome genus which contained species in Europe, Africa, Southeast Asia, and North America. Such a wide range, extending from tropical to mid-latitude climates, would suggest dissimilarity but genus *Natrix* remained intact until rather recently. Since 1960, genus *Natrix* has been examined closely and the result has been an extensive revision. Additional careful research, including an analysis of biochemical characteristics, has resulted in sub-division of the former genus *Natrix*.[6] Refer also to: Figure 23, page 247.

---

6. For an extensive discussion refer to:
Rossman, Douglas A. and W. Gary Eberle.
Partition of Genus Natrix With Preliminary Observations on Evolutionary Trends in Natricine Snakes.

Figure 23. Present Division of Genus *Natrix*.

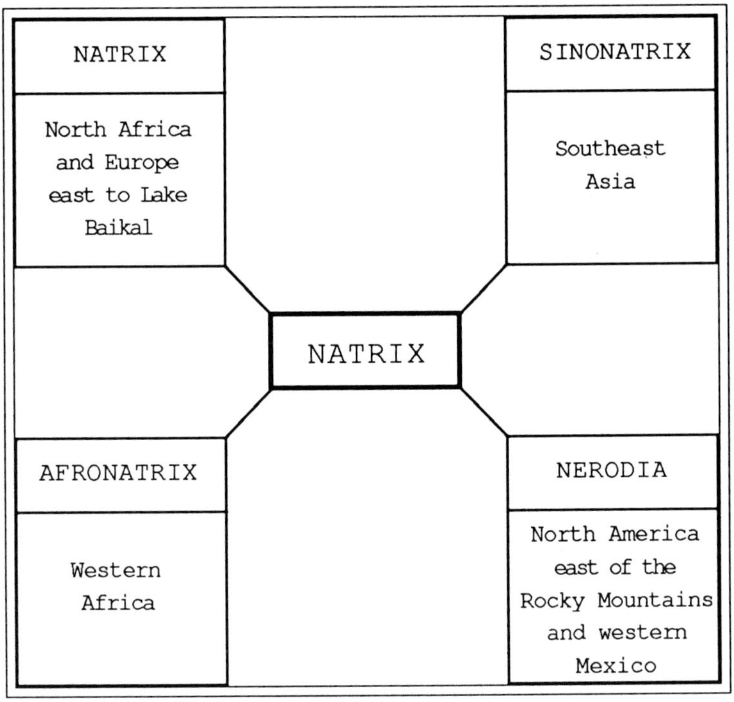

NATRIX		SINONATRIX
North Africa and Europe east to Lake Baikal		Southeast Asia
	NATRIX	
AFRONATRIX		NERODIA
Western Africa		North America east of the Rocky Mountains and western Mexico

Sinonatrix has distinct chromosomal and hemipenal characteristics. The palatine and supratemporal bones are normal. The dorsal scales are in 19 rows at mid-body, there are two pairs of anterior temporals, and the anal plate divided. Each member has more than 30 maxillary teeth which gradually increase in size posteriorly. The internasals become narrow anteriorly and the nasals are located dorso-laterally. Members of genus *Sinonatrix* are oviparous.

Sinonatrix percarinata
Chinese Keelback

งูลายสอเมืองจีน
งูลายสอเกล็ดได้ตาสอง

The body is olive-gray with light-edged, black bars on the sides. The top of the head and the anterior supralabials are also olive-gray, but the remainder of the head is yellowish-white, as is the ventral surface. The posterior portion of the ventral surface, however, has small black spots and specks, whereas, the subcaudals are dark gray with black spots. Juveniles are dark olive-green or gray. The body is fairly stout and the head is noticeably wider than the neck. The nostrils are located dorsolaterally. Adults may reach one meter in length. Females produce from 4 to 12 eggs at a time.

There is one large preocular, one loreal, three postoculars, and a small subocular. Of the eight supralabials, the fifth and sixth border the eye. The body scales are strongly keeled and arranged in rows of 19 at mid-body. The ventrals vary in number from 133 to 157 and the paired subcaudals from 68 to 85. The anal plate is also divided.

This snake is found on the higher elevations within its range. It should be kept a degree or two cooler than the temperatures shown in Appendix 2. In Thailand, it is found in the North in the province of Chiang Mai. It has also been reported in Burma, Laos, Vietnam, and southern China.

Sinonatrix trianguligera   Plates 81 and 82
Triangle Keelback   งูลายสอลายสามเหลี่ยม

This is a large species, attaining a length of 1.2 meters. The head is dark olive and the supralabials yellow with black edges. The vertebral area is black throughout its length. On the anterior portion, the black coloration descends down the sides of the body, forming a series of inverted triangles separated by areas of dark red. This red and black pattern gradually becomes indistinct posteriorly where it becomes nearly solid black. Juveniles have traces of green and yellow superimposed on the anterior pattern. See Plate 82, page 270. The ventral surface is yellowish-white with thin black lines crossing over it from the black areas of the body sides. The red areas of the anterior pattern also intrude a short distance on to the ventral surface.

The scales are strongly keeled and arranged in 19 rows at mid-body. The ventrals number from 134 to 145 and the subcaudals range from 86 to 96. The subcaudals are paired and the anal divided. There is one preocular, one loreal, and three postoculars, the uppermost the largest. Of the nine supralabials, the fourth, fifth, and sixth border the round eye. There are ten infralabials.

The Triangle Keelback is found in Thailand from Ranong Province south to the Malaysian border. Its range continues through Malaysia into western Indonesia. It is also found in Burma.

Genus Xenochrophis

These snakes are of moderate length. The teeth on the upper jaw gradually become enlarged posteriorly. The body scales are heavily keeled; members of this genus are oviparous. They are at home in the water, spending much of their time there. Their nostrils are slightly upturned enabling them to breathe close to the surface of the water without exposing their body.

Snakes of this genus are easy to maintain in captivity. Although they are quick to bite when first caught, they usually adjust to their captive environment rather quickly and make an attractive display. A medium to large cage equipped with hiding places, such as plants, and a large water receptacle should make them comfortable. A diet of fish and frogs will keep these animals in a healthy condition.

Xenochrophis flavipunctata Plate 83
Common Keelback งูลายสอธรรมดา

This species attains a length of 1.2 meters. The head is dark green, but the labials are of a lighter shade. Two dark lines extend back and down from the eye, the space between them being yellowish-green. A black line extends from the base of the jaw on each side of the head, the two lines converging, but not merging, on the back of the neck. Two additional narrow, parallel black lines extend forward from them to the parietal. The body color is dark green with quite large black spots on the anterior which become smaller posteriorly until they are obscure or absent on the tail. The ventral scales are light green with black posterior edges.

Anteriorly, the dorsal scales are weakly keeled but become more strongly keeled posteriorly. These scales are arranged in rows of 19 at mid-body. Ventrals number from 122 to 158 and the paired subcaudals from 70 to 97. The anal plate is divided. There is one preocular, a loreal, three postoculars, and eight supralabials. The fourth and fifth supralabials border the round eye with the sixth and seventh being the largest. There are ten infralabials.

A clutch of 25 eggs hatched in 43 days (February 4 to March 19, 1989) with 100 percent of the eggs hatching. The hatchlings averaged a length of 0.122 m six days after hatching. Shedding began four days after leaving the egg; by the sixth day every hatchling had shed. On the 12th and 13th days after hatching, the neonates began eating small fish and finely sliced pieces of fish. Taylor reports a large female containing 60 embryos being found in Bangkok.[7] Perhaps the short incubation period, large number of eggs, and high percentage of hatchlings are factors which contribute to this species being common in Thailand.

This species is distributed throughout the Kingdom and is very common. It is also found in Burma, southern China, Laos, Vietnam, Kampuchea, and northern peninsular Malaysia.

---

7. Taylor, Edward H.
The Serpents of Thailand and Adjacent Waters.

Xenochrophis piscator
Checkered Keelback

Figure 24
Plate 84
งูลายสอใหญ่

The Checkered Keelback reaches a length slightly more than one meter. It is a colorful snake. The head is brown with a black stripe extending from the eye to between the sixth and seventh supralabials. Another line extends from the postoculars through the anterior temporals onto the eighth supralabial. The labials and chin are whitish-yellow.

Figure 24. *Xenochrophis piscator*.

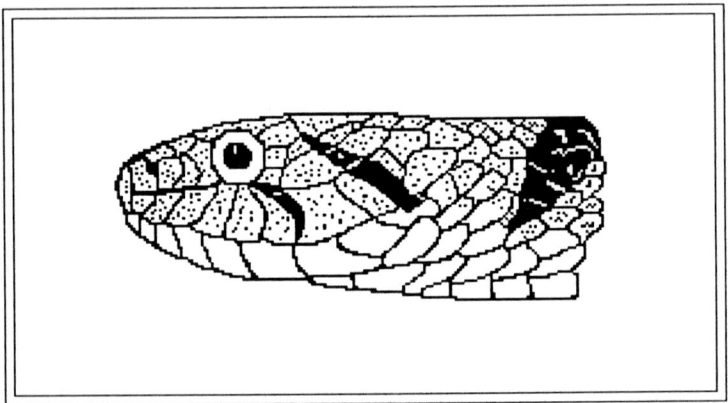

The stout body is olive-brown and marked with a series of dark spots. The spots are arranged in alternating rows creating a checker-board appearance. The pattern is more distinct anteriorly than posteriorly. The ventral surface is yellowish, and each ventral scale has a dark anterior edge. The eyes are round. Females are significantly larger than males.

The scales are heavily keeled and arranged in rows of 19 at mid-body. The ventrals number from 122 to 158, the paired subcaudals from 70 to 97, and the anal divided. One large preocular, a loreal, and three postoculars are present. The lowermost of the postoculars is the smallest. There are ten infralabials, and, of the nine supralabials, the fourth and fifth contact the eye. The internasals become narrow anteriorly with the nostrils located dorsolaterally.

This species is prolific; clutches containing 100 eggs have been recorded. The incubation period ranges from 37 to 51 days and the hatchlings from 0.17 to 0.215 m in length. If the food supply is suitable and adequate, they will easily achieve a total body length of over 0.4 m by the end of the first year.

This is a common snake throughout its range. It is nocturnal and found throughout Thailand in, or near to ponds, streams, and flooded rice fields. It also occurs in India, Bangladesh, Burma, Kampuchea, Laos, southern China, Malaysia, and western Indonesia.

Subfamily Pareatinae

This subfamily includes small snakes which prey primarily upon snails and slugs. Snails are extracted from their shells by use of fairly long teeth and specially adapted jaws. A unique feature of this subfamily is that the chin shields are not separated by a median line nor are they symmetrical. The chin shields are quite asymmetric, and this lack of symmetry somewhat restricts their ability to swallow relatively large prey. See Figure 25.

Figure 25. Heads of Pareine Snakes.

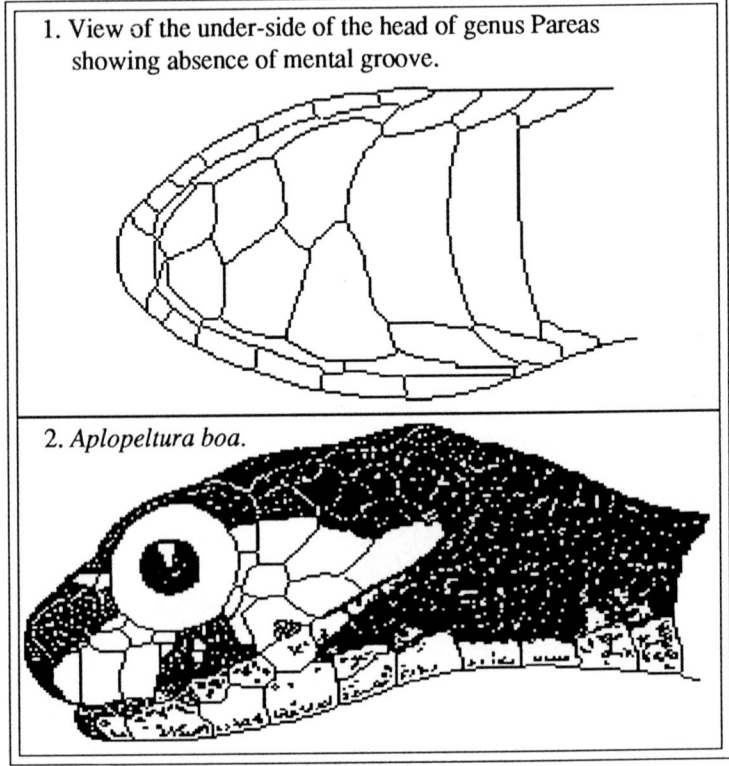

Subfamily *Pareatinae* is divided into the two genera *Aplopeltura* and *Pareas* by the criteria displayed in Figure 26.

Figure 26. Subfamily Pareatinae Criteria.

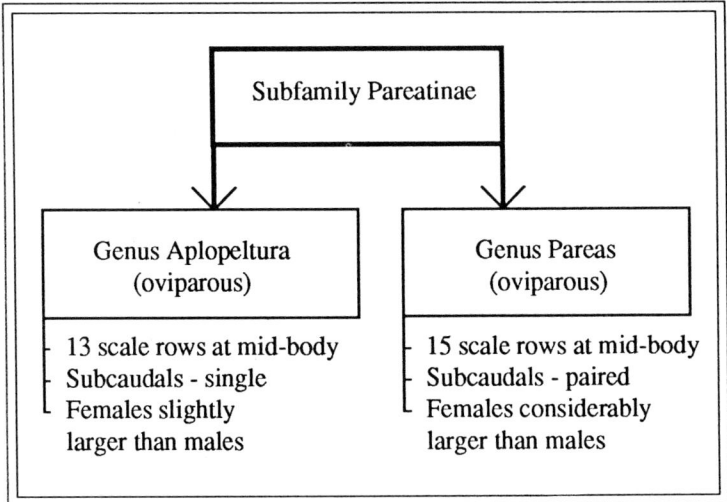

Genus Aplopeltura

In this genus, the head is short, blunt, and quite distinct from the slender neck. The body is slender and strongly compressed vertically; the tail is long and prehensile. There is only one member of this genus.

Aplopeltura are found in humid forests. They require a medium-sized cage; since it is arboreal, the cage is to contain a number of plants that will require daily spraying to provide drinking water. To maintain a high humidity a water dish should also be included. A source of food animals is to be assured before keeping *Aplopeltura* in areas of severe winters.

Aplopeltura boa   Figure 25
Blunt-headed Slug Snake   Plate 85

งูกินทากหัวโหนก

งูบอ

The head is quite distinct from the neck, the eyes are large and bulging, the pupil vertical. The body is brownish-gray with irregular dark markings. On the head, in front of the eyes, are white spots. A black line extends from the eye to the supralabials. This species reaches a mature length of approximately 0.75 meters.

The scales are smooth and arranged in rows of 13 at mid-body. The ventrals range from 148 to 191, and the single subcaudals number from 88 to 127. The anal plate is single. There are two preoculars, two postoculars, and three loreals each of a varying size. Supralabials number nine or ten, and none contact the eye due to the presence of subocular scales. There are 11 infralabials.

This gentle arboreal snake preys on slugs, snails, and perhaps small lizards. It is oviparous, females producing approximately five eggs at a time.

The Blunt-headed Slug Snake is native to the South particularly the provinces of Pattani, Nakhon Si Thammarat, and Chumphon. It is also native to Burma, Malaysia, the Philippines, and western Indonesia.

Genus Pareas

Members of this genus have heads which are short, thick, and distinct from their necks. Their eyes are large to very large and the pupils vertical. Their bodies are vertically compressed; the tails are long, slender, and prehensile. Body scales are arranged in rows of 15 at mid-body, and the subcaudals are paired.

Despite their fierce appearance and the presence of the venom producing Duvernoy's gland, these snakes are docile and not known to bite. Any venom they produce only seems to be effective against their prey and has not yet proven to be harmful to humans. Most are excellent climbers and cages should be provided with plants for them to climb upon. The plants should be sprayed frequently. Their diet consists primarily of slugs and snails. Adult snails are removed from their shells by manipulation of the lower jaw. Young snails are swallowed whole. Captives have eaten large insects and pinky (newborn, hairless) mice as well. Pareine snakes are oviparous, with females producing three to eight eggs at a time.

Pareas carinatus — Plate 86
Keeled Slug Snake
งูกินทากเกล็ดสัน

Adults rarely exceed 0.5 m in length. The head is light brown with a few black dots. The supralabials and the rostral are pale yellow. The body is light brown and crossed by a large number of thin, dark bars. On the back of the neck is a black "X" mark. The ventral surface is yellowish-white with fine dark dots. Females produce from three to eight eggs per clutch that hatch in approximately 60 days.

Infralabials number eight or nine, and there are either seven or eight supralabials, none being in contact with the eye. There is one preocular, two postoculars, and one loreal which is greater in height than length. Mid-body scale rows number 15, the ventrals range from 170 to 184, and the paired subcaudals number from 60 to 88. The anal is divided. The scales are smooth except for the vertebral scales, which are keeled and slightly enlarged.

The Keeled Slug Snake is found throughout Thailand. It has also been reported in Burma, Kampuchea, Laos, southern Vietnam, peninsular Malaysia, and Indonesia.

Pareas hamptoni
Hampton's Slug Snake   งูกินทากลายขวั้น

Adults are less than 0.5 m in length. The body is brown with black bars on the sides. The head is distinct from the neck. Eyes are large with elliptic pupils. A black lines runs from above the eye to the base of the neck and another runs from the eye to the base of the jaw. The top of the head is peppered with black spots, and the ventral surface yellow-white.

The scales are smooth, although those on the vertebral ridge are mildly keeled. There is one preocular, one postocular, one loreal, and seven or eight supralabials. None of the supralabials touch the eye. Mid-body scale rows number 15; ventrals vary from 191 to 196 on males, which also have 93 to 98 subcaudals. Females have 180 to 194 ventrals and 73 to 87 subcaudals. The anal plate is single.

This snake has not been collected in large numbers anywhere in its range.

Hampton's Slug Snake was originally discovered near Pok Loi, N. E. Siam [8] at an elevation at approximately 500 meters. Since then Pok Loi has become part of Laos. However, this snake may occur in adjacent Nan Province. Elsewhere it has been reported in Burma, Vietnam, and southern China.

Pareas laevis งูกินทากเกล็ดเรียบ
Smooth Slug Snake งูกินทากสีน้ำตาล

The Smooth Slug Snake is a relatively small snake reaching a mature length of approximately 0.55 m at maturity. The body is brown with a series of irregular black cross-bands. The sides are dull orange in color. The ventral surface is brownish-yellow with small brown spots. The eye is large and the pupil vertical.

The scales are smooth and arranged in rows of 15 at mid-body. The vertebral row is enlarged. Preoculars are not present, however, one or two postoculars may exisit in addition to an elongated loreal. Of the six supralabials, the third, fourth, and sometimes the fifth have contact with the eye. The sixth supralabial is quite long. The third, of the six infralabials, is relatively large. The ventrals number from 148 to 176, the subcaudals range from 34 to 59, and the anal is single.

This species is rare in Thailand and has only been found on mountains in Nakhon Si Thammarat Province in the South. The range extends across the border into Malaysia and it has also been sighted in western Indonesia.

---

8. Smith, Malcolm A. **The Fauna of British India, Ceylon, and Burma Including the Whole of the Indo-Chinese Sub-Region.**

Pareas macularius
Spotted Slug Snake   งูกินทากจุดดำ

The basic color is light brown. The top of the head is covered with small, dark brown spots. A thin dark line extends from the eye to the base of the jaw. There is a white band around the neck with some dark markings superimposed. The dorsal side of the body is marked with a series of closely spaced black dots creating the appearance of irregular bars. The ventral surface is dull white.

There are seven supralabials, none of which touch the eye. Seven or eight infralabials are present, as well as one preocular, one postocular, and a loreal. Mid-body scale rows number 15. The scales are smooth, except for those on the vertebral ridge. Males have from 148 to 166 ventrals and 40 to 53 paired subcaudals; females have 154 to 165 ventrals and 39 to 55 subcaudals. The anal is single. This species is small, attaining an adult length of only 0.375 meters.

Specimens have been found on Doi Inthanon (Inthanon Mountain), Chiang Mai Province, in the North. This species is also present in northeast India, Burma, Laos, and northern Vietnam.

Pareas malaccanus   Plate 87
Malayan Slug Snake   งูกินทากมลายู

This species averages 0.45 m in length. It is yellowish-brown and has irregular brown bars on its back. A yellow line extends along the vertebral ridge. The sides of the neck are black and the ventral surface white.

There are seven supralabials with the third and fourth touching the eye. There are no preoculars; however, there are two postoculars and a loreal. The loreal is in contact with the eye. The scales are smooth and arranged in rows of 15 at mid-body. There are 154 to 170 ventrals and the subcaudals number 26 to 55. The anal is single.

Within Thailand, the Malayan Slug Snake has only been found in Yala Province in the South. Its range extends through Malaysia into western Indonesia.

Pareas margaritophorus Plate 88
Mountain Slug Snake งูกินทากจุดขาว

The head of this snake is light brown, and there may be a red, or yellow, band on its neck. The body is gray-brown with narrow cross-bars containing black and white spots. The belly is white and covered with small black spots. Adults attain a length of approximately 0.45 meters.

There are seven supralabials, none of which contact the eye. There are also seven infralabials, one triangular preocular, an elongated loreal, and one postocular. The scales are smooth and arranged in rows of 15 at mid-body. The ventrals range from 138 to 159 and the subcaudals number 32 to 56. The anal plate is single.

This species is terrestrial and will not need climbing facilities. Plants are attractive and offer places to hide. Therefore, a small, planted aquarium is recommended for the Mountain Slug Snake.

This terrestrial species is widely distributed in the highlands of the Kingdom of Thailand. It is also found in Burma, peninsular Malaysia, Vietnam, southern China, Kampuchea, and Laos.

Subfamily Pseudoxenodontinae

This newly created subfamily includes members which do have some similarities with members of other subfamilies, but insufficient to classify them elsewhere. Posterior hypapophyses and certain of the hemipenal characteristics are common to members of this subfamily. Future research may possibly alter the composition of this grouping.

Genus Macropisthodon

Some question the inclusion of genus *Macropisthodon* in Subfamily *Pseudoxenodontinae*. For the present, however, this inclusion will be retained as classification elsewhere is considered to be inappropriate. The reader is cautioned that this genus may be listed under Subfamily *Natricinae* in other literature.

This is a genus of small to medium snakes. The last two teeth on the upper jaws are enlarged and strongly recurved, but members of this genus are not considered dangerous to humans. The pupils are round and the scales are keeled. Each species is oviparous, with reports of up to 25 eggs being produced per clutch.

When first captured, these attractive snakes flatten their necks, somewhat like a cobra, and strike. However, they usually settle quickly and thrive in captivity. A medium-sized aquarium will provide adequate housing, but it should contain hiding places. A large water receptacle should also be present. These snakes prey primarily upon frogs, lizards, and tadpoles.

Two species are found in Thailand.

Macropisthodon flaviceps Plate 89
Orange-necked Keelback งูรังแหหัวแดง

The head is light brown and distinct from the neck. A distinctive, black-edged, orange band encircles the neck. The body is black with faint light cross-bands that narrow near to the vertebral ridge but widen on the sides. The ventral surface is either black or dark green with black bands. Adults reach a length of 0.7 meters.

There are five or six infralabials and usually eight supralabials, but on occasions only seven are present. Either the third and fourth or the fourth and fifth supralabials contact the eye. There is one preocular, one loreal, and three or four postoculars. Scales are arranged in rows of 19 at mid-body and are strongly keeled. The ventrals range from 120 to 138 and the paired subcaudals from 49 to 60. The anal is divided. Females seem to achieve greater lengths than males.

The Orange-necked Keelback is a snake of the South, found only as far north as the province of Surat Thani. Its range continues through peninsular Malaysia, and western Indonesia to Borneo.

Macropisthodon rhodomelas Plate 90
Blue-necked Keelback งูรังแหหลังศร

This attractive snake reaches an adult length of about 0.45 meters. The head is light olive-green and distinct from the neck. A wide black chevron appears on the neck and a black line extends from its base along the vertebral ridge to the tip of the tail. The sides are pinkish-brown and are crossed by faint thin lines extending from the black vertebral stripe. The belly is pink or yellow with small black dots near to the edge of each ventral.

The strongly keeled scales are arranged in rows of 19 at mid-body. The ventrals number from 124 to 138, the subcaudals from 42 to 58. Both the subcaudals and anal plate are divided. There is one preocular, one loreal, three postoculars, ten infralabials, and eight supralabials. The fourth and fifth supralabials contact the eye.

This is a snake of peninsular Thailand, found in Nakhon Si Thammarat Province and those to the south. Its range continues in Malaysia and Singapore and parts of western Indonesia.

Genus Plagiopholis

Only one species is found in Thailand. It is a small snake, averaging only 0.47 m in length. The body is cylindrical, and the head is indistinct from the body. The diet consists primarily of frogs, lizards, and perhaps earthworms. Species in this genus are oviparous.

Since this species is small, its cage need not be large. Pieces of bark and other items should be placed in the cage to provide hiding places, and, of course, the cage should be kept warm and humid.

Plagiopholis nuchalis Plate 91
Assamese Mountain Snake งูหัวศร

งูหัวลายลูกศร

These snakes are reddish-brown, and many of the scales have black edges. This results in a banded pattern along the body. The head is dark brown, and there is a dark chevron on the back of the neck, its apex ending at the parietals. Below the eye, there are two dark marks. The chin and throat are yellowish, as are the ventrals. The subcaudals have a series of dark marks. The eyes have vertically elliptical pupils.

The scales are either smooth or keeled posteriorly. Scale rows at mid-body number 15. The ventrals number from 122 to 142, the subcaudals from 23 to 30. The subcaudals are paired, and the anal is single. There is one preocular. The supralabials number six and the third and fourth touch the eye. There are also six infralabials.

This terrestrial species has been found at elevations above 600 m in the northern provinces of Mae Hong Son, Chiang Mai, and Lampang and the western province of Kanchanaburi. It has also been found in Burma.

Genus Pseudoxenodon

In this genus, the head is slightly distinct from the neck. The eye is large and the pupil is round. The body is cylindrical, and the scales are keeled and arranged obliquely. The last two teeth on the upper jaw are abruptly enlarged. Members of this genus are oviparous, and males are slightly larger than females.

Only one member resides in Thailand. It is of moderate length and a medium to large aquarium will comfortably accommodate it. As with most members of this subfamily housing should be equipped with hiding places and a large water receptacle. A diet of frogs and lizards is adequate to maintain good health. Thai specimens have been found at elevations of nearly 2000 m; therefore, they should be kept at temperatures slightly less than those indicated within Appendix 2.

Pseudoxenodon macrops งูลายสาบตาโต
Big-eyed Mountain Keelback งูลายสาบคอบั้ง

The head is brown to gray with a black chevron at the rear of the neck, the apex of which points toward the parietals. The body is also brown to gray with a series of yellowish or reddish spots along the vertebral ridge. Anteriorly, these spots are greater in length than width, but become greater in width than length posteriorly. A series of black spots is present on the sides. The anterior portion of the ventral surface is white with black marks which decrease in size, become less numerous, and light brown towards the middle

Plate 73. *Oligodon inornatus*. Inornate Kukri Snake. งูปี่แก้วสีจาง
Not as colorful as other Thai members of the genus.

Photo: Jarujin Nabhitabhata

Plate 74. *Oligodon joynsoni*. Grey Kukri Snake. งูปี่แก้วใหญ่
Body short and stout. Note: typical thick neck.
Found in the Central Region.

Photo: Suthigit Patramangorn

Plate 75. *Oligodon taeniatus*. Striped Kukri Snake. งูงอด
Attractive, small. Does well in small aquarium.
Makes for an interesting display.

Plate 76. *Amphiesma deschauenseei*. Northern Keelback. งูลายสาบท้องสามขีด
Not common in Thailand. Specimen freshly killed.

Photo: Jarujin Nabhitabhata

Plate 77. (below)

Amphiesma stolata chinensis White-striped Keelback
งูลายสาบคอกหญ้า

Young specimen - loses some bright coloration on maturity.

Photo: Jarujin Nabhitabhata

Plate 78.

Rhabdophis chrysargus Speckle-bellied Keelback
งูลายสาบจุดดำขาว

Eye-catching white supralabials, with black sutures. Specimen not alive.

Photo: Jarujin Nabhitabhata

Plates 79 and 80. *Rhabdophis subminiatus subminiatus*
Red-necked Keelback. งูลายสาบคอแดง

Plate 79.

Attractive, easy to maintain in captivity. Thrives on diet of frogs. Specimen has just eaten a frog.

Photo: Jarujin Nabhitabhata

Plate 80.

Juvenile markings differ from adult. Note: black and yellow neck area on this juvenile.

Plate 81.
Sinonatrix trianguligera
Triangle Keelback
งูลายสอลายสามเหลี่ยม
Attractive, semi-aquatic - can move rapidly on land. When fleeing, has peculiar habit to spring off ground for short distances.

Photo: Jarujin Nabhitabhata

Plate 82.

Sinonatrix trianguligera
Triangle Keelback
งูลายสอลายสามเหลี่ยม

Juvenile.
Colorful youngster photographed in Trang Province. Darkens on maturity.

Photo: Jarujin Nabhitabhata

Plate 83. (above)
Xenochrophis flavipunctata
Common Keelback
งูลายสอธรรมดา
Common and easily maintained in captivity.

Plate 84.
Xenochrophis piscator
Checkered Keelback
งูลายสอใหญ่
Common Thai snake. Feeds readily on fish, frogs and thrives in captivity. Harmless but frequently bites.

Plate 85.
Aplopeltura boa
Blunt-headed Slug
Snake
งูกินทากหัวโหนก, งูบอ

Handsome, arboreal. In group "slug-eating snakes" as prey slugs, snails. Also known as "Chunk-headed Snake" due to head shape.

Photo: Suthigit Patramangorn

Plate 86.
Pareas carinatus
Keeled Slug Snake
งูกินทากเกล็ดสัน

"Slug-eating" or "Chunk-headed". Arboreal, rare on ground. Throughout Thailand.

Plate 87. (below)
Pareas malaccanus
Malayan Slug Snake
งูกินทากมลายู

Very rare in Thailand - few found in Yala Province.

Photo: Suthigit Patramangorn

Photo: Dr. Lim Boo Liat

Plate 88.
Pareas margaritophorus
Mountain Slug Snake
งูกินทากจุดขาว

Terrestrial, contrasting to other Thai genus members. A smaller slug snake.

Plate 89. (below)

Macropisthodon flaviceps
Orange-necked Keelback
งูรังแหหัวแดง

Found in southern lowlands. Not common.

Photo: Dr. Lim Boo Liat

Plate 90. *Macropisthodon rhodomelas*. Blue-necked Keelback. งูรังแหหหลังศร. Beautiful, also found in southern lowlands. Uncommon.

Photo: Dr. Lim Boo Liat

Plate 91.

Plagiopholis nuchalis
Assamese Mountain Snake
งูหัวศร. งูหัวลายลูกศร

Found at high elevations. Specimen captured - Doi Inthanon Mountain, Chiang Mai Province.

Photo: Jarujin Nabhitabhata

Plate 92.

Sibynophis melanocephalus
Malayan Blackhead
งูคอควั่นปลายหัวดำ

Colorful specimen, just killed. Reported to eat grasshoppers and lizards.

Photo: Jarujin Nabhitabhata

Plate 93.

Sibynophis triangularis
Triangle Blackhead
งูคอควั่นหัวลายสามเหลี่ยม

Bright chevron on back of neck - easy to identify.

Photo: Jarujin Nabhitabhata

Plates 94 and 95. *Bungarus candidus*. Blue Krait. งูทับสมิงคลา.
Bands wider than young *L. subcinctus*. (Plate 70). Note: white bands wider posteriorly. Known as Krait - toxic venom - fatal humans.

Photo: Suthigit Patramangorn

Plate 95. Juvenile - has larger white areas than adult.

Photo: Jarujin Nabhitabhata

Plate 96.

Bungarus fasciatus
Banded Krait
งูสามเหลี่ยม

Common and the largest of Thai kraits.
Note: characteristic vertebral ridge and blunt tail.

Photo: Suthigit Patramangorn

of the body. The posterior portion of the ventral surface is dirty white with a few scattered marks. The supralabials are white with narrow black edges; the infralabials, chin, and throat are white. Adults average a little more than one meter in length. A captured and preserved specimen contained ten ovarian eggs.

In this species the eye is large, as is the nostril, and the nasal is divided. There is a single large loreal, a preocular which does not contact the prefrontal, three postoculars, and nine or ten infralabials. Of the eight supralabials, the fourth and fifth contact the eye and the seventh the largest. The body scales, except for the smooth lower row, may be lightly or strongly keeled and arranged in 17 or 19 rows at mid-body. The ventrals range from 151 to 180, and the paired subcaudals from 55 to 80. The anal is paired. On mature males, the keels on the scales in the pubic area develop strong tubercles.

The Big-eyed Mountain Keelback is a snake most often found between 1500 and 2000 m above sea level. Thai specimens have only been found in the northern provinces of Chiang Mai, Loei and Mae Hong Son. Elsewhere, this species has been found in Nepal, northeastern India, Burma, Laos, Vietnam, and peninsular Malaysia.

Subfamily Sibynopheinae

This subfamily is small, containing only three genera. One, *Sibynodontophis*, is found in Central America and a second, *Parasibynophis*, is restricted to the African island nation of Malagasay. The third, *Sibynophis*, is native to Asia.

Genus Sibynophis

The teeth are numerous and closely set on the maxillary, palatine, pterygoid, and dentary bones, but all are of equal size. The dentary is completely detached from the articular, posteriorly. Hypapophyses is present throughout the vertebral column. The tail is long and slender.

These are small to medium-sized terrestrial snakes. The eyes are large and contain round pupils; the head is barely distinct from the cylindrical body. Subcaudals are paired, as is the anal. Although Duvernoy's gland is present, this genus has not yet proven dangerous to mankind. Members are oviparous and reported to produce two to five eggs per clutch.

An aquarium of moderate size is adequate to house members of this genus. This should be well planted to provide numerous hiding places for the captive. *Sibynophis* is reported to subsist on skinks, but it seems likely that they eat frogs as well.

Three species occur in Thailand.

Sibynophis collaris
Common Blackhead

Figure 27

งูคอขวั้นหัวดำ

The head is dark olive to brown with small, darker markings. The eye is fairly large, and the pupil round. There is an obscure bar between the eyes and another on the back of the head. The supralabials are white, bordered with black. The body is brown to gray-brown, with a black stripe or a row of black dots extending along the vertebral ridge. There is a black band, bordered with yellow, on the neck. The ventral surface is white with the darker color of the body sides intruding on the ventrals. Two rows of small black dots are present on the ventrals. This species averages 0.75 m in length.

Figure 27. *Sibynophis collaris.*

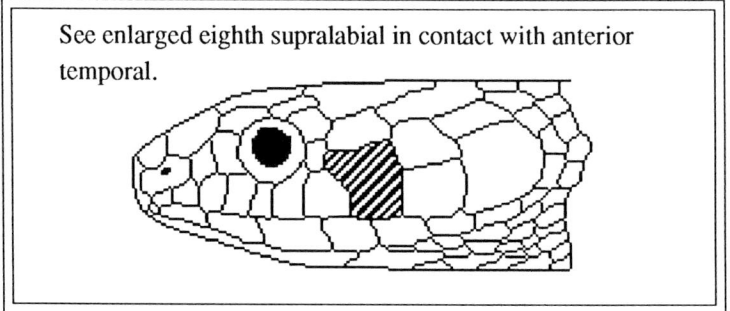

The scales are smooth and arranged in rows of 17 at mid-body. The ventrals number 155 to 186, the paired subcaudals vary between 100 and 125. The anal is single. There is one large preocular, one loreal, two postoculars, and ten supralabials. The eighth supralabial is the largest, and is in contact with the single anterior temporal; the fourth, fifth, and sixth border the eye. There are nine infralabials.

Sibynophis collaris is a mountain snake which is active during the day. It is most often found in the North, especially in Chiang Mai Province. This species has also been recorded in Nakhon Ratchasima (Northeast) and Kanchanaburi (West). It should be kept at temperatures a few degrees cooler than those for lowland forms. Refer to: Appendixes 2, 4, and 6. Its range extends to northern India, Burma, Laos, southern China, and peninsular Malaysia.

Sibynophis melanocephalus   Plate 92
Malayan Blackhead   งูคอชวันปลายหัวดำ

The head is reddish-brown to dark brown, and the supralabials have a black-edged white stripe. The posterior portion of the head changes to black, the neck is pale orange. The body is gray-brown to brown, and there is a black stripe on the vertebral ridge flanked by light yellow or brown stripes. These stripes include a series of dark spots or bars. The entire pattern gradually becomes obscure towards the posterior. The throat, ventrals, and the subcaudals are yellowish-white with the ventrals and subcaudals having black edges. The average length is 0.6 meters.

There is one large preocular, one small loreal, two postoculars, and nine supralabials. The fourth through the sixth supralabials border the eye and the last is the largest. There are nine infralabials. The scales are smooth and in 17 rows at mid-body. The ventrals number 145 to 154 and the subcaudals range from 100 to 120. The anal is divided.

Sibynophis melanocephalus lives alongside streams in the South. It is most common in Narathiwat and Pattani Provinces, but has been found as far north as Nakhon Si Thammarat. The range extends into Malaysia.

Sibynophis triangularis Plate 93
Triangle Blackhead งูคอขวั้นหัวลายสามเหลี่ยม

The head is light brown and crossed by two blackish-brown cross-bars, one passing behind the eyes, the other passing through the temporals. The base of the skull and part of the neck are also brown. A yellow-white line extends from the snout back through the supralabials, gradually widening and eventually merging on the back of the neck to form an inverted chevron. The body is also light brown but becomes darker posteriorly. A series of closely spaced, light brown spots forms a dorsolateral line that becomes obscure posteriorly. The ventral surface is white with a series of black dots on the edges of each ventral and subcaudal. At maturity, this species probably reaches a length of less than 0.7 meters.

Scale rows number 17 at mid-body, the scales being smooth. There are 160 to 189 ventrals and 113 to 124 paired subcaudals. The anal plate is divided. As in the other Thai members of this genus, there is one preocular, one small loreal, and two postoculars. The supralabials number ten with the tenth the largest. The fourth to sixth supralabials are in contact with the eye. There are nine infralabials, the fourth being the largest.

This species is found in both the Northeast and the Southeast, especially in the provinces of Chaiyapum and Chon Buri. It is likely that the potential range extends into Kampuchea.

Subfamily Xenoderminae

In addition to the common characteristics of skull structure and other internal structures, members of this subfamily have heads which are covered by small scales rather than the usual head shields. Members are found in both Southeast Asia and Central America.

Genus Xenodermus

This is a monotypic genus. The internasals and prefrontals are present, but the remaining conventional head scales are replaced by small keeled scales. The nostril is in a single nasal and directed forward.

This species may reach a length of 0.65 m, therefore an aquarium of moderate size will be required for its housing. It is reported to burrow, and, therefore, loose soil should be used as a substrate. It is semi-aquatic and will require a large water receptacle as well as hiding places, for example, potted plants and pieces of bark or wood placed upon the substrate. Frogs are the only known prey of this species, but it may also accept other amphibians or fish. Females are oviparous and reported to produce 2 to 4 eggs at a time.

Xenodermus javanicus   งูออนไม้
Xenodermine Snake   งูท้องขาว

The head is distinct, the eyes of moderate size, and the pupils vertical. The body is slender and slightly compressed; the tail is rather long, comprising approximately 35 percent of the total length. The head is brown with small white marks on the nasals and rostral. The dorsal portion of the body is also brown but the flanks are grayish. The chin and throat are white and the ventral surface gray with each ventral scale having a gray-brown marking along the anterior edge. The subcaudals have brown markings.

The body scales are arranged in three distinct rows. The vertebral row contains alternating large and small scales. Below the vertebral set is another row of scales smaller than those of the vertebral row intermixed with much smaller scales. Scales here do not usually have contact with each other and often interstitial skin is exposed. The lowest row contains small scales which increase in size progressively, as they near the ventral surface. These scales are also not in contact, exposing interstitial skin. Most of the body scales in each row are keeled. The total number of individual scale rows around the body ranges from 40 to 48. Ventrals range from 169 to 185 and the subcaudals from 113 to 147. Both the anal and subcaudals are single. There are 17 or 18 supralabials. Approximately 20 rows of small scales lie between the supraoculars.

This species is reported to be relatively common on Java, but is rare on mainland Southeast Asia. Only two have been found in Thailand and three in Malaysia. Throughout its range it is reported between sea level and 1000 m, and always in well-watered areas. The only Thai specimens have been found in Yala Province bordering Malaysia. It has been reported in lower Burma, however, if this occurence is accurate it would seem likely that specimens will eventually be found in provinces north of Yala. The full range extends throughout Malaysia and into western Indonesia.

FAMILY ELAPIDAE

The latter anterior teeth on the maxillary bones are fixed fangs with enclosed canals (Proteroglypha, Figure 7), rather than the posterior grooved fangs common in some *Colubridae*. The maxilla are moderately elongated and non-rotating. Each family member has specialized venom producing glands. This venom is highly toxic and often fatal to man. Indeed, this family includes species considered to be the most dangerous in the world. Hypapophyses exists throughout the vertebral column. Primitive features, for example, a coronoid bone, pelvic vestiges, and a left lung, are absent.

Subfamily Bungarinae

The gland which keeps the eye moist (Harder's Gland) is confined to the orbit. The palatine and pterygoid bones overlap, causing a slight to extensive erection of the palatine. The teeth on the lower jaw extend farther posteriorly than in other genera.

Genus Bungarus

Members of this genus are referred to as kraits. The head is slightly distinct from the neck and the pupil round. The scales are smooth and the vertebral scales are usually distinct. The loreal is absent.

These are medium to large terrestrial snakes which are generally mild mannered and non-aggressive. Since their venom is extremely potent always approach them with caution. They are assuredly not for amateur keepers. This also applies to the species discussed in the remainder of this text. Members of this genus are nocturnal and their prey usually consists of cold-blooded animals, although some have been induced to eat mice. Normally they eat other snakes, lizards, and frogs. All species are oviparous. In Thai members of this genus, the males are slightly larger than the females.

With few exceptions, kraits do not thrive in captivity. Perhaps greater success would be enjoyed if it was remembered that they are nocturnal and rather timid. Therefore, they should be provided with a large cage and a number of hiding places. The traditional hiding box should be supplemented with hollow logs, pieces of bark, and similar natural hiding places. They should not be disturbed. Observation should be undertaken at night. Also to be considered is the fact that they are native to areas of high humidity and will do well in cages with a high humidity. Keep a large water receptacle with clean water in the cage at all times. This is to be sufficiently large for soaking. Frequent spraying with tepid water also helps to maintain a high humidity. Consult Appendixes 2 to 7 for clear detailing of monthly relative humidity levels in Thailand. With reference to their husbandry, cannibalism has been noted, therefore, it would not be wise to house kraits with other species, nor would it be wise to house large members of the genus with smaller ones.

Bungarus candidus Plates 94 and 95
Blue Krait งูทับสมิงคลา

This is the smallest of the three species of *Bungarus* living in Thailand. The average adult is probably near to one meter long, although larger ones have been recorded. The head is distinct from the neck, the body cylindrical, and the tail pointed. The head is gray-black, but the supralabials are slightly lighter. The tongue is red. A series of black and white bands alternate down the length of the body. The black band on the neck is chevron-shaped with the apex intruding onto the base of the skull. The black bands halt at the ventral scales. The white bands become broader on the sides of the body and fuse with the white ventral surface. Individual black scales are often found on the white bands. Females produce four to ten eggs per clutch. The young, as is true of all venomous snakes, are venomous at birth.

The vertebral scales are enlarged. The scales are smooth and the scale rows at mid-body number 15, although rarely 17 rows are present. Ventrals number 194 to 237, the single subcaudals 40 to 50, and the anal single. The supralabials number seven, with the third and fourth touching the eye, and the fifth and sixth being the largest. There are also seven infralabials, one preocular, and two postoculars.

Bungarus candidus is found throughout Thailand, but is said to be most abundant in the Northeast. It is also found in Vietnam, Laos, Kampuchea, Malaysia, and western Indonesia.

Bungarus fasciatus
Banded Krait

Figure 28
Plate 96
งูสามเหลี่ยม

This is a rather large snake, attaining a length of more than 1.8 m, but averaging around 1.2 meters. The Thai call this species the "triangular snake" because of the very distinct vertebral ridge. A cross section of the body is triangular. The head is distinct from the stocky body, and the eyes are small and round.

Figure 28. Head of Banded Krait, *Bungarus fasciatus*.

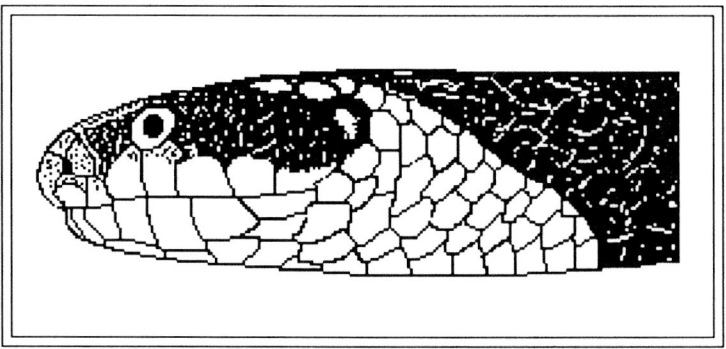

It is an attractive snake with yellow and black bands of almost equal width alternating throughout the length of its body. The bands continue through the ventral surface. The head is predominantly black with yellow supralabials. The tail is blunt, rather than pointed. The Banded Krait lays 4 to 14 eggs at a time. This species thrives better in captivity than the other members of the genus. One specimen survived for 14 years in captivity.

Scales are smooth and arranged in rows of 15 at midbody with vertebral scales enlarged. The ventrals range from 200 to 234, the single subcaudals from 23 to 39, and the anal is single. There are seven supralabials, with the third and fourth contacting the eye, and the sixth the largest. There are also seven infralabials, one preocular, and two postoculars.

This is a common snake and is well known throughout its range, despite being nocturnal. It has a reputation for not biting and being a ravenous consumer of other snakes, including venomous species. Consequently, they are welcomed in much of their range and seldom molested. Banded Kraits bite fewer people within their range than do most other types of poisonous snakes. This is probably due to the fact that they are nocturnal, while most rural Asian people are not. Kraits are very active at night and are likely to strike at any movement in their search for food. Extreme care should be exercised with them at all times, but especially during the night. Although it is true they are extremely docile during the day and do not seem to bite, and some people handle them freely at this time, it is most unwise to take chances with these creatures. After all, it only takes one bite to put a life in jeopardy.

Bungarus fasciatus is found in every part of Thailand, sometimes at high elevations. It is also native to India, Bangladesh, southern China, all of mainland Southeast Asia, and western Indonesia.

Bungarus flaviceps   Plate 97
Red-headed Krait   งูสามเหลี่ยมหัวหางแดง

This is a very beautiful snake. The head and tail are red, and the body black and shiny. Small white specks are present along the vertebral ridge and on the body sides. The ventrals are white, as are the first few body scales. The subcaudals are red. The head is distinct from the neck, and the small black eyes quite prominent on the red head. There is a distinct vertebral ridge, but this is not as pronounced as that of *Bungarus fasciatus*. The tail is pointed. The Red-headed Krait does not exceed two meters in length.

This species has one preocular, two postoculars, six infralabials, and seven supralabials. The third and fourth supralabials touch the eye. There are 13 rows of smooth scales at mid-body. Males tend to be larger than females and have a ventral count ranging from 220 to 236 and a subcaudal count from 47 to 53. The female counts are 193 to 217 and 42 to 54 respectively. In both sexes, the anterior subcaudals are single, but the remainder are paired. The anal is single.

This is a rare snake within its range. It is a snake of the forest and apparently not as fond of watered areas as are the two other Thai members of the genus. Unfortunately, this beautiful animal does not thrive in captivity. Perhaps improved results would be achieved if it is maintained within conditions approximating the natural environment and stress minimized. Keep this species in a fairly large cage with two to three centimeters of soil on the floor. Offer a number of hiding places with pieces of bark strewn on the floor and live plants. A variety of food should be provided, including small snakes, lizards, and frogs. It is known to eat members of genera *Trimeresurus* and *Thamnophis*, in addition to small skinks. Disturb it as little as possible. Perhaps most importantly, keep the humidity and temperature within the limits defined in Appendix 7. A large water dish and frequent spraying will assist in maintaining the humidity within the desired limits. Furthermore, be sure that the animal drinks. For this, it may be necessary to gently guide the snake towards the water dish.

Within Thailand, the Red-headed Krait is found in Ranong Province and south to the Malaysian border, the range extending into Malaysia. *Bungarus flaviceps* has also been found in Burma.

Genus Naja

Figure 29

This is probably the best known and the most feared genus of snakes in the world. The most distinguishing feature is that members have the ability to flatten their necks to form a "hood" and to raise about one third of the anterior body portion from off the ground into a defensive position. This occurs whenever a cobra feels threatened. The defensive position is usually accompanied by a loud hiss. It is a very effective defensive technique and usually makes a potential adversary think twice before pressing on with an attack. A second notable feature of this genus is that some of its species are capable of "spitting" their toxic venom into the eyes of an aggressor. This is possible because, in such species, the hole through which the venom is expelled is located in the front of the fangs rather than the tip. The

Figure 29. Heads of Cobras.

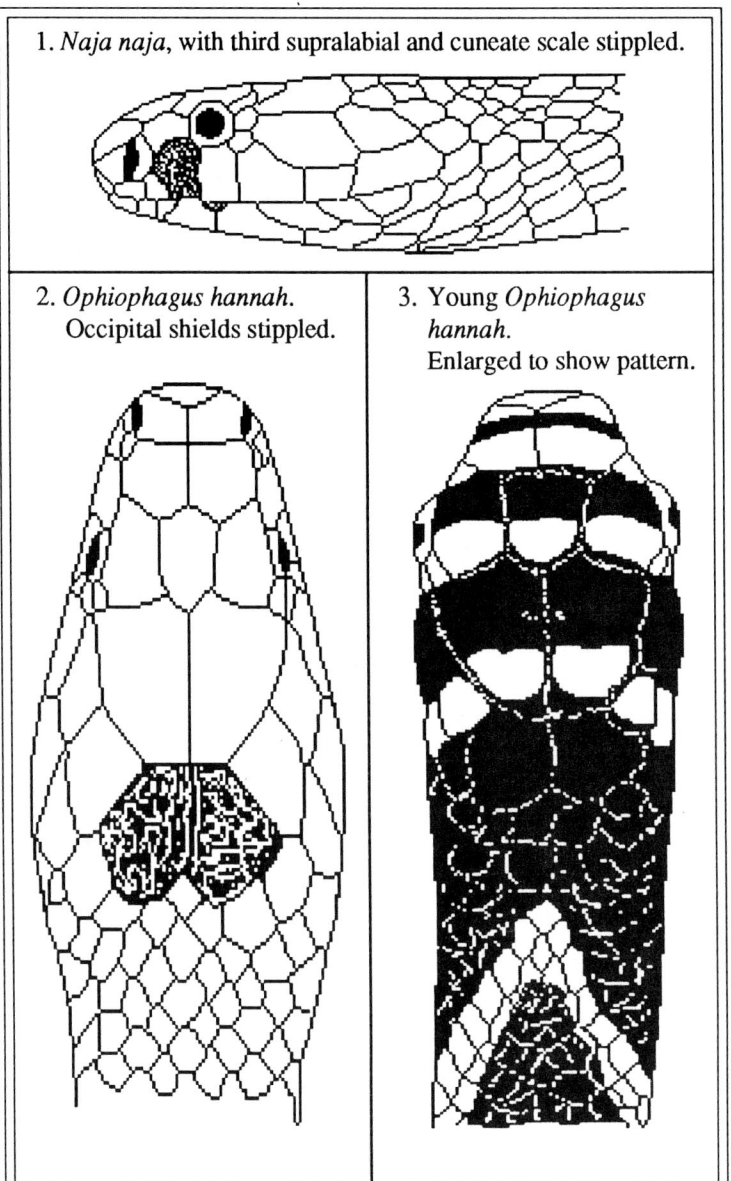

1. *Naja naja*, with third supralabial and cuneate scale stippled.

2. *Ophiophagus hannah*. Occipital shields stippled.

3. Young *Ophiophagus hannah*. Enlarged to show pattern.

spitting cobra rears into a defensive position, opens its mouth, and then forces venom through the fangs toward the eyes of its adversary. They can hit the eyes with great precision within a range of up to two meters. This is a very effective defense as the immediate result is pain and temporary blindness, providing the cobra with time to escape. If the venom is washed from the eyes and they are properly cared for, normal vision returns within a few days. Victims rarely die from cobras spitting venom into their eyes. The bite of any cobra, (and spitting cobras also bite) is quite another matter. The cobra's venom is quite potent and their bites are often fatal if not treated rapidly and correctly. Cobras must always be approached with great caution. Other common characteristics include the lack of a loreal scale, round pupils, barely distinct heads, smooth scales, and oviparity.

The range of genus *Naja* extends from Africa through the Middle East and South Asia into southern China and Southeast Asia. African species are generally larger than those from Asia and have narrower hoods. Furthermore, each species is fairly distinct and there is little controversy regarding their classification. The cobras of Southeast Asia are generally smaller than their African counterparts and can spread their hoods to a greater width. Southeast Asian cobras, however, have few distinct external characteristics. There is considerable variation in both color and pattern, even among individual specimens from the same geographic area. Furthermore, head scalation varies little throughout the region.

Southeast Asian members of genus *Naja*, therefore, pose a taxonomic problem. Presently almost all are classified as *Naja naja* and little progress has been made towards the classification of the various Southeast Asian cobra populations since Linnaeus originally described these forms as *Coluber naja* in 1758. The lack of distinct differences in head scalation and wide variation in coloration and pattern have made it extremely difficult for herpetologists to agree on which populations warrant a distinct taxonomic status. Most likely the best opportunity to solve this problem lies in the research of internal differences, especially the biochemical composition of these snakes. Recent research has led some authors to elevate *Naja naja sumatrana* to species status, *Naja sumatrana*. Others consider that *Naja naja kaouthia* should become *Naja kaouthia*. Work has begun, but the taxonomic problem posed by *Naja naja* is far from being settled.

The problem may be more acute in Thailand than elsewhere for two reasons. Firstly, the Thai people recognize and have given common names to six forms of native Thai cobras. Secondly, unrecognized Latin names have been carelessly applied to members of this genus which have subsequently appeared in international print, thus adding to the confusion. In the descriptions of the cobras of Thailand which follow, I intend to use only those Latin names for which there is substantial agreement within the herpetological community. Other forms will be listed as *Naja* sp., adjacent to the Thai common name. The Latin names which have been applied to these unrecognized forms will be included within the text.

Thai members of Genus *Naja* range from one to two meters in average length. Their heads are round and indistinct from their cylindrical bodies. Their eyes are prominent, the pupils round. Head scalation varies little, therefore it will be discussed here rather than form by form. There is one preocular and three postoculars, but the loreal is absent. The third and fourth of the seven supralabials contact the eye. The third supralabial is large and contacts the posterior edge of the nasal and the eye. There are seven or eight infralabials and often a small triangular scale, the cuneate scale, is present on the edge of the mouth between the fourth and fifth infralabial. The head scalation of *Naja naja* is illustrated in Figure 29, page 289.

Body scalation is more variable and can serve as a taxonomic clue. Unfortunately, such data has not yet been systematically accumulated, correlated and applied and as a potential taxonomic tool it has not been fully utilized. Such data, as is available for Thai forms, will be included in the discussions relating to each form.

Although they are nervous animals, cobras usually do well in captivity. They eat a wide variety of food including other snakes, frogs, lizards, birds, and small mammals. In fact, it is wise to separate cobras during feeding as they become very excited and have been known to eat cagemates. Many have been bred in captivity. They are oviparous, with an average clutch size of 15 to 30 eggs. Incubation periods are usually from 50 to 60 days. There does not seem to be any appreciable sexual size difference with the *Naja* of Thailand.

Thai members of this genus should be given quite large cages with doors always located on the top. This will minimize the chance of the keeper being bitten. Furthermore, the cage should include a hiding box that can be securely closed. This will enable the keeper to chase the cobra into the box, close it, and clean the cage in safety. Such a device will also expose the cobra to a minimum of stress. In addition to these steps, spitting cobras require a few additional security measures. Cages are not to have openings through which they can spit their venom. Also, the eyes of the keeper must be protected at all times when working with them. This may be done by wearing a pair of goggles. Cobras occasionally climb trees in search of food, but they are terrestrial and cages do not require objects for climbing. They like to soak, however, and the cage should contain a large water receptacle. Both the cage and water receptacle are to be cleaned often as cobras assimilate food rapidly, making cleaning a frequent task.

At present, Thai members of genus *Naja* appear to categorize within three groups, *Naja kaouthia* and the Suphan Cobra forming one group. The second group includes three closely related spitting cobras: the Black Spitting Cobra, the Black and White Spitting Cobra, and the Isan Spitting Cobra. The third group is formed of one single species, *Naja sumatrana.*

An additional note concerning the economic value of cobras seems worthy of mention here. In Thailand, many farmers become snake hunters after completing the harvest when large numbers of *Naja kaouthia* and the Black and White Spitting Cobra are captured and then sold in the Central Region. *Naja kaouthia* is sold at a price three times greater than that for the Black and White Spitting Cobra because of the skin quality. The skin of *Naja kaouthia* is manufactured into luxury products; the skin of the Black and White Spitting Cobra is of inferior quality and cannot be used for the same purposes. The Black and White Spitting Cobra is sold only for food.

Subfamily Bungarinae

GROUP 1

This group includes the only two non-spitting cobras of Thailand, *Naja kaouthia* and the Suphan Cobra. Both are rather large cobras and are distinguished by minor differences in scalation, coloration, and the absence of a hood pattern on the Suphan Cobra. An increasing number of herpetologists believe that *Naja naja kaouthia* should be reclassified as *Naja kaouthia*. If this is eventually done, the Suphan Cobra may become one of the subspecies.

Naja kaouthia   Plates 98 to 111
Monocled Cobra
(Siamese Cobra)   งูเห่าหม้อ

This is the most common of the cobras found in Thailand. It is a relatively large snake, attaining an adult length in excess of one meter. *Naja kaouthia*'s color is usually a shade of brown, although this may range from light, almost yellowish, brown to dark, blackish-brown. On most specimens the hood pattern is a single, monocle-shaped mark. This usually has a light brown center encircled by a dark brown edge which, in turn, is encircled by a fairly wide, white area with a dark brown edge. See Plate 98. However, this hood pattern is not constant and a number of variations have been observed. Some specimens are nearly patternless, whereas, others may be elaborately patterned. Some variations are shown in plates 100 to 108. In addition, a patternless white variety (Plate 111) is occasionally found, as are albino specimens (Plates 109 and 110).

The Monocled Cobra has 21, rarely 23, scale rows at mid-body, 27 or more dorsal scale rows around the neck, and 13 or 15 anterior to the vent. There are usually from 170 to 196 ventrals and from 43 to 58 subcaudals. Only rarely are there less than 170 ventral scales.

The Siamese Cobra is most frequently found in low lying, humid areas. It begins to search for food at dusk and, although primarily terrestrial, it will enter water and climb trees in search of food. It is found near human habitation where there is an abundant supply of rodents, ducklings, and chicks. This species lays 15 to 30 eggs at a time, which hatch in approximately 50 days. Neonates weigh roughly 15 grams and average 0.28 meters. Captive raised specimens have reached sexual maturity in three years.

Naja kaouthia is abundant throughout Thailand. This snake has also been found in Burma, Kampuchea, Vietnam, and northern peninsular Malaysia. The range also extends westward through Bangladesh, northeastern India, Sikkim, and Nepal.

Naja sp.   Plate 112
Cream Colored Cobra
(Suphan Cobra)   งูเห่าสีนวล

This form has been previously described under the names *Naja naja kaouthia* (New Species) and *Naja kaouthia suphanensis*. This type may be a color variation of the true *Naja kaouthia*, and, therefore, not a new species or subspecies. It may ultimately prove to be a new subspecies, but currently the name *Naja kaouthia suphanensis* is not scientifically valid.

This is one of the largest forms occurring in Thailand, averaging a length of 1.5 meters. The top of the head is light brown; the labials and lower temporals are white. Both the body and hood are patternless. The dominant color is pale yellow. The scales along the body sides are light yellow, but the scales near to the vertebral area are both equally proportioned light yellow and tan. The interstitial skin is white and the entire vertebral surface yellowish-white. The eyes are reddish-brown, and the tongue is light red. Hatchlings are pinkish-white and are often mistaken for albinos. Scalation is similar to that of *Naja kaouthia*, however the Suphan Cobra may have 23 or 25 scale rows on the neck, whereas *Naja kaouthia* always has 27 or more. The Suphan Cobra has 21 scale rows at mid-body, 15 before the vent, and the subcaudals are paired.

This cobra is found in the lowlands of the Central Region, especially in the provinces of Ang Thong and Suphan Buri. It is not as common as *Naja kaouthia*, nor is it found in areas inhabited by that species. The Suphan Cobra is nocturnal, and its diet is similar to that of the other cobras of Thailand.

GROUP 2

This group includes three spitting cobras: the Black and White Spitting Cobra, the Black Spitting Cobra, and the Isan Spitting Cobra. Each member has 27 or fewer scale rows around the neck and some have a faint "U"-shaped pattern on the hood.

Naja sp. Plates 113 and 114
Black and White Spitting Cobra
(Mottled Spitting Cobra)   งูเห่าด่างพ่นพิษ

In the past this form has been referred to as *Naja naja sputatrix*. However, questions have been raised regarding the status of this form and it will be referred to as the Black and White Spitting Cobra.

This is a relatively small cobra which grows to a length of approximately one meter. The top of the head is black, the supralabials brownish-black, and the infralabials white. The anterior portion of the tongue is black, the posterior portion dark red. The body is mottled very irregularly with black and white. Some specimens are predominantly black, others predominantly white. Still others have irregular bands of black and white. Some have a very thin, white "U"-shaped pattern on the hood, while others do not have a hood pattern. The throat and ventral surface are white with irregular brownish-black blotches or streaks on the ventral surface of the hood. Furthermore, on those with a banded pattern, the black bands continue across the ventral surface where they become brownish-black, rather than black. Hatchlings do not resemble their parents and, in fact, could be mistaken for albinos if it were not for their dark eyes. They possess only the slightest hint of the forthcoming pattern due to the seeming lack of any pigmentation. The black and white patterns develop as they mature. Females are known to produce up to 19 eggs per clutch which hatch in 48 to 58 days.

This form has 25 rows of scales on the neck, 19 to 21 at mid-body, and 13 or 15 before the vent. Ventrals range from 162 to 172 and the subcaudals from 49 to 55. The first four or five subcaudals may be single and the remainder paired.

This is a rather common snake in the West and Central Regions, especially in the provinces of Ang Thong, Suphan Buri, Kanchanaburi, and Tak. It is also found in the Southeast and the South, where its range has recently been extended as far south as Amphoe Tha Chana in Surat Thani Province.

Naja sp.   Plates 115 and 116
Black Spitting Cobra   งูเห่าดำพ่นพิษ

Both the Golden Spitting Cobra, of Group 3, and the Black Spitting Cobra have been described in various publications as *Naja atra* and *Naja naja atra*. *Naja naja atra* has not been recorded in Thailand and is not a spitting cobra.

The dorsal surface of the Black Spitting Cobra is uniformly black, the ventral surface a uniform bluish-black. There is no pattern on the hood. Adults average one meter in length. This form appears to be the most nervous of Thai cobras and sometimes does not do well in captivity. Its diet is similar to other members of the genus.

This form has 27 dorsal scale rows on the neck, 17 at mid-body, and 13 before the vent. There are approximately 170 ventral scales, and 45 to 56 paired subcaudals. Although this form has a higher number of scale rows on the neck and fewer at mid-body than the other members within Group 2, its low number of ventral scales and the anterior location of some of its internal organs seem to indicate that it is more closely related to the Black and White Spitting Cobra and the Isan Spitting Cobra than to the other cobras of Thailand.

The Black Spitting Cobra is reported to be most numerous in the provinces of Chon Buri (Southeast), Suphan Buri (Central Region), and Kanchanaburi (West). In January 1989 it was also discovered in Petchaburi Province (West) where local residents report it to be common. A population was found living in caves in Amphoe Tha Yang, Tambon Klut Muang (12° 53' N, 99° 40' E) at an elevation of 30 m above sea level. Some were found 30 m inside one cave, where the atmosphere was cool and moist.

Naja sp.   Plates 117 and 118
Isan Spitting Cobra   งูเห่าอิสานพ่นพิษ

This form has been referred to as *Naja naja isanensis*, a name not yet officially recognized. The snake is quite large, averaging a length of over one meter. Olive-green specimens of the eastern areas of range tend to be smaller than the grayish-green types of the west. Although this form is usually patternless, some specimens have a very light "U"-shaped hood pattern, a few have a spectacle-shaped hood pattern, and a few have indistinct body cross-bands. The head is greenish-brown, the supralabials light greenish-brown, and the infralabials greenish-yellow. The ventral surface is a dull greenish-yellow. Grayish-green bands extend across the throat under the hood, and widely spaced greenish-brown blotches exist on the ventral body surface.

There are usually 25, rarely 27, scale rows on the neck, 21 at mid-body, and 13 or 15 before the vent. Ventrals number over 170, usually between 180 and 195. The subcaudals are paired and range from 53 to 60.

This form is found in the provinces of the Northeast and Central Region that border the Northeast. It may also occur in neighboring Laos.

GROUP 3

This group contains a single member, *Naja sumatrana*. This species may be referred to as *Naja naja sumatrana* in other literature. It most commonly occurs in a uniformly black color phase throughout most of its range, but the golden phase is the most common form in Thailand.

Naja sumatrana Plates 119 and 120
Golden Spitting Cobra
(Equatorial Spitting Cobra) งูเห่าทองพ่นพิษ

This is another of the spitting cobras inhabiting Thailand. This snake has been described by others with the names of *Naja atra* and *Naja naja atra*. *Naja naja atra* ranges from southern China to northern Vietnam and is not a spitting cobra. Refer to page 298.

The Golden Spitting Cobra is a rather small cobra. averaging a mature length of less than one meter. The body is yellow or yellowish-green and the hood patternless. The head is dark yellow, and the labials light yellow. The eyes are dark and prominent against the yellow background coloration and the tongue entirely pink. The ventral surface is usually pale yellow but many specimens have irregular brown blotches on the ventral surface of the hood. Black forms of this species are reported to inhabit the South, but this presence has not been personally observed.

This form usually has 21 scale rows on the neck, 19, rarely 17, scale rows at mid-body, and 11 before the vent. There may be as many as 188 ventrals; the first four or five of the subcaudals may be single and the remainder paired.

This small cobra has a diet similar to that of other members of the genus native to Thailand. It is the least common of the Thai cobras. The range is restricted to the South, where it has been found in the upland areas of Surat Thani, Nakhon Ratchasima, and Phatthalung Provinces. The range of *Naja sumatrana* continues through peninsular Malaysia (where the same golden phase has also been reported) into western Indonesia. In Indonesia it has been reported on the islands of Sumatra, Bangka, Belitung, the Riau and Lingga island groups, and in Indonesian Borneo.

It is noted that bites from *Naja sumatrana* require treatment with a specific anti-venom. *Naja kaouthia* and *Naja sumatrana* are not conspecific and anti-venom which is effective against the venom of one, is not effective against the venom of the other.

Genus Ophiophagus

This genus contains only one species, the world's largest poisonous snake - the King Cobra. The King Cobra is a truly impressive animal. It averages four meters in length, but the largest ever recorded measured 5.85 m, a greater length than achieved by many pythons. In fact, King Cobras often prey upon quite large pythons.

Although many cobras of the genus *Naja* eat cold-blooded and warm-blooded prey, *Ophiophagus hannah* normally restricts its diet to cold-blooded animals, particularly other snakes. At Siam Farm, adult specimens have eaten *Boiga dendrophila melanota*, *Ptyas mucosus*, and *Python reticulatus* with preference for *Python reticulatus*. They are also reported to eat the following genera: *Ahaetulla, Cerberus, Enhydris, Fordonia, Homalopsis,* and *Trimeresurus*. In the United States they have been fed freshly killed *Agkistrodon piscivorus*. Apparently they are immune to the venom of Asian snakes. Although their venom is potent it works slowly on other snakes. Previously I observed a large specimen struggle with a young *Python reticulatus* for more than two hours before the python finally succumbed to the venom of the King Cobra. Lizards are also eaten, including *Calotes*, *Gekko*, and *Varanus*. Occasionally captive specimens have been trained to eat rats, but this is a very difficult task to accomplish. Usually this is done by transferring the scent of a snake to a rat and offering the dead rat to the King Cobra. Some keepers believe that King Cobras do not thrive on a diet of warm-blooded animals.

This genus is oviparous. The female excavates a shallow depression in the ground prior to filling it with small twigs and dry leaves, most often bamboo leaves as bamboo thickets appear to be a favorite area for nest construction. The nests are circular and may be one meter in circumference. Approximately 20 to 43 eggs are deposited in the nest, and the female remains on, or near the nest to defend the eggs from predators. After 70 to 77 days the hatchlings emerge and maternal responsibility ends. The hatchlings range from 0.5 to 0.535 m in length.

The venom of this snake is extremely toxic. This, coupled with the quantity of venom that can be injected by such a large animal, makes the bite of a King Cobra extremely dangerous. Water buffalo and elephants are known to have died from such bites. The keeping of the King Cobra is not for inexperienced amateurs and they should not be kept within private collections. Safety requires that they be housed in extremely large cages as plenty of space is required to effectively care for them. A hiding box, into which they can be chased and then securely closed, will also help to ensure the safety of the keeper. They are terrestrial, so a climbing apparatus is not necessary. A large container of clean water is required in the cage at all times, and the humidity and temperature maintained within the levels indicated in the Appendixes.

Adults of this genus do quite well in captivity, but their husbandry presents some special problems. Cage requirements have already been mentioned. Another factor to be considered is that these are big animals with big appetites, and a well established captive specimen can and will regularly eat a large number of snakes and/or large lizards. Such animals are not normally available throughout the year in high latitude countries and are usually expensive. The problem of an adequate and reliable supply of food must be considered before a King Cobra is added to any collection.

Most juveniles do not adjust well to the stress of captivity, even if born in captivity. In many cases they refuse food and starve. Force feeding is usually futile and is always very dangerous because King Cobras, as well as other venomous snakes, are toxic from birth.

Ophiophagus hannah Figure 29
King Cobra Plate 121

אורפיעך

The head is fairly blunt and quite distinct from the neck. Considering its great length, the body is relatively slender. It does not have the girth of a large python. Newborn specimens are black with yellow or light yellow bands across the snout, in front of, and behind the eyes. There is a yellow chevron on the neck, and thin bands encircle the body. The ventral surface is yellow or light yellow with black cross-bands. Adults show considerable variations, both in color and pattern. Those from the South tend to be light green-gray. Patterns are not apparent after the juvenile pattern has disappeared, although black skin between the scales can be seen, especially on the tail. King Cobras from the South are the largest in the Kingdom. Those from the Central Region are smaller but slightly more colorful. Some are light green and some orange-yellow. There remains a hint of the juvenile pattern on the body and it is clearly visible on the tail. Those from the West and North are the smallest found in the Kingdom. They are much darker than others and banded. In each case, the top of the heads of adults do not have juvenile markings, but incomplete bands exist on the ventral side of the hood in addition to orange markings. There is no pattern on the hood. The hood of *Ophiophagus hannah* is longer than that of *Naja,* but is narrower.

    The head scales of this large snake are sharply defined. There is one preocular, three postoculars, but the loreal is absent. Of the seven supralabials, the first is in contact with the nasal, and the third and fourth border the eye. There are eight infralabials, the third and fourth are large. The cuneate scale of *Naja* is missing, but there is a pair of large occipital scales posterior to, and bordering the parietals, a distinctive characteristic. The body scales are smooth and arranged in rows of 15 at mid-body. The ventrals vary from 240 to 266 and the subcaudals number from 84 to 106. The first five to seven subcaudals are single but the remainder are divided. The anal plate is single.

King Cobras are most often encountered in forests and on plantations. Reports of unprovoked attacks have not been substantiated. If encountered by man, their first instinct is to flee. If they cannot escape they will defend themselves vigorously. In doing so, the King Cobra is agile and appears fearless. It generally stands its ground and stares intently at the source of provocation. There is little of the nervousness or wasted motion which is often seen in *Naja*. The defensive movements and strike of a King Cobra are usually well measured.

This impressive reptile is found throughout the country, except in heavily populated areas. Elsewhere, it is found in India, Bangladesh, southern China, and all of Southeast Asia.

Subfamily Hydropheinae

Members of this subfamily are common in the waters off the coasts of Thailand. They are quite venomous and potentially dangerous but only account for a few bites or fatalities as contact with humans is infrequent. It appears that they do not molest people who are swimming or wading. Most of the bites and fatalities attributable to members of this subfamily are suffered by fishermen when they inadvertently grasp ensnared sea snakes during removal of fish from the nets.

Hydropheinae have fixed fangs (Proteroglypha, Figure 7) and round pupils similar to those of *Bungarinae*. They differ, however, in that the palatine and pterygoid bones of the roof of the mouth are locked into a horizontal position. Furthermore, members of this subfamily have numerous adaptations to suit them for a totally marine existence. For instance, these sea snakes have tails that are vertically flattened, like the blade of an oar, and their ventrals are either greatly reduced or non-existent. Most have valved nostrils located well forward and on top of the snout. Such a location enables the snake to breathe when only a small portion of the body is out of the water, the valves in the nostrils keeping water out of the respiratory system when it is submerged. *Hydropheinae* also have very short tongues.

All *Hydropheinae* give birth to living young in the water. The only known captive births of Thai specimens in Thailand are attributed to two specimens each of *Lapemis hardwickii* and *Enhydrina schistosa*, which were pregnant when captured in January and February of 1962. The two *Lapemis hardwickii* gave birth to three and six young, respectively. The babies measured from 0.2 to 0.22 meters. The two *Enhydrina schistosa* each gave birth to ten young which measured from 0.25 to 0.27 meters. Unfortunately, the sizes of the mothers were not recorded. The births occurred during March and April of 1962, suggesting a gestation period of two to three months. These figures are probably typical of most sea snakes, and suggest that sea snakes follow a reproductive cycle similar to that of the majority of snakes of Thailand. Copulation occurs during the cold season and birth occurs during the hot or rainy seasons.

Members of this subfamily are rarely kept in captivity as they require spacious and expensive aquariums; furthermore, they suffer high mortality in captivity. Perhaps survival could be improved if they were maintained in accordance with the seasonal fluctuations of water temperature and salinity detailed in Appendix 8, page 461. This appendix reveals that surface temperatures are coolest

Plates 307

Plate 97.
Bungarus flaviceps
Red-headed Krait
งูสามเหลี่ยมหัวหางแดง
Considered as most beautiful Thai snake. Note: coloration of head and tail. Uncommon - does not thrive in captivity.

Plates 98 to 111.
Naja kaouthia
Monocled Cobra
Siamese Cobra. งูเห่าหม้อ

98

Photo: Suthigit Patramangorn

Plate 99. Typical defensive stance. Does not rear up and spread hood when hunting. Note: Throat pattern. Most common Thai Cobra.

Photo: Suthigit Patramangorn

Plates 100 to 103.
Naja kaouthia. Monocled Cobra. Siamese Cobra. งูเห่าหม้อ.
Atypical hood patterns.   Photos: Piboon Jintakune

Plates 104 to 107.
Naja kaouthia. Monocled Cobra. Siamese Cobra. งูเห่าหม้อ.
Atypical hood patterns.   Photos: Piboon Jintakune

310   *Plates*

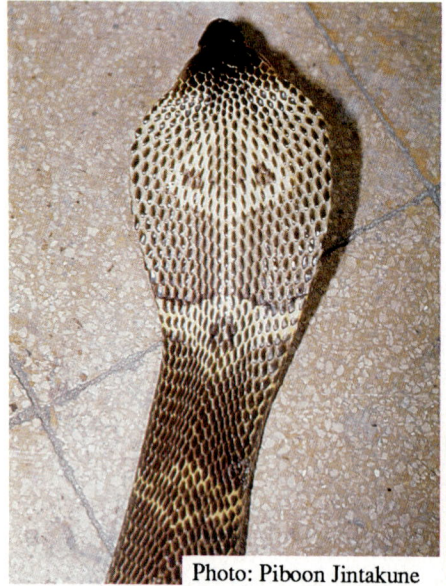

Plate 108. (top)
Atypical hood pattern.

Plate 109. (center)
Dorsal view of albino. Large female - captured in Central Region. Note: pink hood pattern.

Plate 110. (bottom)
Ventral view of albino. Although not common, albino cobras found more frequently than other albinos.

Photo: Piboon Jintakune

Plates 311

Photo: Suthigit Patramangorn

Photo: Suthigit Patramangorn

Plate 112. (above)
Naja sp. Cream Colored Cobra
Suphan Cobra. งูเห่าสีนวล
Not a spitting cobra.

Plate 111. (above)
White Cobra.
Not an albino.
Compare black eyes, black patch on the head with Plates 109, 110. Due to long display at Pata Zoo, Bangkok, specimen rarely hoods.

Plate 113. (right)

Naja sp. Black and White Spitting Cobra
Mottled Spitting Cobra
งูเห่าด่างพ่นพิษ

Dorsal view.
Spitting cobra. Variable pattern.
Note: white "U"-shaped hood pattern - absent on some.

312 *Plates*

115

Plate 114. (above)
Naja sp. Black and White Spitting Cobra
Mottled Spitting Cobra
งูเห่าด่างพ่นพิษ

Ventral view.
White ventral surface on most - on some black bands cross ventral surface. Note: rubbed rostral, result of an inadequate cage.

Plates 115 & 116.
Naja sp. Black Spitting Cobra
งูเห่าดำพ่นพิษ
Plate 115. Dorsal view.
Found in West, Southeast and Central Regions.

Plate 116.
Ventral view. Uniformly black and patternless. Can accurately spit venom - protective goggles required.

116

Plates 313

117. Dorsal view.

118. Ventral view.

Plates 117 and 118. *Naja* sp. Isan Spitting Cobra. งูเห่าอิสานพ่นพิษ
    One of four native Thai spitting cobras. Hood pattern absent or as light "U." Some have "spectacle-shaped" pattern. Ventral surface mottled. Hoods of Asian cobras wider than African.

Plates 119 and 120. *Naja sumatrana*. Golden Spitting Cobra
    Equatorial Spitting Cobra. งูเห่าทองพ่นพิษ.
    Resembles Suphan Cobra (Plate 112) - small, different range. Uncommon. Body patternless, no distinct hood, throat markings.

119. Dorsal view of hood.

120. Ventral view of hood.

Photo: Suthigit Patramangorn

Plate 121.

Ophiophagus hannah
King Cobra

งูจงอาง

Truly impressive. Largest poisonous snake - can inject large quantities of venom each bite. Potentially most dangerous in the world. Juvenile 2 m.

Photo: Dr. W. Dunson

Plate 122.

Acalyptophis peronii
Peron's Sea Snake

งูทากลายท้องขาว

Most sea snakes banded.
Note: oar-like tail and small head.

Photo: Harold K. Voris

Plate 123.

Acalyptophis peronii
Peron's Sea Snake

งูทากลายท้องขาว

View of head.

Characteristic spines on head.

from January to February (24.79° C) and warmest from April to May (30.44° C). Salinity is at a maximum during the cold and hot seasons and at a minimum during the rainy season. If such a regime is not possible then it would be advisable to maintain the water at the average annual values indicated in Appendix 8. It is also important to filter and aerate the water. Fish, eels, and marine invertebrates comprise the bulk of their diet but a recalcitrant eater may be induced to eat frogs. Their care should be similar to that of Thailand's members of *Laticaudinae*, except that *Hydropheinae* are totally marine and will not require dry surfaces.

Genus Acalyptophis

This genus contains a single species not often seen in Thai waters. This variety is distinguished by the presence of spines on the head, the absence of prefrontal scales, and fragmentation of the frontal and parietal scales. The nostrils are located dorsally and the nasals are in contact.

Acalyptophis peronii　　　　　　　　Plates 122 and 123
Peron's Sea Snake　　　　　　　　งูทากลายท้องขาว

This is a snake of moderate length, with adults averaging one meter in length. The head is very small and short, the body thick, and the tail oar-like. The ground color is light brown and a series of dark bands encircle the body. These bands are widest on the vertebral ridge becoming narrower towards the ventral surface. The head is pale brown and remains as such, although the body pattern may become obscure with age. This species feeds primarily upon marine invertebrates; females produce as many as ten living young.

There are tubercles on the supraoculars and postoculars of juveniles which become conspicuous spines on adults. Infralabials number five or six and there are either six or seven supralabials, with the third the largest and the third and fourth touching the eye. The prefrontal scales are absent; both the parietals and frontal are fragmented. There are one or two preoculars and three or four postoculars. Spines are present on the posterior edges of the postoculars and supraoculars on adult specimens. The body scales are keeled and overlap anteriorly but are side by side posteriorly. The scale rows number from 23 to 29 on the thickest part of the body. The narrow ventrals vary from 142 to 206; the anal is single.

This species ranges from the Gulf of Thailand to the waters off the northern coast of Australia.

Genus Aipysurus

Several species belong to this genus but only one inhabits Thai waters. In this genus, the nostrils are situated on the dorsal surface of the head and the nasals are in contact. The head scales may be complete or divided. The ventrals are large and distinct, being at least three times the size of the adjacent body scales. There are six or more supralabials and the posterior chin shields are usually reduced and separated by small scales.

Aipysurus eydouxii Plate 124
White-spotted Sea Snake งูทะเลจุดขาว

This is a slender snake which rarely exceeds one meter in length, most averaging 0.6 meters. The heads of adults are dark olive, those of juveniles tend to be black. The dorsal surface of the body is brownish or olive and crossed by bands of yellow, black-edged scales. These bands do not often extend across the vertebral ridge and do not cross the yellow ventral surface. The species is reported to subsist on fish eggs. An average litter consists of four or five living young.

The body is only slightly compressed and is nearly uniform in diameter throughout its length. There is a large frontal and usually two prefrontals but occasionally four are present. Except for the occasional presence of four prefrontals, the head scalation of the White-spotted Sea Snake is not fragmented. One preocular and two postoculars are present as well as six infralabials. Of the six supralabials, the second does not contact a prefrontal, the fourth touches the eye, and the sixth is usually the longest. The body scales are smooth, overlap, and in 17 rows on the thickest part of the body. The ventrals vary between 129 and 144, the single subcaudals from 23 to 35. The anal is divided.

In Thailand, this species seems to be restricted to the northern portion of the Gulf of Thailand. It has also been found along the shores of the countries bordering the South China Sea and is also known to inhabit the Timor, Java, Flores, and Arafura Seas, in addition to the waters of northern Australia.

Genus Astrotia

This is yet another monotypic genus, its sole member being a large, bulky species. Some species exceed it in length, but none can match its girth. The ventrals are the distinguishing characteristic of this genus. With the exception of those on the juveniles and throats of adults, they are divided into pairs of elongated, pointed scales which form a distinct ventral keel. The head shields are large and unfragmented, the nostrils located dorsally, and the nasals in contact. The body scales overlap, each scale having either a median keel or tubercle.

Astrotia stokesii Plates 125 and 126
Stokes' Sea Snake งูทากลาย

This snake has a very large body with a girth sometimes in excess of 0.25 meters. Its length is usually about 1.2 m, and it has a very bulky appearance. The body color ranges from black to dull yellowish-white and there may, or, may not be a series of narrow cream bands encircling the body. The head is only slightly distinct from the neck. This species feeds primarily upon marine invertebrates and females are considerably larger than males.

The body scale row count is high due to the large body girth, varying from 45 to 63. The ventrals are small, divided, poorly developed, and very irregular. They form a distinct median ventral keel and number from 226 to 286. The anal is divided, and the body scales are keeled. There is one preocular, two postoculars and eight supralabials. The second supralabial is the largest and the third, fourth, and fifth border the eye. The last three are small. There are eleven infralabials; the nasals are in contact.

Astrotia stokesii has been found near to Sri Lanka as well as throughout the waters of Southeast Asia and those off the northern coast of Australia.

Genus Enhydrina

This is yet another monotypic genus but its sole member is rather abundant in Thai waters. Its head shields are enlarged and regular, and the enlarged mental scale protrudes back into the mental groove. Females are slightly larger than males and produce unusually large litters for sea snakes. Litters average 18 young ranging from 0.25 to 0.27 m in length. It has been reported that newborn specimens reach maturity in their second year.

Enhydrina schistosa	Figure 30
Beaked Sea Snake	งูคออ่อนปากจะงอย
	งูคออ่อนหัวโต

This snake's common name has arisen because the rostral resembles the beak of a bird and projects down below the level of the supralabials. It is a rather large snake, attaining a length of approximately 1.4 meters. The dorsal side is greenish-gray and crossed by dark, narrow cross-bands. The head is somewhat darker and the posterior supralabials are cream as is the ventral surface. This species feeds primarily on fish.

320   *Subfamily Hydropheinae*

Figure 30. *Enhydrina schistosa.*

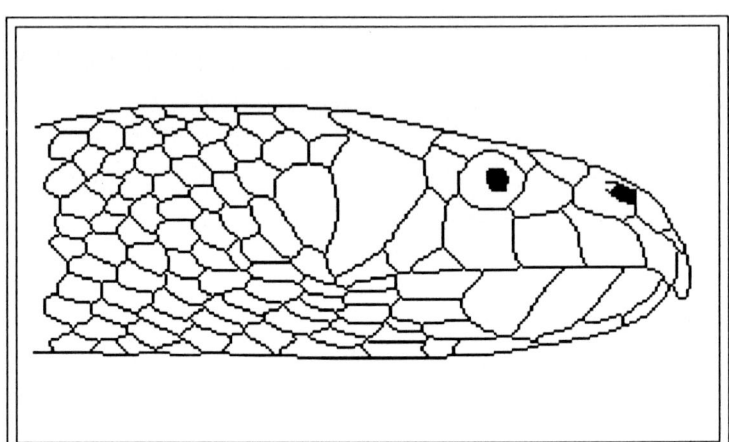

The body scales are keeled, overlap, and number 49 to 66 at the thickest section of the body. The narrow but distinct ventrals number from 236 to 322 and though distinct, are barely wider than the adjoining body scales. Females are larger than males. Of the seven or eight supralabials, the third and fourth or only the fourth touch the eye and the last two or three are extremely small. There are 11 infralabials, one preocular and one postocular, although occasionally specimens with two postoculars are found. The nasals are in contact, the nostrils dorsal. The venom of the Beaked Sea Snake is highly toxic. Fortunately few bites occur. Litters contain 4 to 10 babies, the average length being 0.28 meters.

Enhydrina schistosa is found in the waters off both the eastern and western coasts of Thailand. The range extends from the Persian Gulf through Southeast Asia to the waters off northern Australia.

Genus Hydrophis

This genus contains more species than any of the other genera of sea snakes and it is well represented in Thai waters. As one would expect in a genus containing a large number of species, individual species may vary considerably but each has a number of characteristics in common. For instance, in each species the head shields are enlarged and regular, the nostrils located dorsally, the nasals in contact, the anterior chin shields large, and the mental shield triangular. In nearly every species, the ventrals are scarcely larger than the body scales and almost never divided. The scales may be arranged side by side or overlap.

Hydrophis brookii   Plate 128
Brooke's Sea Snake   งูแสมรังท้องเหลือง

Brooke's Sea Snake has an exceptionally small head, the anterior portion of the body is quite slender, and the posterior portion relatively compressed. The dorsal surface is gray with dark gray bands encircling the anterior portion of the body. Posteriorly, however, the bands become narrow on the flanks and often do not encircle the body. The ventral surface is yellowish-white. The head is blackish or grayish with yellow markings. Young specimens have a yellow mark across the snout that continues along the sides of the head. Brooke's Sea Snake grows to a length of about one meter and its diet consists of eels and marine invertebrates. Significant differences are not apparent between the length of the male and the female. A litter contains an average of seven young.

The keeled scales overlap throughout the body. Scale rows on the thickest part of the body number from 37 to 45. The small but distinct ventrals number from 328 to 414. There is one preocular, one postocular, seven or eight infralabials, and six supralabials. Occasionally specimens may have two postoculars or five supralabials. The second supralabial contacts the prefrontal, the fourth borders the eye, and the sixth is very small. On some specimens the single large anterior temporal contacts the eye.

This species is common in the Gulf of Thailand and is also found off the west coast. Its range extends into the coastal waters of Kampuchea, Malaysia, and Indonesia.

Hydrophis caerulescens Plate 129
Dark Blue-banded Sea Snake งูแสมรังลายเยื้อง
งูแสมรังเกล็ดหยาบ

At maturity, this snake reaches a length of 0.8 meters. It is blue-gray and the back is crossed by wide, dark bands that narrow towards the belly and do not cross the ventral surface. The tail is also banded. These bands become indistinct with age. The ventral surface is pale yellow. The head of an adult is gray and may, or, may not have a yellow mark behind its eye. It eats eels and marine invertebrates. Males are larger than females, which give birth to an average of six young per litter.

The scales are heavily keeled and overlap. There is one preocular, two postoculars, and eight supralabials. The second supralabial touches the prefrontal and the third and fourth border the eye. The infralabials number nine; the last five are small. Scales are arranged in rows of 37 or 39 on the thickest part of the body. The ventrals are very small and number from 266 to 287. The head is fairly small, the anterior portion of the body slightly slender, and the posterior portion compressed.

Hydrophis caerulescens is found in the waters off both coasts of Thailand in addition to the coastal areas of India, Bangladesh, the remainder of Southeast Asia, and the Gulf of Carpentaria in Australia.

Hydrophis cyanocinctus Figure 31
Blue-banded Sea Snake Plate 130
งูแสมรังเหลืองลายคราม
งูแสมรังลายฟ้า

This species may grow to a length of 1.7 meters. Its head is indistinct from its cylindrical neck. The body, however, thickens rapidly and becomes appreciably thicker than the neck. Coloration and pattern vary considerably from specimen to specimen. The color is usually olive or yellow with transverse markings. The markings may, or, may not encircle the body and may be broken or incomplete. The head is black and the ventral surface is yellowish. This species also preys primarily upon eels and marine invertebrates. Females produce an average of ten young each litter and are much larger than males.

The scales are keeled, overlap throughout the body, and in rows of 37 to 47 on the thickest part of the body. The ventrals are small, distinct, and number from 290 to 390.

Figure 31. *Hydrophis cyanocinctus.*

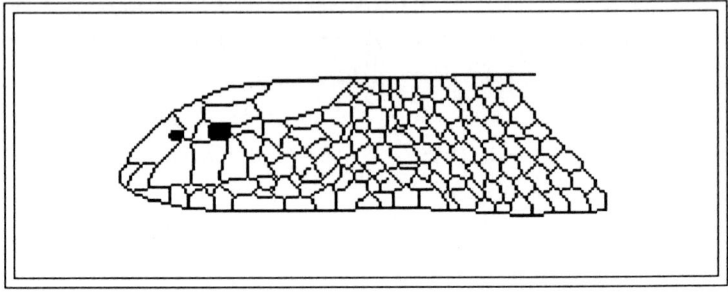

There are seven or eight supralabials with the third, fourth, and fifth, or only two contacting the eye. The second contacts the prefrontal. Nine or ten infralabials are present in addition to one preocular and two postoculars.

The Blue-banded Sea Snake is widely distributed. In Thailand, it is found in the Gulf of Thailand and along the west coast. Its range is from the Persian Gulf through the waters of Southeast Asia and north to Japanese waters. The distribution within this range, however, is irregular and there are locations where it has not been reported.

Hydrophis fasciatus atriceps
Banded Small-headed Sea Snake   งูแสมรังปล้องลายแถบ

This subspecies has a small head which is indistinct from its neck. Anteriorly, the body is very slender but becomes thicker and extremely compressed posteriorly and the tail, of course, is compressed vertically. The head is shiny black, the body dark olive, and there are pale yellowish oval spots on the sides of the body. On some specimens these oval spots are connected and form dorsal bars. Eels comprise the bulk of its diet, but occasionally some marine invertebrates are eaten. Males are larger than females; each litter averages two or three young.

The keeled scales overlap and are arranged in rows of 39 to 49 on the thickest part of the body. The usual number is 43 or 45. The ventrals are small but distinct and range from 323 to 452. There are six or seven supralabials, the second being the largest, the third and fourth contacting the eye, and the last the smallest. There are seven or eight infralabials, one preocular, and usually one, but rarely two postoculars. Often there is a small scale lodged between the third and fourth infralabials.

This subspecies has been reported along the Thai coast of the Gulf of Thailand. It has also been found in the waters off Kampuchea, Vietnam, the Philippines, Indonesia, New Guinea, and northern Australia near Darwin.

Hydrophis fasciatus fasciatus Plate 131
Striped Sea Snake งูแสมรังลายแถบ

This species grows to a length of approximately 1.1 meters. The head is very small and barely distinguishable from the neck. The anterior portion of the body is quite slender, but increases in girth posteriorly and becomes very compressed. The head is shiny black. A series of alternating black and light yellow bands begins at the base of the head. Initially they are distinct, with the light yellow bands narrower than the black ones. As the bands proceed posteriorly and the body increases in girth, they become wider and less distinct. Small black specks begin to appear on the light yellow bands. Furthermore, the black bands retain their width on the vertebral ridge, but narrow ventrally and become indistinct, leaving the ventral surface entirely yellow. The light yellow and black bands, however, remain evenly spaced and distinct on the tail.

Scales are keeled, overlap, and are arranged in rows of 47 to 58 on the thickest part of the body. The small but distinct ventrals range from 414 to 514. There is one preocular and usually one, but occasionally two, post-oculars. Either six or seven supralabials are present, with the second contacting the prefrontal and the third and fourth bordering the eye. The last one or two supralabials are small. Usually there is a small scale between the third and fourth infralabials.

Hydrophis fasciatus fasciatus has not yet been reported in Thai waters. It is included, however, because its confirmed range makes it very likely that this species exists off the west coast of Thailand. It has been found in the waters off India, Bangladesh, and Burma. The Striped Sea Snake has also been reported in the Straits of Malacca as far south as Singapore.

Hydrophis klossi
Kloss's Sea Snake   งูผักมะรุม

This is a medium-sized sea snake which attains a length of about 1.3 meters. Its head is quite small and indistinct from the neck, the body long and slender anteriorly but thicker and compressed posteriorly. It is greenish-gray and has a series of broad, dark bands. The head and neck are entirely black in young specimens, but the head of older animals is a lighter shade of black and there is often a semicircular mark on the top of the head. The chin and throat are yellow.

Hydrophis klossi has one preocular, one postocular, and five, rarely six, supralabials. The second supralabial is the largest and touches the prefrontal, the third and fourth border the eye, and the sixth, if present, is small. Five infralabials are present. The scales overlap throughout the body and are smooth or faintly keeled and in rows of 31 to 37, rarely 39, on the thickest part of the body. The ventrals are very small but distinct and number from 360 to 430. The snout projects slightly over the chin.

Kloss' Sea Snake is found in the waters along the east coast of peninsular Thailand. It also has been recorded off both the east and west coasts of peninsular Malaysia.

Hydrophis lapemoides
Persian Gulf Sea Snake   งูทะเลอ่าวเปอร์เชีย

This species was recently discovered off the west coast of Thailand, in the vicinity of Phuket Island, and was subsequently reported near the southern end of the Straits of Malacca off the west coast of peninsular Malaysia. The head is of moderate size and the body is rather robust, being two to three times the diameter of the neck. Juveniles are considerably more slender than adults. Males are slightly larger than females, achieving an adult length slightly in excess of one meter; females are slightly less than one meter. As is often the case with sea snakes, juveniles are more vivid in color than adults. They are white or yellow and have a series of 33 to 43 black cross-bars, wider dorsally than ventrally, along the length of their body. The head is black with a curved yellow mark that has its apex at the nostrils, and conclusion at the temporals. With maturity the color and markings become dull and the yellow mark on the head and the ventral portion of the cross-bars may disappear entirely.

The body scales number 43 to 51 rows at the thickest part of the body, with 29 to 35 on the neck. The posterior body scales are somewhat hexagonal or quadrangular in shape, juxtaposed, or slightly imbricate, and are slightly keeled. The ventrals are distinct throughout the body. They are double the width of adjacent body scales anteriorly, narrow posteriorly, and range in number from 314 to 372. The head has one preocular and either two or three postoculars. Of the eight supralabials, the second is usually in contact with the prefrontal and either the third and fourth or the third, fourth, and fifth contact the eye. There is a series of small scales at the oral margin posterior to the second infralabial.

This snake is not plentiful anywhere within its range. It has been found in the Persian Gulf and off the coasts of Pakistan, India, and Sri Lanka and off the west coasts of Thailand and peninsular Malaysia.

Hydrophis ornatus ornatus
Reef Sea Snake

Plate 132

งูแสมรังหางขาว

Adult specimens average one meter in length. The head of the Reef Sea Snake is barely distinct from the neck, the body is stocky with the maximum girth being achieved posteriorly. The body is gray to olive-green and is crossed by dark bands. These bands are separated by areas of white on the tail. The ventral surface is yellowish-white. Males are larger than females.

This snake has one preocular and two or three postoculars. Of the seven or eight supralabials, the second contacts the prefrontal and both the third and fourth border the eye. The body scale row count varies with sex; males have 33 to 45 on the thickest part of the body and females have 39 to 55. These scales overlap throughout and may be either smooth or faintly keeled. Although they become narrower posteriorly, the ventrals are distinct throughout. Males have from 209 to 260 ventrals, females from 236 to 312.

This is not a common snake in Thai waters; only a few specimens have been found in the Gulf of Thailand. Its range extends from the Persian Gulf through the Arabian Sea and the Bay of Bengal into the waters of Southeast Asia.

Hydrophis spiralis   Plate 133
Yellow Sea Snake   งูแสนรังเหลือง

The record length for this species is 2.745 m, but the average is closer to two meters. The head is distinct and the body not really slender, reaching its maximum diameter posteriorly. The basic color is yellowish-green. The body scales have black edges and 35 to 50 narrow black bands are present on the body, some encircling it. Dark spots are occasionally present on the yellow or yellowish-green surfaces. The ventral surface is yellow. Juveniles frequently have a black line extending along the ventral surface. Adults have yellow heads but the heads of juveniles are black with a yellow, horseshoe-shaped mark. Adult females are larger than males and give birth to 5 to 14 young per litter. The young are 0.2 to 0.4 m in length at birth.

    Scales slightly overlap and are usually smooth, but may be slightly keeled. They are arranged in rows of 33 to 38 on the thickest part of the body. Supralabials number from six to eight. The second contacts the prefrontal and either the third through the fifth, the third and fourth, or the fourth and fifth contact the eye. A single large anterior temporal scale sometimes contacts the supralabials and there is often a small scale on the lip after the third or fourth infralabial. Ventrals number from 295 to 362. They are distinct, being approximately double the size of the adjoining body scales.

    The Yellow Sea Snake is found off the west coast of Thailand but has not been found in large numbers. The range extends from the Persian Gulf through the Arabian Sea and Bay of Bengal into the Straits of Malacca in the vicinity of Singapore.

Hydrophis torquatus aagaardi
Lower Gulf Black-headed Sea Snake

งูแสมรังเทาอ่าวไทยตอนล่าง

The head is elongated, moderate in size, and distinct. The anterior portion of the body is moderately thick and cylindrical, but increases in girth posteriorly and also becomes vertically compressed. Dorsally, the body is gray or greenish-gray and is crossed by a series of dark gray or black cross-bars which may encircle the body; the ventral surface is yellowish-white. The head is dark olive to black with yellow markings across the snout and along the sides. Some specimens have yellowish spots on the frontal and parietals. These head markings fade and become less distinct with age but are usually retained. This subspecies may attain an adult length of 1.1 meters.

Head shields are large and well developed. The rostral is wide and visible from above; the nasals are large and partially divided by a suture extending to the second supralabial. There is one preocular and one postocular. The supraocular is large, only slightly narrower than the frontal. Of the eight supralabials, the second is the largest and in contact with the prefrontal, the third and fourth border the eye, and the last four are small. The first four infralabials are large; three of them border the chin shields. The three subspecies of *Hydrophis torquatus* are distinguished by the number of scale rows on the thickest part of the body and the number of ventral scales. In this subspecies, the ventrals range from 276 to 325, the average being 297. The ventrals are narrow, but distinct. However, on some specimens they may be divided into groups posteriorly. The body scales may be keeled or have tubercles; if tubercles are present, there are usually two of unequal size. Dorsal scales are in rows of 39 to 47 on the thickest part of the body and slightly overlap.

Hydrophis torquatus aagaardi has been found in the Gulf of Thailand as far north as the vicinity of Songkhla. It has also been found off the east coast of peninsular Malaysia and off the west coast of Borneo.

Hydrophis torquatus diadema
Upper Gulf Black-headed
Sea Snake   งูแสมรังเทาอ่าวไทยตอนบน

The head is moderately large, elongated, and distinct. The anterior portion of the body is of moderate girth. The girth increases posteriorly until this body portion is double the thickness of the anterior, in addition to being vertically compressed. Coloration and pattern are nearly identical to *Hydrophis torquatus aagaardi*. The head is black to dark olive with yellow markings on the snout and sides. On some specimens, the yellow coloration is predominant. Dorsally, the body is gray or greenish-gray and crossed by a series of dark gray or black cross-bars, some of which encircle the body. The ventral surface is yellowish-white. The color and pattern lose definition with age as does that of the previously mentioned subspecies but both retain more in maturity than does *Hydrophis torquatus torquatus*. Adult specimens of *Hydrophis torquatus diadema* achieve a length of approximately one meter.

Head scales are fairly large, well defined, and generally similar to those of *Hydrophis torquatus aagaardi*. However, *Hydrophis torquatus diadema* has one preocular and one, or two postoculars. Furthermore, there are usually seven and sometimes eight supraoculars. The second is in contact with the prefrontal, the third and fourth border the eye, and the last three to four are quite small. This subspecies has the highest average number of ventrals, 301, of any of the three subspecies living in, or near Thai waters. The range is from 271 to 343; the ventrals are narrow, less than double the width of the adjoining body scales, but are distinct. Conversely, this subspecies has fewer scale rows on the broadest part of its body, from 35 to 42, than the other two subspecies under discussion. The body scales, which slightly overlap, have either a central tubercle or a short keel and are usually in 35 to 42 rows at the thickest part of the body.

This subspecies is restricted to the extreme northern portion of the Gulf of Thailand. It has been found at 13° N near Ampheo Mekong on the western side of the Gulf and in the waters surrounding Chang Island at 12° N on the eastern side.

Hydrophis torquatus torquatus   Plate 134
West Coast Black-headed Sea Snake   งูแสมรังเทาหัวดำเหลือง

This snake is the smallest of the subspecies of *Hydrophis torquatus*, specimens averaging approximately 0.8 m in length. The body structure is similar to the two other subspecies; the head is moderate in size, the anterior portion of the body is not slender, girth increasing posteriorly and the body becoming vertically compressed. The young have bright colors and a distinct pattern. The body is white and crossed by a series of black bands which usually do not encircle the body. The head is black with whitish or yellow marks across the snout and the sides. With maturity the colors become dull and the pattern becomes indistinct when the dorsal surface changes and becomes pale gray with poorly defined gray bands. On some specimens, the bands almost disappear leaving a nearly uniformly gray dorsal surface. The ventral surface becomes yellowish-white and the head gray with more extensive but less clearly defined yellow markings.

The head scales are fairly large and clearly defined. One preocular and one or two postoculars are present. Seven or eight supralabials may exist; the second contacting the prefrontal, the third and fourth touching the eye, and the remainder small. Usually there are 43 to 49 scale rows on the widest section of the body but occasionally there are only 41. Members of this subspecies probably will have more scale rows than those of the other two subspecies. The individual scales may have either a central tubercle or a short keel and slightly overlap. The narrow but distinct ventrals range from 242 to 306 with an average of 278, the lowest range of the three subspecies.

Hydrophis torquatus torquatus has not yet been found in Thai waters. It is included here, as it is quite likely that it will be found off the west coast of Thailand. It is known to exist in the Straits of Malacca, off the west coast of peninsular Malaysia, approximately in the area between 2° N and 5° 30' N., bringing it very close to Thai territorial waters.

Genus Kerilia

Of the two members in this genus only one, *Kerilia jerdonii siamensis*, occurs in Thai waters. The snout slopes abruptly down and is quite narrow anteriorly. The nasals are in contact and the nostrils on the dorsal side of the head. The large head scales are well defined without small scales; the body scales are large, keeled, and overlap. The ventrals are distinct and slightly wider than the adjacent body scales.

Kerilia jerdonii siamensis งูแสมรัง
Jerdon's Sea Snake งูชายธงลายข้าวหลามตัด

This subspecies may reach a length of 0.9 meters. The head is scarcely distinct from the body, which is fairly consistent in diameter throughout. The top of the head is bluish-gray and the chin area bluish-white. The body is blue-gray at the vertebral area, but becomes yellowish on the body flanks and the ventral surface. The body is crossed by approximately 40 black bands which are at their widest on the vertebral ridge, becoming narrower on the sides without crossing the ventral surface. Males and females grow to similar adult sizes.

The scales are keeled, overlap, and arranged in rows of 21 or 23, rarely 19, at the thickest part of the body. The ventrals are noticeable and number from 225 to 253. There is one preocular, one postocular, and six supralabials. The third and fourth supralabials contact the eye, the last is often in contact with the single anterior temporal. There may be from six to eight infralabials.

This snake has only been found in the Gulf of Thailand close to the southern provinces of Songkhla and Pattani. It has not been recorded outside of Thai territorial waters.

Genus Kolpophis

In this genus, the nostrils are situated on the dorsal side of the snout. The head shields are irregular and some fragmented. The ventrals are small but distinct. The supralabials are heavily fragmented and the infralabials are often separated from the oral cavity by small scales.

Only one species of this genus exists and it has been found in Thai waters.

Kolpophis annandalei   งูชายธงหัวโต
Big-headed Sea Snake   งูกะรังหัวโต

This is a short, relatively stout snake, thicker posteriorly than anteriorly, and usually less than one meter in length. Its olive head is distinct from its neck. The body is a yellowish color and is crossed by a series of pale gray bands that are quite broad. These fade with age, becoming indistinct in older specimens. The belly is white.

The body scales are quite small, slightly keeled, and arranged in rows which vary from 74 to 93 at the thickest part of the body. The ventrals are small and number from 320 to 403. The labials are quite variable. The supralabials vary from 9 to 13, the infralabials from 9 to 14. The head scales are either fragmented into small scales, or they may be irregular. The frontal is partly fragmented and the parietals are separated by small scales, making this species particularly easy to identify. There is a large supraocular.

Fragmentation of the labials has resulted in both the variability of their numbers and the presence of three suboculars. There is one preocular, three or four loreals, and two postoculars.

The Big-headed Sea Snake has been taken mostly from the waters off Songkhla and Pattani in the South. It has also been found in the waters off Kampuchea and Vietnam as well as off the coasts of Malaysia and the Indonesian island of Java.

Genus Lapemis

Of the two recognized species in this genus, only one, *Lapemis hardwickii*, occurs in Thai waters. It consumes a variety of marine life, including eels, invertebrates, and a variety of fish.

The nasals are in contact and the nostrils located dorsally. The head shields are unfragmented, although the parietals may be divided. The head is fairly large and distinct from the short, stout body, the posterior portion being approximately twice the diameter of the neck. Body scales are nearly square or hexagonal, the first several rows being the larger. The ventrals are small but usually distinct anteriorly; they are vestigial or absent posteriorly. On *Lapemis hardwickii*, the parietals are entire and the ventrals extremely small or absent on the posterior three quarters of the body.

Lapemis hardwickii
Hardwicke's Sea Snake

Plates 135 and 136

งูใอ้ฮัว

This is a short, stout snake which reaches a mature length of about 0.86 meters. The relatively large head is pale olive to black and the body is yellow with dark olive bars that normally taper to a point at the sides. These bars may encircle the ventral surface or may merge on the vertebral ridge. The tail is black and the ventral surface white. Litters of 3 to 6 have been recorded, the offspring being approximately 0.21 m long at birth.

There is one loreal, one preocular, one or two postoculars, and seven or eight supralabials, with the third and fourth bordering the eye. Either 10 or 11 infralabials may be present. The nearly square body scales are keeled. In males they are in rows of 25 to 37 at the thickest part of the body; in females the rows number from 33 to 41. In each sex, the lower three or four rows of body scales are larger than the others and those of males are usually quite spiny. The ventral scales are not bigger than the body scales and become indistinct posteriorly. They range in number from 114 to 186 on males and from 141 to 230 on females.

Hardwicke's Sea Snake is fairly common in Thai waters and is found along the peninsular coast of the Gulf of Thailand. It is also reported to be very common along the coasts of peninsular Malaysia. The range also includes the waters of Kampuchea, Vietnam, the Philippines, Indonesia, and the tropical waters of Australia.

Genus Microcephalophis

As the name implies, members of this genus have very small heads. The head shields are entire, the nasals contact each other, and the nostrils are located dorsally. The anterior portion of the body is very long and slender, the posterior portion is much thicker and vertically compressed. The ventrals are entire and distinct anteriorly; posteriorly, they are separated into halves by a longitudinal fold. Females are slightly larger than males.

Although not a common snake, one genus member is found Thai waters. Eels are its primary prey.

Microcephalophis gracilis gracilis [9]　　　Plate 137
Needle-headed Sea Snake　　　งูคออ่อนหัวเข็ม

The head of this snake is long, narrow, very small, and almost indistinct from the neck. The snout projects beyond the lower jaw. The slender anterior portion of the body gives way to a much thicker and vertically compressed posterior portion. The head is shiny black; distinct bluish-black bands, separated by bands of dark gray, appear on the neck and continue posteriorly but gradually fade and become indistinct. At the point where the body begins to thicken, the bluish-black bands merge at the vertebral ridge and become gray. They narrow ventrally and appear as inverted triangles. The former dark gray areas become yellowish-white. At the thickest part of the body, the

---

9. Sometimes presented as *Hydrophis gracilis gracilis.*

vertebral area remains gray but the lower portions of the "inverted triangles" almost disappear and the flanks and ventral area are yellowish-white. The gray cross-bands gradually reappear and become closer together posteriorly until they merge, making the tail entirely dark gray. Males average 0.95 m in length and females 1.025 meters.

On Thai specimens, the scales are keeled and in 29 to 37 rows at the thickest part of the body. There is one preocular, one postocular, eight infralabials, and six supralabials. The second supralabial touches the prefrontal, the third and fourth contact the eye.

Although this snake is uncommon it has been observed in the Gulf of Thailand. Elsewhere, its range extends from the Persian Gulf through the Arabian Sea and the Bay of Bengal into the Straits of Malacca. Its range continues north through the South China Sea and east through the waters of Indonesia.

Genus Pelamis

This another monotypic genus. The head shields of its sole member are regular and enlarged, but the mental grove is obscured by the presence of small scales between the chin shields. The body is extremely compressed and the ventral scales are very small.

Pelamis platurus Plates 138 to 140
Yellow-bellied Sea Snake งูชายธงหลังดำ

This is a medium-sized snake with a mature length of 0.9 meters. The head is large, somewhat narrow, and with an elongated snout; the head is distinct from the neck and the body is greatly compressed vertically. Patterns are highly variable (Refer to: Plate 139, page 360) and, although most are yellow and black, some have brown coloration. Unlike the majority of sea snakes, however, *Pelamis platurus* are not banded. In the most common form, the top of the head is black but both the rostral and labials are yellow. The dorsal body surface is black and the ventral surface yellow. Variations might include vertebral stripes, either distinct, sinuous, or broken. Some specimens may have black spots on the flanks or ventral surface. It is not surprising that a snake with such an extensive range would be so variable in pattern. Most Thai specimens have a narrow black vertebral stripe extending the length of the body from the top of the head to the tail tip. The black vertebral stripe often becomes sinuous or divides into spots posteriorly. The body sides and the ventral surface are usually yellow and completely free of black spots or stripes.

There are one or two preoculars, two or three postoculars, and 10 or 11 infralabials. Of the seven or eight supralabials, the second is large and contacts the prefrontal; the fourth and fifth are usually separated from the eye by suboculars, but on some specimens they do contact the eye. The sixth and seventh supralabials are small. There are 49 to 67 scale rows on the widest part of the body. The scales are hexagonal or nearly square and contiguous. The lowermost scales have small tubercles which are well developed and the most noticeable on adult males. The ventrals, which are the same size as the body scales, number from 264 to 406 and are often divided by a longitudinal furrow or fragmented and indistinguishable from the body scales.

Other species of sea snakes occasionally enter the mouths of rivers but this species has not been reported in such an environment. Its range extends out into the sea and it is truly pelagic. In addition to both Thai coasts, the Yellow-bellied Sea Snake ranges from the cold waters off the coast of Siberia to southern Australia. Its east-west range extends from the warm waters of east Africa through South Asia into Southeast Asia and continues through the vast Pacific Ocean to the coasts of Central and South America. It has the most extensive range of any sea snake.

Genus Praescutata

In this genus the nasal scales are in contact and the nostrils located on the dorsal surface of the snout. The head scales are large and regular, without separation by smaller scales. The ventrals are wider anteriorly than posteriorly.

There is only one species in this genus with its range extending into Thai waters.

Praescutata viperina
Viperine Sea Snake

Plate 141
งูชายธงท้องขาว

The head is short, depressed, and distinct from the neck; the body is relatively stout. The top of the head is black but mottled with lighter colors; the labials and chin are white. The body is gray and a series of moderately large black spots extend along the back from head to tail. These spots become indistinct with age. The tail is black and the ventral surface is white. Males attain approximately one meter in length and females 0.9 meters.

The body scales are keeled and arranged in rows of 37 to 50 on the thickest part of the body. The ventrals number from 226 to 274, are distinct, but narrow posteriorly. There is one large preocular, one small pre-subocular, a large supraocular, one large postocular, and eight or nine infralabials. Of the seven to nine supralabials, the second is the largest. The third through the fifth, or only two, contact the eye.

Prey consists primarily of marine invertebrates, but eels are occasionally eaten. Thai specimens have been found in the Gulf of Thailand and have recently been discovered on the west coast in the vicinity of Phuket. Elsewhere, the known range extends from the Persian Gulf to the Java Sea and north into the South China Sea.

Genus Thalassophis

In this genus, the nostrils are supra-lateral and the nasals separated by a pair of elongated internasals. The ventrals are distinct but not wider than the body scales.

This is another monotypic genus.

Thalassophis anomalus
Anomalous Sea Snake   งูเสมียนรังหัวสั้น

This snake has a short head, short, stout body, and strongly compressed tail. Its adult length is less than one meter. The body is light gray with broad dark bars. The bars often connect at the vertebral ridge or become narrower on the sides but sometimes they encircle the body. The ventral surface is white, as are the throat, labials, and chin. The top of the head is black. Males are larger than females; an average of five young are born at a time.

The scale rows range from 31 to 35 at the thickest part of the body, and the scales strongly keeled. The ventrals, barely wider than the body scales, number from 210 to 256. The frontal is small and partly or completely divided on some specimens. There is one preocular and one or two postoculars. Of the seven or eight supralabials, the third, fourth, and fifth border the eye. The second usually touches the prefrontal.

In Thailand, the Anomalous Sea Snake is found in the northern part of the Gulf of Thailand. It is also found off the coast of Kampuchea. It is reported to be uncommon in Malaysian waters but is found in the South China, Sulu, Java, and Flores Seas.

Subfamily Laticaudinae

Although certainly at home in the water, members of Subfamily *Laticaudinae* are also comfortable on land and are often found on coral reefs, on shore, in mangrove swamps, and sometimes rather far inland. Indeed, these snakes are oviparous and leave water to lay their eggs on land. Unlike completely marine sea snakes, they have relatively wide ventral scales - at least one half the body width - to facilitate movement on land, as well as vertically compressed tails to facilitate locomotion in water. Furthermore, the nostrils are located laterally, on the sides of the snout, and the nasals separated by internasals. They also differ from *Hydropheinae* in that the palatine and pterygoid bones of the roof of the mouth are not completely locked, allowing vertical movement. Other characteristics of this subfamily include smooth, imbricate scales, the posterior of one overlapping the anterior of another. As is the case with all members of Family *Elapidae*, members of this subfamily have highly toxic venom and can inflict fatal bites.

Members of Subfamily *Laticaudinae* generally do not thrive in captivity. They require a sufficiently large aquarium for free swimming. The water is to be filtered, aerated, and well stocked with fish and/or eels. Appendix 8, page 461, indicates the prevailing surface conditions of the Gulf of Thailand. The water is at its coolest from January to February, averaging 24.79° C, and warmest during the hot season of April and May, averaging 30.44°C. Salinity is at its maximum during the cold and hot seasons and at a minimum during the rainy season. The ideal situation for maintaining Thai sea snakes is to vary the water temperature and salinity according to the seasonal patterns indicated within Appendix 8. Such a regime should optimize the conditions for good health, stimulate feeding, and enhance conditions for captive breeding. If such a seasonal regime is not possible, attempt to maintain the average annual temperature and salinity levels displayed in Appendix 8. *Laticaudinae* require a dry surface area upon which to crawl and, hopefully, deposit eggs. Courtship and mating take place on land, generally at night. Fish and eels comprise the bulk of their diet with some eating eels, exclusively. Reluctant eaters may be induced to eat frogs.

A single genus and two species are found in Thailand.

Genus Laticauda

Species in this genus have fairly large, distinct head shields, the nostrils are situated on the sides of the snout, and the nasals separated by internasal shields. The body scales are smooth and overlap. The ventrals are relatively large, at least half the body width. Adult females are larger than adult males.

Laticauda colubrina Figure 32
Amphibious Sea Snake Plate 142
(Yellow-lipped Sea Krait) งูสมิงทะเลปากเหลือง

The head is small and barely distinct from the neck, the body nearly cylindrical, and the tail vertically compressed. The head is black with a snout and labials of yellow on young specimens. This yellow coloration, however, fades with maturity, becoming dull yellowish-white or gray on mature specimens. The body is light or dark blue-gray and encircled by black bands. Some bands narrow on the sides and become incomplete but most encircle the body. The ventral surface is yellow between the black bands. This species subsists primarily on eels. Females are longer and bulkier than males; females average 1.5 m and males one meter. Mating occurs on land at night. Eggs are deposited on land, but clutch sizes and the incubation period are unknown.

Figure 32. *Laticauda colubrina.*

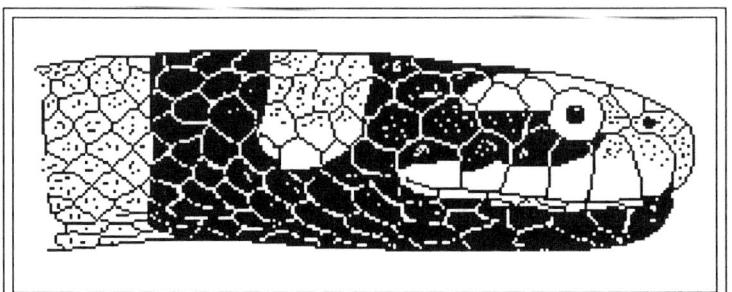

The head scales are not fragmented and are regular with one exception. In most specimens, a single large scale separates the prefrontals. There is one preocular, two postoculars, and seven, occasionally eight, supralabials. The third and fourth supralabials border the eye. Either eight or nine infralabials are present. The smooth scales overlap and are in rows of 21 to 25 at the thickest part of the body. The ventrals number from 213 to 246 and are at least one half the width of the body. The number of paired subcaudals varies with sex, males having 37 to 47 and females 29 to 35. The anal is divided.

This is not a common Thai snake, its only known sighting being in the vicinity of Phuket. It is, however, reported to be common in peninsular Malaysia. Its range extends from the vicinity of Calcutta through Southeast Asia and the waters of northern Australia into the southwest Pacific Ocean.

Laticauda laticaudata
Black-Banded Sea Krait   งูสมิงทะเล

This is a fairly small snake, averaging only one meter in length. The head is barely distinct from the neck, the body nearly cylindrical, the tail vertically compressed and oar-shaped. A yellow mark on the top of the head extends forward to cover the snout and reaches down behind the eye to the labials. The remainder of the head is black. The body is light bluish-gray and crossed by black bands that are nearly uniform in width and may, or, may not encircle the body. The ventral surface is white. This species holds the captive longevity record for the genus, a mere five years and three months.

A single scale does not separate the prefrontals, as in the case of *Laticauda colubrina*. There is a single preocular, two postoculars, and seven supralabials, the third and fourth bordering the eye. There are usually seven infralabials. The smooth body scales are in 19 rows at the thickest part of the body. Ventrals range from 225 to 245; the subcaudals are paired, males having from 36 to 50, females from 25 to 35. The anal is divided.

This species is found in the Gulf of Thailand and on the west coast near Phuket, but not in large numbers. It is not a common Thai snake. It has an extensive range, having been found in the Bay of Bengal, the waters of Southeast Asia, the East China Sea, the Yellow Sea, off northern Australia, and in the southwest Pacific Ocean.

Subfamily Maticorinae

This subfamily contains small, but sometimes long and slender, snakes. Although their venom is very toxic, few bites and even fewer fatalities occur because of their secretive nature and small size. The gland which keeps the eye moist (Harder's gland) is not entirely within the orbit and the dentitional arrangement of the lower jaw is not extended beyond the splenial-angular hinge.

Two genera are found in Thailand.

Genus Calliophis

These are small to medium-sized snakes whose short, rounded heads are only slightly distinguishable from their cylindrical bodies. The eyes are small, the pupils round, the body scales smooth, and the tail short. The loreal is absent and the subcaudals paired, as are the nasals.

Genera *Calliophis* and *Maticora* both contain snakes usually referred to as Asian Coral Snakes. Color patterns differ significantly, but their body structures are quite similar. The principal difference lies in the structure and placement of the venom glands. In *Calliophis* they are of normal size and entirely within the head; in *Maticora* they are elongated and extend from the head far into the body cavity.

Members of *Calliophis* are mild mannered and do not often make attempts to bite. When disturbed, they will often raise and coil their tail, exposing a colorful subcaudal area, in an effort to distract an enemy from their head. As is always the case with poisonous snakes, caution should be exercised; bites from this genus have been fatal. Specimens are rare and seldom seem in Thailand, perhaps because they are small, hide during the day, and only active at night.

A cage of small to medium size is adequate for these snakes. They are shy and secretive, so it will be necessary to provide them with adequate hiding places. Forested areas are their natural habitat, where they are most often found under fallen logs or pieces of bark, between rocks, or even burrowed under loose soil. Similar places should also be provided for captive specimens. Place a few centimeters of soil on the floor of the cage with some stones and pieces of bark or wood upon it. Plants also provide them with desirable hiding places. Place a water dish in the cage and recess it into the soil almost to its rim and spray the plants frequently. Be careful to maintain the temperatures and relative humidity displayed in the Appendixes. Food may be a problem in certain parts of the world. *Calliophis* preys upon small snakes and lizards, but may be induced to eat small frogs as well. Members of the this genus are egg layers. Sexual size differences vary from species to species.

Calliophis gracilis   Plate 143
Gray Coral Snake   งูปล้องหวายเทา

This is a small slender snake whose head is barely distinct from its neck. The length averages only 0.35 meters. The eyes are small and contain round pupils. The body is light or brownish-gray with a narrow black line extending its full length at the middle of the back. There is a black girdle on the neck as well as the tail, and a series of black spots extends along the body sides. Alternating black and white bars cover the belly, but the ventral side of the tail is pink.

This species has six supralabials, with the third and fourth bordering the eye. The first is the smallest, and the last is the largest. There are seven infralabials, one preocular, and two postoculars. The fourth infralabial is the largest. There are 13 scale rows at mid-body. The ventrals number from 303 to 320, and the paired subcaudals range from 21 to 28. The anal is divided.

The Gray Coral Snake is found only in the southern provinces bordering Malaysia. It is also found across the border in Malaysia.

Calliophis macclellandii macclellandii
McClelland's Coral Snake

งูปล้องหวายลายชั้นดำ
งูธิดาพระอาทิตย์

This attractive snake is not common in Thailand. Its head is short and rounded, its body cylindrical. In Thai specimens, the snout is black with the remainder of the head cream. The body is red with a series of narrow black bands crossing over it, each band has a light border. The belly is yellow. Females produce 6 to 14 eggs per clutch and are significantly larger than males, reaching approximately 0.75 m in length. Males probably average 0.65 meters.

The scales are smooth and in rows of 13 at mid-body. There is one preocular which is in contact with the posterior of the two nasals, two postoculars, and six infralabials. Of the seven supralabials, the third and fourth border the eye and the fifth and sixth contact the anterior temporal. On males, the ventrals range from 182 to 212 and the subcaudals from 28 to 36. Females have ventrals ranging from

208 to 244 and from 25 to 33 subcaudals. The subcaudals are normally paired but rarely specimens are found with unpaired subcaudals. The anal is divided.

Calliophis macclellandii macclellandii is found in the North and Northeast but it is far from common. It lives under loose soil and pieces of fallen vegetation on forested hills at elevations up to 1800 meters. Its primary prey consists of snakes but probably would also accept lizards. It is native to northeastern India, Nepal, Bangladesh, Burma, Vietnam, Laos, and southern China.

Calliophis maculiceps hughi
Cochran's Coral Snake   งูปล้องหวายหางแหวน

A total length of 0.26 m has been recorded. Its dorsal color is reddish-brown, the ventral surface lighter. The subcaudal area is light blue with black spots, and the throat bluish-gray. The top of the head is entirely black with a black band running across the nape of the neck that ends at the mouth level. A black ring encircles the body at the base of the tail.

There is one preocular, two postoculars, a single temporal, and seven supralabials, the third and fourth contacting the eye. The dorsal scales are in 13 rows at midbody, there are 186 ventrals, 27 subcaudals, and the anal is divided.

A single specimen has been collected on Tao Island, approximately 10° N, 99° 30' E, in the Gulf of Thailand. None have been found outside of Thailand.

Calliophis maculiceps maculiceps Plate 144
Small-spotted Coral Snake งูปล้องหวายหัวดำ

This small, slender, cylindrical snake is colorful and quite attractive. The head is brownish-black, the body dark orange or reddish-brown with a slight, variable pattern. Some specimens have widely separated dots on their sides, others have irregular large dots whilst others have a black, twisted, segmented pattern on the vertebral ridge. A black girdle exists around the base of the tail with another around the end of the tail; the ventral surface is pink. Females attain an adult length of 0.48 m, adult males may reach 0.435 meters. A captive female of 0.312 m produced a clutch of two large, elongated eggs.

The single preocular contacts the posterior nasal; there are two postoculars, a single temporal, and seven supralabials. Of the seven supralabials, the third and fourth border the eye and the fifth, sixth, and seventh touch the temporal. There are also seven infralabials. the smooth dorsal scales are in 13 rows at mid-body. Ventrals range from 174 to 186 on males and from 189 to 203 on females. Males have from 25 to 31 subcaudals, females from 21 to 25. Both the subcaudals and anal plate are divided.

Calliophis maculiceps maculiceps is the member of the genus most often encountered in Thailand. It has been recorded in every part of the Kingdom, although not in large numbers. The secretive nature of this genus is probably the reason that so few specimens are captured. The Small-spotted Coral Snake is commonly found under vegetation, logs, stones, or similar items. It is mild mannered and preys upon small snakes, particularly *Typhlops*, and lizards. It is found throughout Southeast Asia.

Calliophis maculiceps smithi
Stripe-backed Coral Snake   งูปล้องหวายหลังเส้น

This subspecies is usually around 0.33 m in length. The head is dark green and the body is yellowish-green. A black line extends along the middle of the back from the head to the tip of the tail. Black bands are present on the occiput, neck and base and end of the tail. The ventral surface is light pink but becomes darker posteriorly; the subcaudal area is blue-black with some bluish-white spots.

The description provided by Taylor indicates that the scalation of *Calliophis maculiceps smithi* falls well within the range of *Calliophis maculiceps maculiceps*. Information regarding scales is fragmentary as *Calliophis maculiceps smithi* is rare and few specimens have been studied. The Stripe-backed Coral Snake has one preocular and one postocular, the loreal is absent. The mid-body scale row count is 13 and the scales smooth. There are roughly 184 ventrals, 21 subcaudals, and the anal is divided.

Only two specimens have been found in Thailand, one in Saraburi Province of the Central Region, the other at Nong Kan Ploi in the Southeast. None have been reported outside of Thailand. This subspecies is synonymous with *Calliophis maculiceps malcolmi*.

Plate 124. *Aipysurus eydouxii.* White-spotted Sea Snake. งูทะเลจุดขาว.
  Small. Found in north of Gulf of Thailand. Diet - fish eggs.

Photo: Dr. W. Dunson

Plates 125 and 126. *Astrotia stokesii.* Stokes' Sea Snake. งูทากลาย.
  Light color phase. Not unusual - light, dark phases - pattern varieties.

Photo: Dr. W. Dunson

Plate 126. Melanistic form - no pattern. Large, bulky.

Photo: Dr. W. Dunson

356   *Plates*

Photo: Harold K. Voris

Plate 127.

Enhydrina schistosa
Beaked Sea Snake
งูคออ่อนปากจะงอย
งูคออ่อนหัวโต

Large, heavy-bodied.
Common - both Thai coasts.
Venom - toxic.

Photo: Harold K. Voris

Photo: Dr. Lim Boo Liat

Plate 128. (above)
Hydrophis brookii
Brooke's Sea Snake
งูแสมรังท้องเหลือง
Regular, alternating bands.

Plate 129.
Hydrophis caerulescens
Dark Blue-banded Sea Snake
งูแสมรังลายเยื้อง
งูแสมรังเกล็ดหยาบ
Found - both Thai coasts. Prey - eels, marine invertebrates.

Plate 130. *Hydrophis cyanocinctus*. Blue-banded Sea Snake.
งูแสมรังเหลืองลายคราม. งูแสมรังลายฟ้า. Found off both Thai coasts.

Photo: Harold K. Voris

Plate 131. *Hydrophis fasciatus fasciatus*. Striped Sea Snake. งูแสมรังลายแถบ.
Not yet found in Thai waters. Confirmed range indicates a likely presence off the west coast of Thailand.

Photo: Harold K. Voris

Photo: Dr. W. Dunson

Photo: Harold K. Voris

Plate 132. (above) *Hydrophis ornatus ornatus* Reef Sea Snake
งูแสมรังหางขาว
Not common. Found Gulf of Thailand. Spends life in water, as do all *Hydropheinae*.

Plate 133. *Hydrophis spiralis* Yellow Sea Snake
งูแสมรังเหลือง
First discovered in Thai waters, 1977.

Plate 134. *Hydrophis torquatus torquatus*. West Coast Black-headed Sea Snake. งูแสมรังเทาหัวดำเหลือง. Not yet been confirmed in Thai waters.

Photo: Harold K. Voris

Plates 135 and 136.

Lapemis hardwickii
Hardwicke's Sea Snake
งูใฮ้ข้าว

Found off both coasts of peninsular Thailand. Short, stout snake with a relatively large head.

Photo: Dr. W. Dunson

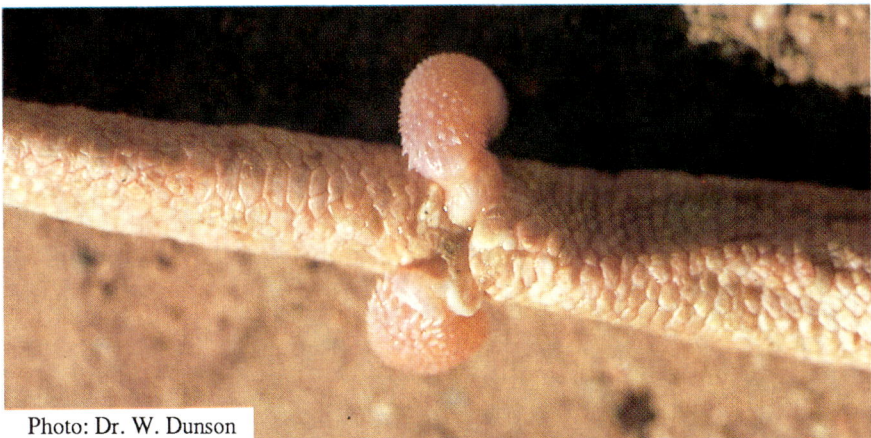
Photo: Dr. W. Dunson

Plate 136. Male sex organs (hemipenis) are located internally, posterior to the anal plate. They are paired and during breeding a hemipenis exits the anal opening of the male and is then inserted into the female's sex organs through the anal opening. The plate shows the male sex organs - forced out of their normal position for examination.
Note: very small, almost indistinguishable, ventral and subcaudal scales, typical of Subfamily *Hydropheinae*.

Plate 137. (left) *Microcephalophis gracilis gracilis* Needle-headed Sea Snake งูคอย่อนหัวเข็ม Very small head. Rare in Thai waters.

Plates 138 and 139. *Pelamis platurus* Yellow-bellied Sea Snake. งูชายธงหลังดำ. Note: elongated head.

Photo: Harold K. Voris

138

Photo: Dr. W. Dunson

Plate 139.

Greatest range of the sea snakes.

Various patterns. Coloration same for each in this species - patterns often variable - true of many species.

Photo: Dr. W. Dunson

Plate 140.

Pelamis platurus
Yellow-bellied
Sea Snake
งูชายธงหลังดำ

Mating snakes.

Photo: Dr. W. Dunson

During copulation, the male and female entwine as the male inserts one of two hemipenis through anal opening into female's sex organs. Fertilization is internal. Method followed by all snakes excepting one known parthenogenetic species, *Ramphotyphlops braminus*.

Plate 141.

Praescutata viperina
Viperine Sea Snake
งูชายธงท้องขาว

Note: subocular scales prevent supralabials contacting the eye.

Photo: Harold K. Voris

Plate 142. *Laticauda colubrina.* Yellow-lipped Sea Krait. งูสมิงทะเลปากเหลือง.
As all sea snakes, has fatal toxic venom. Most often encountered coastal areas - basks on coral reefs, tidal flats.

Photo: Dr. W. Dunson

Plate 143. *Calliophis gracilis.* Gray Coral Snake. งูปล้องหวายเทา.
Uncommon, beautiful. Found in South.

Photo: Dr. Lim Boo Liat

Plate 144. *Calliophis maculiceps maculiceps.*
Small-spotted Coral Snake. งูปล้องหวายหัวดำ.
Pretty, widely distributed - rarely found due secretive nature.
Note: tail position - typical defensive position.

Photo: Raynoo Cox

Genus Maticora

This is the second genus containing Asian Coral Snakes. Genus *Maticora* is represented by two species in the Kingdom. They have smooth scales, small eyes with rounded pupils, and loreal scales are absent. They are oviparous and prey primarily upon snakes and lizards.

It is unfortunate, but the two Thai representatives of this genus do not thrive in captivity. Perhaps they may do better if kept on soil in very large cages. They are timid and secretive; therefore, they should be given numerous hiding places and disturbed as infrequently as possible. Of course, they should be kept within the guidelines for temperature and humidity offered in the Appendixes.

Few people have been bitten by *Maticora*, probably because their secretive habits keep them removed from human contact. This is fortunate because their venom is toxic and those bitten have suffered extreme difficulties. In this genus, the venom producing glands extend into the body cavity. Thus, members are named Long-glanded Coral Snakes.

Maticora bivirgata flaviceps   Plate 145
Blue Long-glanded Coral Snake   งูพริกท้องแดง

This is a very beautiful snake. The head and tail are both red and the body a very dark blue. The scales are iridescent. A narrow, gray-white line extends along the lower body scales from the neck to the tail. The ventral surface is red. The head is barely distinct from the neck, and the body is quite slender. Although slender it is also quite long. Measurements of 1.4 m are common.

The Blue Long-glanded Coral Snake has one large preocular and two small postoculars, but is without loreal scales. There are six infralabials and six supralabials. The third and fourth supralabials touch the eye and the sixth is the largest. Scale rows number 13 at mid-body; the scales are smooth. The ventrals number from 244 to 295, the paired subcaudals from 34 to 53, and the anal plate is divided.

This snake lives in forested areas and is most often found under rocks, logs, and fallen vegetation. It is known to eat other snakes and probably preys upon lizards and frogs. If disturbed, it will raise and coil its tail like a corkscrew, showing the red under surface.

The species does not thrive in captivity. Captives should be kept in cages with numerous items which they can hide under. They should not be disturbed often. They are nervous and probably vulnerable to stress. Furthermore, they seem to be a snake which does not drink water voluntarily, for they seem prone to dehydration. Their cage should contain a recessed water dish. If they are not observed drinking voluntarily, gently place a hook under the anterior portion of their body and "lead" them to water. Place the head in the water dish and they will probably drink.

In Thailand the Blue Long-glanded Coral Snake is native to the provinces south of Songkhla. It also occurs in Burma, peninsular Malaysia, and western Indonesia. It is not a common snake anywhere within its range.

Maticora intestinalis lineata   Plate 146
Brown Long-glanded Coral Snake   งูพริกสีน้ำตาล

This is a small, slender snake, usually under 0.5 m in length. The head is small, flat, rounded, and not clearly distinguishable from the cylindrical body. The head is bright brown, the labials olive, and the chin gray-black. A series of thin lines extend from the head through the body to the tail. The vertebral line is pale red. A sequence of black, yellow-brown, black, cream, and black lines proceeds down the sides of the body from the pale red line on the vertebral ridge. The ventral surface is alternately black and white. The subcaudals are red and crossed by one black bar.

The body scales are smooth and arranged in 13 rows at mid-body. There is one preocular, two postoculars, six infralabials, loreals being absent. Of the six supralabials, the third and fourth are in contact with the eye. The ventrals number from 197 to 273, the divided subcaudals range from 15 to 33, and the anal plate is single. It has been reported that females produce two to four eggs per clutch which require approximately 80 days to hatch.

Like *Maticora bivirgata flaviceps*, the Brown Long-glanded Coral Snake is most likely to be found under rocks and vegetation in forested areas. It has been found at elevations of 1100 meters. When disturbed, the tail is raised to display the red and black ventral surface. This species does not thrive in captivity and should be housed and cared for in the same manner recommended for the Blue Long-glanded Coral Snake.

This snake is not common in Thailand and has only been found in the provinces south of Surat Thani. It is also native to peninsular Malaysia and western Indonesia.

FAMILY VIPERIDAE

Members of this family normally have stout bodies and triangular heads. The pupil is vertical; the scales are keeled. They are venomous, possessing a specialized venom producing gland located behind the eye - completely separate from the superior labial gland. Most species are viviparous.

The distinguishing feature of members of this family is the delivery system for venom injection into prey or adversaries. This differs from other poisonous snakes. Each has a pair of long, tubular, hinged, recurved fangs located on the upper jaw at the front of their mouth (Solenoglypha, Figure 7). When the mouth is closed these fangs are folded back to lie comfortably between the upper and lower jaws. When the mouth is opened, the fangs swing forward into a position perpendicular to the upper jaw for ease of penetration into the flesh of an animal and for conducting venom deep into the wound. These fangs are encased in a membrane when not in use and are also "shed", or replaced, periodically. Unlike the rear-fanged snakes and most *Elapidae* or *Hydropheinae*, these snakes strike their prey and, usually, do not hold on. They strike, withdraw, and wait for their prey to retreat and die. The prey is followed and, when found dead, swallowed. An exception might occur if a large snake has selected a small animal as prey. In such a case where the prey does not represent a threat, the animal may be bitten, held until death, and then swallowed. The large fangs are also useful in the process of swallowing food.

Family *Viperidae* is divided into Subfamily *Crotalinae* and Subfamily *Viperinae*.

Subfamily Crotalinae

In addition to the efficient venom delivery system mentioned above, evolution has favored subfamily *Crotalinae* with yet another useful device to help them in their struggle for survival. On each side of the head, approximately halfway between the eye and the nostril, there is a relatively large noticeable loreal pit. This open pit leads into a small inner chamber housing heat sensitive nerve endings. Thus, members of this subfamily are able to prey upon warm-blooded animals during the night or in dark burrows quite effectively by following the body heat that is emitted. The presence of this pit has given the species in this subfamily the name "Pit Viper".

Two genera and a total of thirteen species and subspecies are native to Thailand.

Genus Calloselasma

This genus has been separated from genus *Agkistrodon* rather recently. The distinctive characteristics are: the loreal pit is separated from the supralabials; the dorsal scales are smooth and the subcaudals paired; the braincase is broad with the supratemporal not extending or barely extending beyond it; the slender ectoptergoid is hooked, not strongly curved; the angular and splenial are fused.

Calloselasma contains two genera, both oviparous. *Calloselasma rhodostoma* is widespread on mainland Southeast Asia and is native to Thailand. The validity of the second genus, *Calloselasma annamensis* of Vietnam, has been questioned. Some consider it is an aberration of *Calloselasma rhodostoma*.

Because of its size, it can be housed in a fairly small cage. The cage should open from the top for added safety as this species can strike with great speed and accuracy. It is terrestrial and rarely climbs. The cage should include a hiding place as well as a constant supply of clean, fresh water. Rodents are the staple of the diet but this may be varied by offering birds and/or frogs occasionally. *Calloselasma* usually does well in captivity.

Calloselasma rhodostoma Figure 33
Malayan Pit Viper Plates 147 to 149

This is an attractive little snake whose head is triangular and quite distinct from the neck. The snout is pointed and slightly turned up at the tip. Adults grow to a maximum length of 0.8 m, with females being larger than males. Their bodies are of moderate girth and a vertebral ridge is noticeable. This snake is capable of spreading its ribs and making its body very flat. The tail is rather thin and short. The eyes are of moderate size and have elliptical pupils.

The body is pinkish to grayish-brown and there is a series of light-edged, black triangular markings on the back. The head is dark with a light line extending from the snout, passing above and through the eye terminating at the base of the jaw. The pupils are dark and the labials are white. The ventral surface is dull white with small brown specks. The newborn are very dark but have white tails. When hungry, juveniles elevate the white tail in order to lure prey within striking distance.

Figure 33. *Calloselasma rhodostoma.*

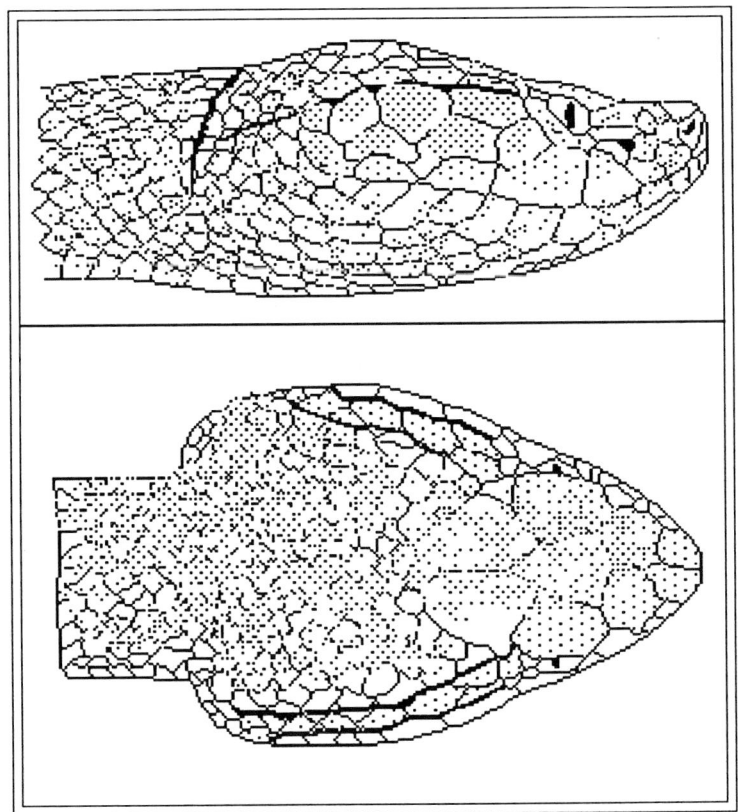

The scales are smooth and in 23 rows at mid-body. The ventrals vary with sex; males have from 148 to 154, females from 156 to 166. The subcaudals are divided and also vary with sex. Males have from 45 to 52, females from 35 to 46. The anal is single. There are three preoculars, a loreal, and one large postocular. The normal head shields are not separated by small scales. There are seven supralabials, the third and fourth bordering a large subocular scale. Eleven infralabials are present.

This snake is terrestrial and is often found coiled between rocks or in vegetation in forested areas, though it is frequently found near human habitation. It is somewhat sluggish and often does not retreat when approached. This viper seems to prefer relying upon its protective coloration to avoid detection. When confrontation is inevitable, however, they strike swiftly and vigorously. Because of the tendency to lie still in order to avoid detection and its very wide range, *Calloselasma rhodostoma* accounts for more snakebites than any other single species in both Thailand and Malaysia. The bites are extremely painful, but not usually fatal if correctly treated. Many of these bites occur near human habitation.

The Malayan Pit Viper is a nocturnal species that reproduces by laying eggs, females producing from 13 to 30 eggs at a time. Females coil about the eggs but are not capable of regulating the temperature of the clutch. They do, however, vary the exposure of the eggs to the air, thus exercising a control over the exposure of the clutch to variations in relative humidity. The brooding female also defends the clutch. The incubation period lasts for roughly 30 to 47 days. The young are very small, approximately 0.08 to 0.16 m. in length, when hatched and probably subsist on insects and small lizards until they are capable of catching and swallowing larger, more substantial meals. Refer to: Plate 149, page 397.

This species is found throughout Thailand. It is also native to Kampuchea, Laos, Vietnam, northern peninsular Malaysia, and western Indonesia.

Genus Trimeresurus

This is a genus of small to medium-sized snakes. Most of them are primarily arboreal and produce living young; however, at least one is basically terrestrial and oviparous. Some others are primarily arboreal and oviparous. In each species the eye has vertical pupils; the head is distinct and is covered with small scales rather than conventional head shields. A loreal pit is present and the small nostril is contained within a single nasal scale. Scale rows at midbody number from 17 to 31 (19 to 27 in Thai specimens) and the subcaudals are almost always paired. The tail is moderate to short in length and prehensile in most. Most species also have two or three postoculars and a single anal plate. Females are usually much larger than males.

Many members of this genus are distinct, such as *monticola meridionalis* and *wagleri*, and easy to identify, but this is not always the case. In Thailand, the species *albolabris*, *hageni*, *macrops*, *popeorum*, *stejnegeri*, and *sumatranus* are often confused and have been incorrectly identified in some publications. The confusion arises from the color as all of these genera are a shade of green, often a very similar shade of green, and do not have consistent adult body patterns to differentiate them. This problem is compounded because juveniles and subadults also have similar coloration and markings. Thus, neither coloration nor pattern is an effective tool by which these "green" pit vipers can be identified. Scalation, however, is an external characteristic which is used to differentiate some of these genera. For instance, on *albolabris* and *macrops*, the first supralabial and nasal are entirely or partially fused; in *popeorum* and *stejnegeri* they are separated. External characteristics, however, cannot be used as a means of identification in every case. Some genera are so closely related that only internal characteristics, for example, dentition and hemipenis structure, can be used for positive identification. Even this may not provide all of the answers.

Some publications have reported the presence of *Trimeresurus erythrurus* in the Kingdom of Thailand, but evidence has not yet conclusively proven its presence east of Burma.

In general, this genus does well in captivity. Most species should be housed in fairly tall cages that include plants and branches upon which to climb. The cage should be fairly tight to keep the humidity at an acceptably high level. Daily spraying with tepid water will maintain a high humidity and provide drinking water for the captive. Many specimens, especially the young, obtain most of their water intake by drinking from the leaves of bushes and trees. A water dish should also be included in the cage. Mice, birds, lizards, and tree frogs are the primary prey of most members of this genus. The babies of some species are so small that it might be necessary to offer them insects or very small lizards for their first few meals.

These are venomus snakes, quick to strike, and some species are abundant within a wide range. Therefore, they account for a large number of the snakebites recorded in Thailand and other parts of Asia. Although their bites are painful and often cause great swelling, fatalities are rare. Nonetheless, care should be taken in dealing with each member of this genus.

Twelve species and subspecies have been recorded in Thailand.

Trimeresurus albolabris albolabris Figure 34
White-lipped Pit Viper Plates 150 and 151
งูเขียวหางไหม้ท้องเหลือง

The head is fairly long, roughly triangular, and very distinct from the neck. The eyes are yellow, the neck slender, the body relatively stout, and the prehensile tail fairly short. The head and dorsal surface are green, the ventral surface is greenish or yellowish-white. The supralabials are pale green, yellow, or white, as are the chin and throat. On males a white stripe begins at the neck and extends the length of the body on the first dorsal scale row. The ventral edge may be dark on young animals. This stripe is indistinct or totally absent on females. The tails of both sexes are reddish-brown. Females average 0.75 m in length, males 0.47 meters.

The small scales on the head are smooth, although the temporal scales may be feebly keeled. The internasals may be in contact or separated by one to three small scales. There are three preoculars, no loreal, and two small postoculars. The supraoculars are undivided and separated by as many as 8 to 13 scale rows. The subocular is separated from the supralabials by one or two rows of small scales. There are usually from 9 to 12 supralabials but there may be as few as 7 and as many as 13. The first supralabial is wholly or partially fused with the nasal. If there is a partial separation, it may be fairly wide. The third supralabial is the largest. Infralabials are usually in the range of 11 to 13, but there may be as few as 6 and as many as 16. The keeled body scales are in rows of 21, rarely 23, at mid-body. In males the ventrals range from 151 to 169, the subcaudals from 61 to 78. Females have from 149 to 173 ventrals and from 48 to 67 subcaudals. In both cases, the subcaudals are paired, the anal single, and the tail prehensile.

Figure 34. *Trimeresurus albolabris albolabris*.

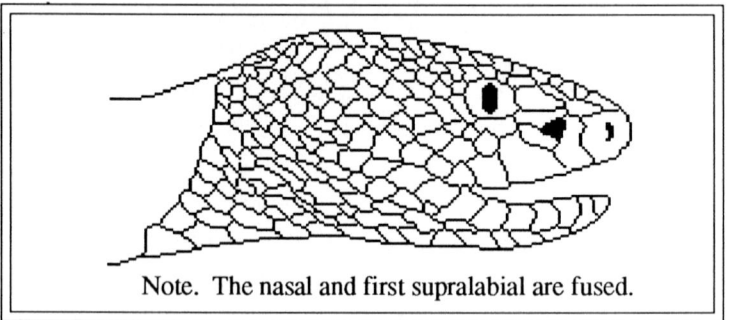

Note. The nasal and first supralabial are fused.

This is a nocturnal snake that is usually quick to strike. It hunts at night on the ground and in bushes and trees, seeking mice, birds, lizards, or tree frogs. It is often found near human habitation, including homes in Bangkok, and is responsible for a large number of bites. Fortunately, its venom is not usually fatal. It gives birth to from 7 to 15 young at a time. The young are only about 0.1 m at birth and it is often difficult to find small food for them. Their first food might well be extremely small lizards or insects.

The White-lipped Pit Viper is widely distributed in Thailand but may be most abundant in the Central Region, especially in the vicinity of Bangkok and the provinces of Ayutthaya and Nakon Pathom. Elsewhere, it is found in India, Nepal, Bangladesh, southern China including Hainan, and all of mainland Southeast Asia as far south as Penang in peninsular Malaysia. In Indonesia it is found on Sumatra, Java, Madura, Borneo, and Celebes. It is also found on the Nicobar Islands.

Trimeresurus hageni
Hagen's Pit Viper

Plate 152

งูเขียวหางไหม้สุมาตราหัวเขียว

This species may reach one meter in length. It is light to dark green. The dorsal body scales have narrow black edges that may form thin longitudinal lines which are too thin to be called cross-bands. A light line, bordered ventrally by a dark line or a series of dark spots, extends along the first and second dorsal scale rows. Some specimens have a series of pink spots along the body sides that ultimately merge with the light red tail. The head is green, the labials, chin, and throat a lighter shade of green. Some specimens have a light streak extending from the eye through the temporals to the corner of the mouth. The ventral surface is light green; the ventrals of some specimens have dark edges.

The body scales are feebly keeled and in 21 rows at mid-body. Ventrals number from 177 to 197, the paired subcaudals from 63 to 85, the anal single, and the tail prehensile. The smooth head scales are in rows of 4 to 9 between the supraoculars. The internasals are in contact, or rarely, separated by a small single scale. There are usually 9 to 11 supralabials. Usually only the third or the third and fourth supralabials contact the subocular.

This species has often been confused with *Trimeresurus sumatranus*, to be discussed later. Table 7 displays the characteristics of the two species.

This is a snake of the lowlands and forested hills, being found at elevations of 600 meters. This species has only recently been added to the herpetofauna of Thailand, specimens being found in the southern provinces of Surat Thani and Trang. This snake is also native to peninsular Malaysia and western Indonesia.

Table 7. Characteristics of *Trimeresurus hageni* and *Trimeresurus sumatranus*.

Characteristic	*Trimeresurus hageni*	*Trimeresurus sumatranus*
supralabials contacting subocular	one (#3) or two (#'s 3 & 4)	three (#'s 3, 4, & 5)
supranasals	in contact, rarely separated	separated, rarely in contact
head and neck scales	without broad black borders	with broad black borders
dorsal pattern	may, or, may not have narrow black lines	very distinct black cross-bands

Trimeresurus kanburiensis Plate 153
Kanburee Pit Viper งูหางแฮ่มกาญจน์

The head is flat, triangular, and distinct from the neck and slender body. The eye is dark brown, the iris black; the head is purplish-brown with an abundance of small, olive-green blotches. The dorsal surface of the body has a pattern of alternating, zigzagging bands of olive-green and purplish-brown, most of the bands being approximately two scales wide. The labials are olive-green, the chin and throat pale green, and the ventral surface yellowish-green. A white stripe, broken by intrusions of purplish-brown which continue a short distance on to the ventrals, becomes prominent in the area of the fifteenth ventral. This continues along the edge of the ventrals and the first body scale row to the vent and beyond, to become obscured in the colors of the tail. The anterior tail portion is mottled olive-green and purplish-brown; it is entirely purplish-brown posteriorly. This species averages 0.8 m in length.

Separating the supraoculars are 8 to 12 small scales; the internasals are approximately double the size of the adjoining small head scales and separated by one scale. Of the ten supralabials, the first is separated from the nasal, the second borders the anterior of the loreal pit, the third is the largest. Supralabials are separated from the narrow, elongated subocular by one row of scales. The first of the 12 infralabials are in contact behind the mental, separating it from the anterior chin shields. There is a single preocular and three postoculars, the lowest being the largest. The body scales are in 19 rows and, except for the lowest, are keeled. The ventrals number 155 to 159, and the paired subcaudals 42 to 74. The anal plate is single and the tail prehensile.

This species was first described by Malcolm Smith in 1943. The specimen, a female, was found in hilly country in the province of Kanchanaburi. Unfortunately, various publications have confused this species with *Trimeresurus purpureomaculatus purpureomaculatus*. The publications have appeared in Thailand, Europe and the United States. These animals have also been exported from Thailand with an incorrect name. The two species are easily distinguished by the number of scale rows at mid-body, the general build of the head and body, and by coloration. Compare the descriptions of the two species as well as their photographs. Refer to: Plates 153 and 160, pages 398 and 401.

The Kanburee Pit Viper has been found in evergreen forests in the provinces of Kanchanaburi and Surat Thani. It has not yet been reported outside of Thailand, although it seems likely that its range would extend across the border into the extreme south of Burma.

Trimeresurus macrops Plate 154
Big-eyed Pit Viper งูเขียวหางไหม้ตาโต

The head is somewhat short and rounded but still triangular and distinct from the neck; the body is slender. Its color distinguishes it from other closely related members of the genus. The top of the head and dorsal surface of the body are bluish-green, the labials are light bluish-green. The chin and throat are bluish-white, and the ventrals bluish, with the ventral scales becoming bluish-green on their lateral margins. The bluish-green of the margins of the ventrals intrudes onto the first few rows of body scales. The tail is reddish-brown, but the subcaudals may retain a hint of the bluish color of the ventrals. The eyes are golden yellow. A bluish lateral stripe along the first dorsal scale row may, or, may not be present. Young specimens may have a white postocular stripe. The stripe is always present on the male and infrequently on the females. Females are slightly larger than males and give birth to from 6 to 12 young at a time, each a miniature replica of an adult. The maximum length recorded for a male is 0.62 m and 0.71 m for a female.

The body scales are keeled, except for the first row, and arranged in 21 rows at mid-body. Rarely only 19 are present. The ventrals range from 160 to 174 on males, with females having from 161 to 175. The subcaudals are paired with males ranging from 62 to 74 and females from 49 to 63. The anal is single, the tail prehensile. The internasals are usually in contact, but occasionally are separated by one to three scales. The supraoculars are wider than the internasals and separated by three scales; the supraoculars are separated by eight scale rows at the narrowest point. The temporals are keeled. From 10 to 13 infralabials are present, the first are in contact and also separate the mental from the anterior chinshields. Of the 9 to 12 supralabials, the first is entirely or partially fused with the nasal, the second forms the anterior edge of the loreal pit, and the third the largest. All of the supralabials are separated from the fragmented subocular by three rows of small scales.

Trimeresurus albolabris albolabris and *Trimeresurus macrops* are closely related and their ranges overlap. The distinguishing characteristics of the two are displayed in Table 8, page 380.

Trimeresurus macrops was proposed as a new species in 1977 by Eugen Kramer.[10] Regretably, some publications have confused this species with *Trimeresurus popeorum popeorum*. Coloration and the fusion of the nasal and the first supralabial clearly distinguish *Trimeresurus macrops* from *Trimeresurus popeorum popeorum*.

---

10. In: **Zur Systematik der Grunen Grubenottern der Gattung Trimeresurus.** 163-205.

Table 8. Characteristics of *Trimeresurus albolabris albolabris* and *Trimeresurus macrops*.

Characteristics	*albolabris albolabris*	*macrops*
pterygoid teeth	12 or less	13 or more
mandibular teeth	13 or less	13 or more
supraocular shield	narrow	wide
temporal scales	smooth	keeled
labial color	light green or white	blue-green
ventral color	clearly lighter than dorsum	slightly lighter than dorsum

Although primarily arboreal and nocturnal, the *Trimeresurus macrops* may often be found at dusk or in the early morning hours on the ground hunting for food. Prey consists of small rodents, birds, tree frogs, and lizards.

This is a common snake in the southern portion of the Central Region, often found within the city limits of Bangkok. The range extends eastward through Kampuchea into southern Vietnam. It may also occur in southern Laos.

Trimeresurus (Ovophis) monticola meridionalis
Mountain Pit Viper Figure 35
Plates 155 and 156
งูทางแฮ่มภูเขา

As was earlier stated, *monticola meridionalis* is distinct and easily recognizable. In fact, it has recently been proposed that *Trimeresurus monticola* and its subspecies be removed from genus *Trimeresurus* and used to comprise a new genus, *Ovophis*. This new genus would contain four species, one of them with three subspecies. This new classification is based upon the following distinctive characteristics of *monticola*: the subcaudals are both single and paired; the border of the maxillary cavity has a rounded projection forming two distinct curvatures; the ectopterygoid bone has a truncated dorsolateral projection; the pterygoid teeth extend nearly to the posterior margin of the ectopterygoid bone; the basal portion of the pterygoid is shorter than the ectopterygoid; the splenial and angular are fused. Furthermore, *Trimeresurus* has a shorter series of pterygoid teeth than does *Ovophis*. *Ovophis* is also oviparous, in contrast to most of the species presently included in genus *Trimeresurus*. Although the argument is convincing, *monticola* remains, for the present, within *Trimeresurus* as it is not yet known if the proposed reclassification has been accepted by the International Commission on Zoological Nomenclature.

This is a short, stout snake, much stouter than other members of *Trimeresurus*. The head is broad and triangular, the eyes relatively small, and the pupils vertical. Specimens from northern Thailand are generally darker than those from the South. The top of the head is black, but

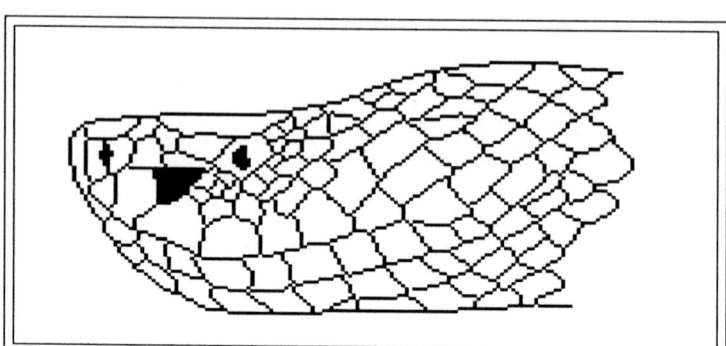

Figure 35. *Trimeresurus (Ovophis) monticola meridionalis.*

a brown stripe passes above the the eyes and through the temporals. The labials are dark brown. A series of large, irregularly shaped black blotches, smaller anteriorly, extend along the vertebral ridge and halfway down the flanks. Smaller, brownish-black blotches extend along the lower half of the body flanks. The remaining dorsal surface is dark brown. The ventral surface is light with small brown spots. The top of the head of southern specimens is usually blackish-brown, the temporal area and sides of the head are dark brown. A series of roughly rectangular, brownish-black marks extend along the sides of the vertebral ridge; they alternate and do not meet at the vertebral ridge on the anterior one-third of the body. Posteriorly most merge, but a few do not. Smaller brownish-black blotches extend along the lower sides of the body near the ventrals. The rest of the dorsal surface is a lighter shade of brown, sometimes brownish-yellow. The ventral surface is nearly yellow and has a dusting of brown. Juveniles have similar patterns and coloration, but their colors are more brilliant. Adult females attain lengths of 1.1 m, males about 0.5 meters. This species does not have a prehensile tail.

Thai specimens usually have the following scalation. The head is covered with unequal, smooth, slightly overlapping scales. The large, undivided supraoculars, are separated by five to nine small scales. The internasals are large and, although occasionally in contact, are usually separated by one or two scales. There are three preoculars, two loreals, the upper the larger, and three small postoculars. The seven to ten supralabials are separated from the fragmented subocular by two to four rows of small scales. The first supralabial is entirely separated from the nasal, the second is sometimes detached from the scale forming the anterior portion of the pit, and the third is the largest. There are usually 11 infralabials. Mid-body scale rows number 23 or 25, but as few as 21 and as many as 27 have been observed. The scales are smooth except for the third through the sixth rows, which are faintly keeled. Ventrals range from 138 to 145 on males, their subcaudals numbering from 44 to 50. Females have 144 to 152 ventrals and 33 to 40 subcaudals. The subcaudals are paired, although the last few rows may be single, as is the anal.

This is a terrestrial animal which lives at elevations above 1000 meters. It is important to remember that they live in cooler environments than do most tropical snakes and do not do well in a hot environment. Consequently, they should be maintained at daytime temperatures of 24^0 C to 27^0 C and at nighttime temperatures which are proportionately lower than those displayed in Appendixes 6 and 7. The Mountain Pit Viper is usually found between rocks or other objects on the ground, or crawling about in the evening searching for their food of small rodents.

This subspecies is not abundant anywhere within its range. In Thailand, it has been found in the North, West, and the South. Elsewhere, it has been recorded in Nepal, the Himalayan areas of India, central and southern China, and mainland Southeast Asia including peninsular Malaysia.

Trimeresurus popeorum popeorum [11]
Pope's Pit Viper

Figure 36
Plate 157
งูเขียวหางไหม้ท้องเขียว

The head is relatively short and rounded but retains a triangular shape. The eyes are large, the pupils vertical and yellow, and the neck thin. The coloration is very similar to that of *Trimeresurus stejnegeri stejnegeri*. The top of the head and dorsal surface are medium green, the ventral surface pale green. A white stripe extends the length of the body along the first dorsal scale row and the ventral half of the second dorsal row. On some specimens, the white stripe has a brown ventral border. Some males have a white postocular stripe but none appear on females. The labials are pale green and the tail is red. This species grows to a length of approximately one meter.

Except for the first row, the body scales are keeled. They are arranged in rows of 21 at mid-body. The ventrals range from 156 to 169 on males while the subcaudals number from 64 to 76. Females have ventrals numbering from 155 to 168, their subcaudals ranging from 52 to 66. The subcaudals are paired, the anal single, and the tail prehensile. The upper head scales are small, unequal, and smooth. The supraoculars are narrow and separated by 9 to 13 scale rows. The elongated subocular is separated from

---

11. An early taxonomic error may have made this species synonymous with *Trimeresurus gramineus*.

the supralabials by a single row of scales. Temporal scales are small and lightly keeled. The internasals are large, elongated, and separated by one or two scales. Of the 9 to 11 supralabials, the first is entirely separated from the nasal and the third is the largest. There are 12 to 14 infralabials, three elongated preoculars, two small loreals, and two postoculars.

Figure 36. *Trimeresurus popeorum popeorum.*

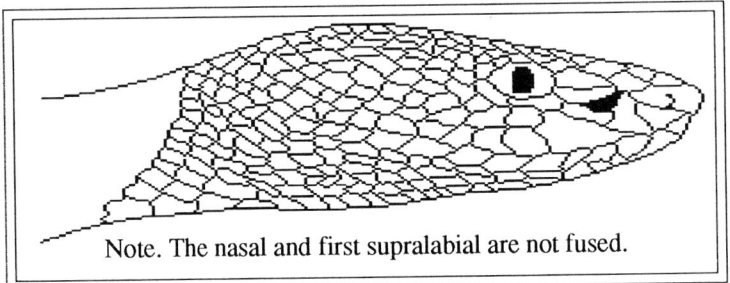

Note. The nasal and first supralabial are not fused.

Trimeresurus popeorum popeorum and *Trimeresurus stejnegeri stejnegeri* are closely related. A comparison of the descriptions of each will reveal that there are no significant external characteristics which distinguish them. The only known difference between them is internal - the hemipenis of males are dissimilar. Thus, females are indistinguishable, creating a frustrating taxonomic problem.

Pope's Pit Viper is nocturnal and frequently found on low bushes and shrubs although it probably seeks prey on the ground as well. Prey is similar to that of other members of the genus, consisting of mice, birds, frogs, and lizards. It is reported to be viparous and common on elevations above 1000 meters.

This snake is found throughout the Kingdom. In addition to Thailand, it is found in northeastern India, Burma, Laos, Kampuchea, Vietnam, and in peninsular Malayasia as far south as Kuala Lumpur.

Trimeresurus puniceus puniceus Plate 158
Flat-nosed Pit Viper
(Brown Flat-nosed Pit Viper) งูปาล์มแดง

This is a fairly small snake, females averaging 0.595 m in length, males about 0.43 meters. The head is flat, triangular, and quite distinct from the neck. The snout is spatula-shaped, slightly upturned, and protrudes beyond the chin. There is an elevated ridge starting above each eye and extending to the rostral. The eyes are pale yellow and the pupil vertically elliptical, and the body short and stout. Both the head and body are reddish-brown. When examined closely, an uneven light pattern is seen throughout the body. The ventral surface is mottled brown and darker than the dorsal surface. Juveniles are a lighter shade of brown and have a more distinct pattern than adults.

The head is covered with small, smooth scales. The temporals may be smooth or slightly keeled. The supra-oculars are separated by 9 to 14 scale rows. Two to four postoculars may be present. The subocular may be entire or divided and is separated from the supralabials by two or three scale rows. Of the 10 to 12 supralabials, the second sometimes borders the loreal pit and the third is the largest. The body scales are feebly keeled and arranged in rows of 19, 21, or 23 at mid-body. The ventrals range from 144 to 176 and the paired subcaudals from 38 to 65. The anal is single and the tail prehensile.

This is a snake of the forests that has been found at elevations up to 1500 m above sea level. As such, its should be kept at temperatures slightly cooler than those shown in the appendixes. It is nocturnal and preys upon mice, birds, and tree frogs. Unlike most members of this genus, this species is oviparous, females laying 7 to 12 eggs at a time. Some reports indicate that females incubate their eggs, the incubation period being approximately 60 days.

Only a few specimens have been found in Thailand. Personally I have only seen one and preserved specimens are not included in the national collection. The few specimens collected in Thailand have been found in the provinces bordering Malaysia. Its range continues through peninsular Malaysia to the island of Borneo. It is also native to western Indonesia, where it has been found on the islands of Sumatra, Java, and the Natuna Islands. It is reported to be rare throughout its range.

Trimeresurus puniceus wiroti   Plate 159
Siamese Palm Viper
(Wirot's Pit Viper)   งูปาล์ม

This snake was first found in 1979 in the province of Nakhon Si Thammarat in southern Thailand. Two were captured in a dense, humid forest. The Siamese PalmViper is a small serpent. The largest one found to date, measured 0.558 meters. These snakes have a triangular head which is distinct from the neck and a spatula-shaped snout protruding forward above the mouth and chin. The prominent colors are light and dark brown. The head is dark brown with a light brown line passing through the upper half of the eye which continues to the base of the jaw. The labials have numerous small, light brown blotches. The eyes are yellowish-brown, and the vertically elliptical pupils are dark. The body is predominantly light brown, however, there are numerous blotches of dark brown. Their locations are irregular, although they are more frequent along the vertebral ridge than on the sides. Their sizes also vary but those on the sides are usually larger than those in the vertebral area. Areas of greenish-brown or pinkish-brown also exist. The ventral surface is mottled light or dark brown. The tails of juveniles are yellow but become dark brown at maturity.

The supralabials are separated from the eye by a row of small scales. There are 11 to 14 infralabials and 9 to 11 supralabials, the fourth being the largest. There are two or three elongated preoculars, each bordering the loreal pit, in addition to three small postoculars. The head scales are small and keeled. The supraoculars are fragmented and separated by approximately eight small scales. The keeled body scales are arranged in 21 rows at mid-body. There are between 148 and 176 ventrals, the paired subcaudals range from 46 to 58, and the anal single. The tail is prehensile.

The Siamese Palm Viper has been previously presented in publications appearing in Thailand, Germany, and The United States as *Trimeresurus wiroti*. When studied, however, the minor differences between *wiroti* and *puniceus puniceus* do not appear to warrant separate species status. For comparative purposes the characteristics are displayed in Table 9.

With such minor differences of scalation, it seems more suitable to regard the Siamese Palm Viper as a subspecies of *Trimeresurus puniceus puniceus* rather than a separate species. Furthermore, *Trimeresurus puniceus wiroti* has a separate geographic range. Specimens have been found in the South on the Nakhon Si Thammarat Range, to the north of the Semgalasiri Range where the few Thai specimens of *Trimeresurus puniceus puniceus* have been found. The

Table 9. Characteristics of *Trimeresurus puniceus puniceus* and *Trimeresurus puniceus wiroti*.

Characteristic	*Trimeresurus puniceus puniceus*	*Trimeresurus puniceus wiroti*
head scales	small, smooth	small, slightly keeled
supraoculars	narrow or divided into pointed, erect scales	fragmented
scale rows between supralabials & suboculars	2 or 3	one
dorsal scale rows	19, 21, or 23	21
ventrals	144 - 176	148 - 176
subcaudals	38 - 65	46 - 58
supralabials	10 - 12	9 - 11

reader is cautioned that both the designations *Trimeresurus wiroti* and *Trimeresurus puniceus wiroti* have not yet received widespread recognition in the herpetological community.

The Siamese Palm Viper is a rather mild mannered, attractive little snake. It is primarily arboreal and seldom found on the ground. In captivity it seems to prefer a diet of small mice and house geckos, but in the wild its prey probably includes tree frogs as well and small birds. *Trimeresurus puniceus wiroti* should be housed in a small aquarium that includes a number of plants for climbing and hiding. Frequent sprayings are recommended. As with all snakes from the South, the temperature and humidity should be kept within the ranges described in Appendix 7. The Siamese Palm Viper is oviparous, as is *Trimeresurus puniceus puniceus*. Females are larger than males and produce 8 to 14 eggs per clutch. It is reported that females guard the eggs until the hatchlings emerge. Hatchlings average 0.15 m in length.

This is a snake of the highlands, specimens usually being found between elevations of 500 and 1200 meters. In Thailand, its range is limited to these elevations in, and around the southern province of Nakhon Si Thammarat. There are no reports of this snake being found outside of Thailand.

Trimeresurus purpureomaculatus purpureomaculatus
Mangrove Pit Viper   Plate 160
(Shore Pit Viper)   งูพังกา
                                                  งูเขียวม่วงหางไหม้

The head is triangular and distinct from the neck and the body is relatively stout compared to most other members of the genus. Adults of this species average one meter in length, with females being considerably larger than males. The top of the head is blackish-brown, the labials light yellow but speckled with small black dots. A series of rather large black blotches extends along the vertebral ridge, some having two or three brown scales in or near the center.

Smaller, irregularly shaped blotches extend along the sides of the body. The remaining areas on the dorsal side of the body are covered with yellow or yellowish-white, black-edged scales. There may, or, may not be a white line extending along the first row of body scales. The ventral surface is mostly white but the ventral scales have black edges and a few specimens have random black blotches of varying size. The subcaudals are predominantly black. This species is viviparous. Newborn have a pattern of pinkish-brown blotches alternating with areas of greenish-yellow. A second color variety, apparently more common in peninsular Malaysia, is uniformly purplish-brown dorsally with large, irregularly placed dark blotches. A white dorsolateral line may, or, may not be present. The head and body are the same color; the ventral surface is brown.

The head scales are tuberous and keeled, as are the temporals. The internasals are relatively large and usually separated by one to three scales. The supraoculars are narrow, may be broken, and separated by 12 to 16 small scales. Of the 11 to 13 supralabials, the first is partially or entirely fused with the nasal, the second forms the anterior edge of the loreal pit, and the third is the largest. Two or three rows of small scales separate the supralabials from the elongated subocular. There are from 11 to 13 infralabials, one large preocular, and either two or three postoculars. With the exception of the first row, the body scales are strongly keeled; there are 23 or 25 rows at mid-body on males, 25, 27, or 29 on females. Males have ventrals ranging from 160 to 175 in number, and subcaudals number from 66 to 79; on females ventrals vary from 152 to 170, the subcaudals 54 to 67. The subcaudals are paired the anal single, and the tail prehensile.

As the common name implies, this snake is found in mangrove forests and along canals which connect to the sea, as well as on offshore islands. Although often found on the ground, it is more often observed on low bushes. It is nocturnal and preys upon frogs, lizards, and small rodents. The bites of the Mangrove Pit Viper are painful, but not usually lethal. This snake does very well in captivity, where it has bred. Litters average 7 to 14 babies which are large enough to take pinky mice for their first meal.

Past publications have confused this species with *Trimeresurus kanburiensis*. The two species are easily distinguished, however, by careful examination of the head and body scales and noting the body structure and coloration. Compare the descriptions and photographs of each. Refer also to: Plate 153, page 398.

Thai specimens are found in the provinces located on the Straits of Malacca in addition to Surat Thani Province. Its range continues through peninsular Malaysia to Singapore. It is also found in Burma south of 17°N and on Sumatra, as well as many of the offshore islands in the Straits of Malacca. A subspecies, *Trimeresurus purpureomaculatus andersoni*, occurs on the Andaman Islands and less commonly on the Nicobar Islands.

Trimeresurus stejnegeri stejnegeri
Bamboo Pit Viper
(Chinese Green Tree Viper)   งูเขียวไผ่

This species has a thick, triangular head that is distinct from its slender neck and body. The eye is reddish-brown and the top of the head and dorsal surface medium green. The ventral surface is pale or whitish-green. The labials are light green or white. Males usually have a white postocular stripe, but these do not frequently occur on females. A white stripe, bordered ventrally with dark red or orange, extends along the first dorsal scale row from the vicinity of the eighth ventral into the subcaudals. This stripe is particularly distinct on youngsters. The dorsal surface of the tail is reddish-brown, the ventral surface light green. Males may reach 0.85 m in length, females one meter.

The head scales are small, unequal, and smooth. The supraoculars are narrow, with 9 to 13 scale rows separating them. The temporals are small and smooth. The elongated subocular is separated from the supralabials by one to four scale rows. Supralabials number 9 to 12, the first being completely separated from the nasal, the third being the largest. From 11 to 15 infralabials are present, in addition to two preoculars and two or three postoculars. Except for the first row, the body scales are keeled; scale rows numbering 21 at mid-body. Males have 156 to 172 ventral scales and 62 to 77 subcaudals; on females, ventrals range from 150 to 172 and the subcaudals from 59 to 69. The subcaudals are paired, the anal single, and the tail prehensile.

Trimeresurus stejnegeri stejnegeri and *Trimeresurus popeorum popeorum* are closely related and, externally, indistinguishable. For additional details, refer to the section *Trimeresurus popeorum popeorum*, page 384.

As with most members of the genus, *Trimeresurus stejnegeri stejnegeri* becomes active at night, searching for prey on the ground and in low trees or bushes. The diet includes mice, birds, tree frogs, and lizards. It is viviparous, litters averaging 7 to 10 living young.

Within Thailand, this species is fairly abundant in the mountains of the West, Northeast and Southeast. It is also native to upper Burma, Kampuchea, Vietnam, southern China, including Hainan Island, and Taiwan.

Trimeresurus sumatranus
Sumatran Pit Viper   งูเขียวหางไหม้สุมาตราหัวดำ

The snout is relatively short, the head is elongated and distinct from the neck and body. The scales on the top of the head and neck have broad black edges, surrounding green centers. Some scales are completely black. The labials are yellowish-green. Young specimens may have white post-ocular stripes. The dorsal surface is light green with distinct black cross-bands. On some, a yellowish-white line extends the length of the body along the first two dorsal scale rows. The ventral surface is yellowish-green, with the ventral scales having black edges. The tail is green anteriorly but reddish-brown posteriorly. A few have a yellowish tail with reddish-brown cross-bars. Adults average over a meter in length.

Plate 145. *Maticora bivirgata flaviceps*. Blue Long-glanded Coral Snake. งูปริกท้องแดง. Long, slender, beautiful - very difficult to maintain in captivity. Found in South - not common in range. Venom toxic.

Plate 146. *Maticora intestinalis lineata*. Brown Long-glanded Coral Snake. งูปริกสีน้ำตาล. Less common than Blue Long-glanded Coral Snake. Range also restricted to South.

Photo: Dr. Lim Boo Liat

Plate 147. *Calloselasma rhodostoma*. Malayan Pit Viper. งูกะปะ.
Specimen from the South. Abundant and widespread in Thailand. Coloration and pattern provides excellent camouflage - difficult to see. Responsible for large number of bites annually.

Photo: Dr. Roger Conant

Plate 148. *Calloselasma rhodostoma*. Malayan Pit Viper. งูกะปะ.
Freshly killed specimen from Nakhon Ratchasima, Northeast. Typical light coloration of northeastern specimens; southern are darker. Compare to southern specimen Plate 147.

Photo: Jarujin Nabhitabhata

Plate 149.

Calloselasma rhodostoma Malayan Pit Viper
งูกะปะ

Hatchlings began emerging from this clutch of eggs August 14, 1988. Incubation lasted 33 days.

Photo: Raynoo Cox

Plate 150. *Trimeresurus albolabris albolabris*. White-lipped Pit Viper. งูเขียวหางไหม้ท้องเหลือง. Commencement of shedding.

Photo: Suthigit Patramangorn

Shedding begins at the snout - old skin peels back over the body to tail end. Skins sometimes shed intact but often break. Compare color of newly exposed skin on the snout with color of unshed body skin. Dull body color, opaque eyes indicate imminent shedding.

Plate 151. *Trimeresurus albolabris albolabris* White-lipped Pit Viper งูเขียวหางไหม้ท้องเหลือง Very common in and around Bangkok. Primarily arboreal, many encountered at dusk and dawn foraging on ground.

Photo: Corey Blanc

Plate 152. *Trimeresurus hageni*. Hagen's Pit Viper. งูเขียวหางไหม้สุมาตราหัวเขียว. Recent addition to the fauna of Thailand. Range restricted to South.

Photo: Dr. Lim Boo Liat

Plate 153. *Trimeresurus kanburiensis*. Kanburee Pit Viper. งูหางแฮ่มกาญจน์. Slender, attractive - not yet found outside Thailand.

Plate 154. *Trimeresurus macrops*. Big-eyed Pit Viper. งูเขียวหางไหม้ตาโต. Common in Bangkok and southern portion of Central Region. Shares range with *Trimeresurus albolabris albolabris*.

Photo: Corey Blanc

Plates 155 and 156. *Trimeresurus (Ovophis) monticola meridionalis* Mountain Pit Viper งูหางแฮ่มภูเขา

Dead specimen from Chiang Mai. Typical coloration and pattern of northern specimens.

Plate 156. Lighter southern specimen, live.

155

Photo: Jarujin Nabhitabhata

156

Photo: Don Wells

Plate 157. *Trimeresurus popeorum popeorum*. Pope's Pit Viper.
งูเขียวหางไหม้ท้องเขียว. Relatively large member of genus - quite common at higher elevations throughout country.

Plate 158. *Trimeresurus puniceus puniceus*. Flat-nosed Pit Viper.
Brown Flat-nosed Pit Viper. งูปาล์มแดง.
Unusual species, rare in Thailand - few found extreme south.

Photo: Suthigit Patramangorn

Plate 159. *Trimeresurus puniceus wiroti*. Siamese Palm Viper.
Wirot's Pit Viper. งูปาล์ม.
New subspecies - quite rare, more found than *T. puniceus puniceus*.
Range the South - north of range of Flat-nosed Pit Viper.

Photo: Suthigit Patramangorn

Plate 160. *Trimeresurus purpureomaculatus purpureomaculatus.*
 Mangrove Pit Viper. Shore Pit Viper. งูฝังกา. งูเขียวม่วงหางไหม้.
 Heavy-bodied, strikes quickly. Thrives in captivity. Live offspring,
 large - eat newborn mice first meal.

Photo: Corey Blanc

Plates 161 and 162. *Trimeresurus (Tropidolaemus) wagleri.*
 Wagler's Pit Viper. Temple Pit Viper. งูเขียวตุ๊กแก. งูกะปะเสือ.
 Adult color, pattern different from juveniles. Arboreal.
 Diet - lizards, some mice.

Photo: Corey Blanc

Plate 162. *Trimeresurus (Tropidolaemus) wagleri.*
Four-week juvenile - first shed and meal (house gecko) completed. Compare coloration with adult Plate 161.

Photo: Suthigit Patramangorn

Plate 163.

Vipera r. siamensis
Siamese Russell's Viper
งูแมวเซา

Thai subspecies - smaller, less color than Indian. When disturbed coils defensively, loud hiss. Quite hardy - thrives in captivity.

Plate 164.
Vipera r. siamensis and *Vipera r. russelli*
Siamese Russell's Viper and Russell's Viper
งูแมวเซาและงูแมวเซาอินเดีย

Right: *V. r. russelli* Indian form - bright markings.
Left: *V. r. siamensis* Attractive. Both prolific - large litters.

The head scales are smooth and fairly large. The infranasals are separated and only rarely in contact. The supraoculars are large with only 5 to 9 scales between. The temporals are smooth. The subocular is in contact with the third, fourth, and fifth supralabials. Of the 7 to 11 supralabials, the first is not fused with the nasal, the second borders the loreal pit, and the third is the largest. There are two preoculars and either two or three postoculars. The Sumatran Pit Viper has high ventral counts, usually over 180, with males having a count ranging from 179 to 185, and females from 178 to 200. Males have from 71 to 82 subcaudals, females from 61 to 67. The subcaudals are paired, the anal single, and the tail prehensile. This species is often confused with *Trimeresurus hageni*. The reader is referred to page 376 for a discussion concerning the variations.

The Sumatran Pit Viper is found in cultivated areas and forested hills. It has been observed at elevations 1000 m above sea level. Prey includes mice, birds lizards, geckos, and frogs. It accounts for a large number of painful bites within its range, but, fortunately, fatalities are few.

Only a few specimens have been reported in Thailand, those being observed in Pattani Province and those provinces bordering Malaysia. It is reported to be more common in Malaysia and western Indonesia. There are also reports of its occurence on the Philippine island of Palawan.

Trimeresurus (Tropidolaemus) wagleri
Wagler's Pit Viper   Plates 161 and 162
(Temple Pit Viper)   งูเขียวตุ๊กแก
     งูกะปะเสือ

Trimeresurus wagleri has been recently reclassified and assigned to genus *Tropidolaemus*. Unfortunately, the basis for this reclassification has not been clarified and it is not known whether the International Commission on Zoological Nomenclature has accepted this reclassification.

Wagler's Pit Viper has a particularly large head that is quite distinct from the neck. The head is triangular, body moderately heavy, and the tail short. This species is very attractive. The vertebral area is black with a series of relatively narrow yellow cross-bars. A number of white or light colored scales may intrude into the yellow cross-bars. White and pale green scales often intrude onto the black areas. Green, black-edged scales cover the sides. The top of the head is black with a number of irregular green markings. The pupils are vertical and the eyes yellow, as are the labials and chin. The ventral surface is white, but each ventral scale has a black edge. The tail is solid black. Juveniles are marked quite differently. Both the head and body are light green. A red line extends from the rostral through the eyes to the base of the jaw. There is a pattern of regularly spaced red and white spots along the body. The white-tipped tail is light brown. Juveniles acquire an adult pattern at two years of age. Adults attain a length of about one meter without any significant difference in the sizes of mature males and females. Compare plates 161 and 162, pages 401 and 402.

The head is covered with small, overlapping keeled scales. There are three preoculars, the lower one bordering the sensory pit. Three postoculars, one subocular, and two small loreals are present. Supralabials number from eight to ten. The first is not connected to the nasal nor do the remaining ones border the sensory pit or the subocular. There are either 11 or 12 infralabials. The body scales are keeled and arranged in mid-body rows of 21, 23, or 25. The ventrals number 127 to 154, the divided subcaudals range from 45 to 56, and the anal is divided. The tail is prehensile.

This is, by far, the most docile of the genus. Wagler's Pit Vipers are nocturnal and very sluggish during the day becoming active only when searching for mice, birds, and lizards. This is the snake which frequents the Snake Temple in Penang, Malaysia. Although it is true that this snake seems reluctant to bite and rarely does, the bite does cause swelling and pain but it does not seem to be lethal. The habitat is lowland and mangrove forest. It is arboreal and rarely found on the ground. Females give birth to 11 to 18 babies per litter. Newborn average a length of 0.15 meters.

This species generally does not do well in captivity. The major problem is that most do not defacate. This problem may be solved by weekly soakings. Place specimens which are having difficulty into a container of water at 25^0 C and allow them soak for about an hour. Most snakes defecate readily in water. Once this problem has been solved, captive breeding might be achieved. To date, there have not been reports of successful captive breeding.

In Thailand, this snake is relatively rare and only found in the South, as far north as Surat Thani Province. It is common in peninsular Malaysia and also native to Sumatra, the island of Borneo, Celebes, and the Philippines.

Subfamily Viperinae

Members of this family have keeled scales, are heavy-bodied, and bear living young. They lack the loreal pit of Subfamily *Crotalinae* and are generally referred to as "True Vipers".

Genus Vipera

The head is distinct from the neck and is of moderate size. The snout is rather short and rounded with a protruding upper edge. Traditional head scales, especially the parietals and frontal, are often replaced by small, irregular scales. The eye is of moderate size and the pupil vertically elliptical.

Only one member of this genus is found in Thailand, *Vipera russelli siamensis*. This snake does well in captivity, if given correct care. It is a rather large snake that requires a large cage with a top opening. Being terrestrial, it does not need climbing objects. The cage should be provided with a hiding box and a large water dish. A diet of mice, rats, and birds will keep specimens in good health. They breed well in captivity, the young being miniature replicas of the adults. They are large enough to eat pinky mice for their first meal. Always remember that this is a **dangerous** reptile. Although long term captives appear sluggish, they can always strike unexpectedly, swiftly, and accurately. Their venom is quite toxic and potentially lethal. Mortality from their bites is fairly high. In some respects, they are more dangerous than cobras.

Vipera russelli siamensis Plates 163 and 164
Siamese Russell's Viper   งูแมวเซา

This snake reaches a length of 1.5 m, although the average adult length is nearer to one meter. Its head is roughly triangular and very distinct from the neck. The body is slightly stout and the tail short. It is an attractive snake with a light brown body and a series of multicolored blotches covering the entire length. The blotches vary in shape and size and some merge. All, however, are dark brown with a black inner edge and a white outer edge. The same blotches, although smaller, appear on the top of the head. The ventral surface is dirty white and covered with small black spots.

The scales are heavily keeled and arranged in rows of 27 to 33 at mid-body. The ventrals number from 153 to 180, the paired subcaudals range from 41 to 46, and the anal is single. The head is covered with small scales; nostrils are exceptionally large. There are 12 or 13 infralabials, and 10 to 12 supralabials separated from the eye by numerous small scales. The eyes are surrounded by 10 to 15 small scales.

Siamese Russell's Viper inhabits dry, grassy plains and hills, where it likes to lie coiled between rocks, in cavities in the earth, and in clumps of grass. When threatened, it makes a very loud, continuous hissing sound that can be quite alarming. It gives birth to an average of 30 to 45 babies at a time. It is prolific. Over 60 young have been born in some litters. The young, of course, are venomous and equipped to take care of themselves at birth. They are approximately 0.2 m long at birth.

This snake is not evenly distributed within its range. For instance, in Thailand it is only found in the provinces of Nakhon Nayok, Lop Buri, Saraburi, and Chai Nat in the Central Region; in the province of Prachin Buri in the Southeast; and in the province of Nakhon Ratchasima in the Northeast. The Siamese Russell's Viper has also been reported in Amphoe Phap Sakae of Prachuap Khiri Khan Province and Amphoe Lang Suan of Chumphon Province. This area is astride the border of the West Region and the South Region. The Javan subspecies is also only found in specific locations within its range. *Vipera russelli siamensis* is also found in Burma and southern China and probably Laos and Vietnam as well. Other subspecies are found in Pakistan, India, Bangladesh, Sri Lanka, Indonesia, and Taiwan.

Appendixes

Appendix 1

World Herpetological Societies and Organizations

This worldwide list of herpetological societies and organizations was originally compiled by the 1981-82 Regional Herpetological Society Liaison Committee of the Society for the Study of Amphibians and Reptiles (SSAR). I am very grateful to SSAR for permission to include their original and amended compilation in this publication. Their original work has been updated to reflect the names of officials and addresses current as of June 1989. In some cases current addresses are not known, therefore, only the society name has been included. As a comprehensive list, it is always under constant change.

Please send additions or address changes to:
   Society for the Study of Amphibians and Reptiles
   For the attention of:
   Stephen H. Hammack
   Reptile Dept., Dallas Zoo
   621 East Clarendon Drive
   Dallas, TX 75203, U.S.A.

ARGENTINA

Asociacón Herpetológica Argentina
Dr. Jorge D. Williams
Secretary
Museo de Ciencias Naturales
Cassilla 745
1900 La Plata.

Asociacón Latino-Americana Ictiologo y Herpetologo
Dr. Marcos A. Freiberg
Museo Argentina Ciencias Naturales
Avenida Angel Gallardo 470
Buenos Aires.

AUSTRALIA

Australasian Affiliation of Herpetological Societies (AAHS)
Dr. Harry Ehmann
Convenor
School of Biological Sciences
Sydney Technical College
Broadway
N. S. W. 2007.

Australian Herpetological Society
(member society of AAHS)
Dr. Tony Sheargold
President
P.O. Box R79
Royal Exchange
Sydney
N. S. W. 2000.

Australian Herpetologist's League
G.P.O. Box 864
Sydney, N.S.W. 2001.

Australian Society of Herpetologists, Inc.
Dr. Gilliam P. Courtice
Secretary
School of Physiology
University of New South Wales
P.O. Box 1
Kensington
N. S. W. 2003.

North Queensland Herpetological Society
Mr. A. Taplin
25 B Stanton Terrace
North Ward
Townsville
QLD 4810.

South Australian Herpetology Group, Inc.
(member society of AAHS)
c/o South Australian Museum
North Terrace
Adelaide
South Australia 5000.

The Australian Society of Herpetologists
John Wombey
Secretary/Treasurer
CSIRO Division of Wildlife and Rangelands Research
P.O. Box 84
Lyneham ACT 2602.

Victorian Herpetological Society
(member society of AAHS)
c/o Brian F. Barnett
16 Suspension St.
Ardeer, Victoria 3022.

Western Herpetological Group
(member society of AAHS)
Address not known.

BELGIUM

Belgische Bond Voor Aquariumen-en Terrariumkunde
c/o G. Rens
Musstraat 55
3530 Houthalen-Helchteren.

Campaigne Protection Tortues
c/o Mr. J. Bouvry
Rue du Pot d'Etain 13
7500 Tournai.

Centre d'observation Belge des Reptiles et Amphibiens
Mr. A. Goethals
President
Avenue General Medecin Derache, 153
B-1050 Bruxelles.

Terra, Herpetology Society
Redaction and
Foreign Correspondence
F. Vanderstraeten
Wolterslaan 93
B. 9110 Gent/Sint-Amandsberg.

Verenigen Voor Terrariumkunde en Herpetologie
Address not known.

CANADA

Canadian Amphibian And Reptile Conservation Society
c/o Barbara Fromm
101 Roehampton Avenue
Apt. 1706
Toronto
Ontario M4P 2W2.

Canadian Association of Herpetologists
Dr. David M. Green
President
Redpath Museum
McGill University
859 Sherbrooke Street West
Montreal
Quebec H3A 2K6

Canadian Herpetological Society
P.O. Box 130
Station "G"
Toronto
Ontario M4M 3E8

Venomous Animal Society of Canada
Address not known.

CZECHOSLOVAKIA

Czechoslovak Zoological Society
Herpetological Section
Petr Roth
Czechoslovak Academy of Sciences
Institute of Physiology and Genetics
CS - 27721 Libechov

DENMARK

Nordisk Herpetologisk Forening
c/o Ulf Olsen
Ornevej 6
4040 Jyllinge.

FRANCE

Société Batrachologique de France
(Société pour l'étude et la Protection Des Amphibiens)
Dr. J. J. Morère
President
Dr. A. Dubois
General Secretary
Laboratorie des Reptiles et Amphibiens
Museum Nationale d'Histoire Naturelle
25 rue Cuvier
Paris 75005.

Société Herpétologique de France
Dr, Jean Lescure
President
Labratoire des Rept. et Amphib.
Muséum National d'Hist. Naturelle
25 rue Cuvier
75005 Paris

GERMANY (DRG)

Kultursbund der DDR
(Zentraler Fachausschuss Terraristik)
P S F 34
DDR-1040 Berlin.

Zag Schildkröten/ Panzerechsen
c/o H.W. Rudloff
Strasse der Neuerer 201
DDR-4602
Wittenberg-Piesteritz.

GERMANY (FRG)

Deutsche Gesellschaft Für Herpetologie Und Terrarienkunde E.V.
Dr. W. Böhme, President
Zoologisches Forschungsinstitut und Museum A. Koenig
Adenauerallee 150-164
D-5300 Bonn 1.

Herpetofauna
Postfach 1110
Stuttgarter Strasse 35
D - 7056 Weinstadt 1.

**Interessen-Gemeinschaft
Schildkrötenschutz**
Address not known.

**ISIS Gesellschaft Für
Bilogische Aquarien Und
Terrarienkunde**
Joseph Woolmann
Schwanthalerstrasse 123
D 8000 München 2.

**Societas Europaea
Herpetologica**
Dr. Ulrich Joger
General Secretary
Hessisches Landesmuseum
Friedensplatz 1
D-6100 Darmstadt.

**Verband Deutscher Vereine
Für Aquarien Und
Terrarienkunde**
Ewald Somann
Burscheider Weg 11C
D 1000 Berlin 20.

HUNGARY

**Herpetological Congress of
The Socialist States**
Dr. O. Gy. Dely
President
Zoological Department
Hungarian Natural History
Museum
Baross u. 13
H-1088 Budapest.

**Herpetological and
Terraristic Faculty of the
Society of Natural Sciences**
Tit Természettud
Bocskai ut 37
Studio 1113 Budapest.

INDIA

Indian Herpetological Society
Neelimkumar Khaire
Secretary
Poona Snake Park (Katraj)
Pune 411046.

Madras Snake Park Trust
Guindy Deer Park
Madras 600 022.

ISRAEL

**Israel Herpetological
Information Center**
Amos Bouskila
Director
Hazeva Field Study Center
86815 Mobile Post Arava

ITALY

**Associazione Piemontese
Erpetologiae Acquariologia**
c/o Luciano Mariotto
Secretray A.P.E.A.
Via Leoncavallo 57/C
10154 Torino.

Unione Erpetologica Italiana
Dr. S. Bruno
Centro del Parco Nazionale
d'Abruzzo
"L'Aquila", 67032 Pescasseroli.

JAPAN

Herpetological Society of Japan
Dr. Richard C. Goris
Secretary
Sugao 9480 - 7

The Japan Snake Institute
Dr. Y. Sawai, M. D.
Director
Yabuzuka-honmachi, Nittagun
Gumma Prefecture 379-23.

MEXICO

Comité Herpetológico Nacional
MS Zeferino Uribe-Peña
President
Departamento de Zoologia
Instituto de Biologia, UNAM
Apartado Postal 70 - 153
Mexico, D. F.

NETHERLANDS

Chelonian Documentation Center
PO Box 125
Workumertrekweg 17
8700 AC Bolsward.

Nederlandese Bond "Aqua Terra"
c/o H. M. A. van Lier
V. Riebeecklaan 23
2024 AE Haarlem.

Nederlandse Doelgroep Slangen
(affiliated with NVHT)
Ruud Verbeek
Secretary
Alholm 76
2133 DD Hoofddorp.

Nederlandse Schildpadden Vereniging
Headquarters
Papelaan 18
2522 EJ Voorschoten.

Nederlandse Studiegroep Anolissen
c/o Frats van Leeuwen
Secretary
2e Boerhavestraat 5hs
1091 AK Amsterdam

Nederlandse Vereniging Voor Herpetologie en Terrariumkunde (NVHT)
c/o E. F. Elzenga
Secretary
Burg. H. van Konijnenburglaan 46
3925 XB Scherpenzeel.

The Dutch Turtle and Tortoise Foundation
P.O. Box 125
8700 AC Bolsward

NEW ZEALAND

New Zealand Herpetological Society Inc.
Robert Porter
Secretary
28 Spinella Drive
Glenfield, Auckland.

PEOPLE'S REPUBLIC OF CHINA

Board of Acta Herpetologica Sinica
Zhao Ermi, Chief Editor
Chengdu Institute of Biology
Academia Sinica
P.O. Box 416
Chengdu, Sichuan.

Chinese Society For The Study of Amphibians And Reptiles
(Chinese SSAR, or SSAR-China)
Zhao Ermi
President
Chengdu Institute of Biology
Academia Sinica
P.O. Box 416
Chengdu, Sichuan.

POLAND

Polish Zoological Society Herpetological Section
Prof. Dr. Leszek Berger
Secretary
Plac Wielkopolski 2/55
61 - 746 Poznan

PUERTO RICO

Chelonia-Sociedad Para el Estudio de los Quelonios
c/o Jorge Luis Piñero
President
P.O. Box 22061
UPR Station
San Juan 00931.

SOUTH AFRICA

Herpetological Association of Africa
Dr. J. H. vahn Wyk
Chairman
Nasionale Museum
Postbus 266
Bloemfontein 9300.

SPAIN

Associacion Herpetologica Espanola
Mrs. M. V. Vives-Balmana
Vice-President
Museo Naacional de Ciencias Nat.
Jose Gutierrez Abascal 2
28006 Madrid.

Associacion Iberica de Herpetologica
Carlos Perez-Santos
Avenida de Logrono 23
Madrid 22.

Grupo de Estudio de los Anfibios y Reptiles Ibericos
c/o Juan Pablo Marinez, or
Dr. E. Balcells
R. Centro Pirenalco de Biologia Experimental
Apartado 64
Jaca (Huesca).

Societat Catalana d'Ictiologia i Herpetologia (S.C.I.H.)
S. Gimenez i Lopez,
Vice-Secretary General
Museu de Zoologia de Barcelona
Apartat de Correus, 593
08080 Barcelona.

SWEDEN

Eskilstuna Terrarieförening
c/o Mats Olson
Rosstorpsvägen 34
633 53 Eskilstuna, 016-13 14 23.

Göteborgs Herpetologiska Förening
(member society of SHR)
Address not known.

Lunds Terrarieförening
(member society of SHR)
c/o Leif Aman
Dr. Strömas väg 3R
241 00 Eslöv, 0413-163 95.

Örebro Terrariekluubb
(member society of SHR)
c/o Häkan Molin
Drahenbergsgaten 61 702 19
Örebro.

Stockholms Herpetologiska Förening
(member society of SHR)
c/o Mikael Norström
Mickelbergsvägen 78
126 63 Hägersten, 08-97 41 72.

Swedish Herpetological Society
Dr. Goran Nilson
Department of Zoology
University of Gotenborg
Box 25059
S - 40031 Gotenborg.

SWITZERLAND

IUCN/SSC Amphibia/Reptilia Group
Mr. R. E. Honegger
Vice-Chairman
Zoo Zurich
Zurichbergstrasse 221
CH-8044
Zurich.

Korrdinationsstelle für Amphibien Undreptilienschutz in Der Schweiz
Dr. Kurt Grossenbacher
Naturhistorisches Museum
Bernastrasse 15
CH-3005 Bern.

Schildkroten-Informationsdienst
c/o H. H. D. Falk
Bachserstrasse 10
CH - 8174 Stadel b.
Niederglatt / ZH

Société Erpétologique De
Genéve
Case postal 20
CH-1211 Geneva 2 Gare.

UNION OF SOVIET
SOCALIST REPUBLICS

**All-Union Herpetological
Committee**
Dr. I. S. Darevsky
Chairman
Zoological Institute
Academy of Sciences
Leningrad V-164.

UNITED KINGDOM

**Association for the Study of
Reptilia and Amphibia**
c/o Cotswold Wildlife Park
Burford
Oxon
OX8 4JW.

British Chelonia Group
Miss Fiona McGrattan
Corresponding Secretary
10 Clyde Park, Redland
Bristol
BS6 6RR.

British Herpetological Society
Headquarters
c/o Zoological Society of
London
Regent's Park
London NW1 4RY.

**Conservation Committee of
Societas Europaea
Herpetologica**
Mr. K. F. Corbett
Chairman
136 Estcourt Road
Woodside
London SE25 4SA.

**International Herpetological
Society**
A. J. Mobbs
Secretary/Treasurer
27 St. Thomas Close
Dartmouth Avenue
Walsall
West Midlands
WS3 1SZ

**IUCN/SCC
Tortoise Group**
Dr. Ian R. Swingland
Chairman
School of Continuing Education
and Biological Laboratory
Rutherford College
University of Kent
Canterbury
CT2 7NX.

**South Western Herpetological
Society**
c/o Frank B. Gibbons
Secretary
Acanthus
59 St. Marychurch Rd.
Torquay
Devon
TQ1 3HG.

Thames And Chiltern Herpetological Group
Headquarters
The Youth Club
Narcot Lane
Chalfont St. Giles
Buckinghamshire.

UNITED STATES OF AMERICA

All Florida Herpetological Conference
c/o Walter Auffenberg
Florida State Museum
Gainesville
FL 32611.

American Federation of Herpetocultursalists
P.O. Box 1131
Lakeside
CA 92040.

American Society of Ichthyologists and Herpetologists
c/o National Marine Fisheries Service Systematics Laboratory
National Museum of Natural History
Washington
D.C. 20560.

Arizona Herpetological Association
1433 Huntington Drive
Tempe
AZ 85282.

Arkansas Herpetological Society
Perk Floyd, Secretary
16 Lakeside Drive
Hensley, AR 72065.

Association for the Conservation of Turtles and Tortoises
c/o Sandra Jordon
Secretary
RFD #4, Box 291
Sussex
NJ 07461.

Bay Area Amphibian and Reptile Society
Palo Alto Junior Museum
1451 Middlefield Road
Palo Alto, CA 94301.

Bay Area Turtle and Tortoise Society
Address not known.

California Turtle and Tortoise Club
P.O. Box 8952
Fountain Valley
CA 92728.

Central Florida Herpetological Society
P.O. Box 3277
Winter Haven, FL 33881.

Central Illinois Herpetological Society
1125 West Lake
Peoria, IL 61614

Central Ohio Herpetological Society
c/o Mary E. Garrett
436 Canyon Drive North
Columbus, OH 43214.

Chicago Herpetological Society
c/o Chicago Academy of Sciences
2001 North Clark Street
Chicago, IL 60614.

Colorado Herpetological Society
P.O. Box 15381
Lakewood, CO 80215.

Connecticut Herpetological Society
George Whitney, DVM
Whitney Clinic
860 Oakwood Road
Orange, CT 06477.

Dallas Herpetological Society
P.O. Box 153672
Irving, TX 75015.

Delaware Herpetological Society
c/o Ashland Nature Center
Brackenville and Barley Mill Rd
Hockessin, DE 19707.

Desert Tortoise Preserve Committee Inc
P.O. Box 453
Ridgecrest, CA 93555.

East Texas Herpetological Society
P.O. Box 1561
Trinity, TX 75862.

Eastern Seaboard Herpetological League
c/o Ray Logue
3456 Aldine Street
Philadelphia, PA 19136.

El Paso Herpetological Society
7505 Dempsey
El Paso, TX 79925.

Florida West Coast Herpetological and Conservation Society
c/o John Lewis
1312 South Evergreen Avenue
Clearwater, FL 33516.

Gainesville Herpetological Society
P.O. Box 7104
Gainesville
FL 32605-7104.

Georgia Herpetological Society
c/o Reptile House
Atlanta Zoological Park
800 Cherokee Avenue, S.E.
Atlanta, GA 30315.

Gopher Tortoise Council
c/o Patricia Ashton
611 NW 79th Drive
Gainesville, FL 32607

Great Lakes Herpetological Society
c/o Jeff Gee
Treasurer
4308 N. Woodward
Royal Oak, MI 48072.

Greater Cincinnati Herpetological Society
Cincinnati Museum of Natural History
1720 Gilbert Avenue
Cincinnati, OH 45202.

Greater Dayton Herpetological Society
c/o Dayton Museum of Natural History
2629 Ridge Avenue
Dayton, OH 45414.

Greater San Antonio Herpetological Society
c/o W. Rowe Elliot, III
134 Aldrich, San Antonio
TX 78227.

Herpetologists' League
Department of Biology
University of Miami
Coral Gables, FL 33124

Herpetological Society of Southwest Michigan
c/o Scott Averill
Secretary
2557 Bristol, N.W.
Grand Rapids
MI 49504.

Hoosier Herpetological Society
c/o Dennis Brown
2906 South Taft
Indianapolis, IN 46241.

Idaho Herpetological Society
P.O. Box 6329
Boise, ID 83707.

Inland Empire Herpetological Society
c/o San Bernadino County Museum
2024 OrangeTtree Lane
Redlands
CA 92373.

International Society for the Study of Dendrobatid Frogs
c/o Dale Bertram, M. D.
One Virginia Terrace
Madison
WI 53705.

Iowa Herpetological Society
P.O. Box 23035
Des Moines, IA 50322.

Kansas Herpetological Society
Museum of Natural History
The University of Kansas
Lawrence
KS 66045.

Kaw Valley Herpetological Society
Rt. 1, Box 298
Eudora
KS 66025.

Kentucky Herpetological Society
John MacGregor
102 Fourth Street
Nicholasville, KY 40358.

Lehigh Valley Herpetological Society
c/o Leonard Knapp
215 Lawn Avenue
Sellersville, PA 18960.

Long Island Herpetological Society
117 East Santa Barbara Road
Lindenhurst, NY 11757.

Lubbock Turtle and Tortoise Society
c/o Joe Cain
5708 - 64th Street
Lubbock, TX 79424.

Lynchburg Herpetological Society
(local chapter of Virginia H. S.)
c/o S. Whitt
Lynchburg College
Lynchburg, VA 24501.

Maryland Herpetological Society
Natural History Society of Maryland, Inc.
Editor
Herbert S. Harris, Jr.
2643 North Charles Street
Baltimore
MD 21218.

Massachusetts Herpetological Society
P.O. Box 1082
Boston, MA 02103.

Michigan Society of Herpetologists
321 West Oakland
Lansing, MI 48906.

Mid-Mississippi Valley Herpetological Society
c/o Mike Ladato
925 Park Place Drive
Evansville
IN 47715.

Minnesota Herpetological Society
J. F. Bell Museum of Natural History
10 Church Street
Minneapolis
MN 55455-0104.

Mississippi Herpetological Society
Address not known.

National Turtle and Tortoise Society
P.O. Box 9806
Phoenix, AZ 85068 - 9806.

Nebraska Herpetological Society
c/o James D. Fawcett
Department of Biology
Univ. of Nebraska at Omaha
Omaha, NE 68182.

New Mexico Herpetological Society
Department of Biology
University of New Mexico
Albuquerque, NM 87131.

New York Herpetological Society
P.O. Box 1245
Grand Central Station
New York, NY 10017.

New York Turtle and Tortoise Society
c/o Suzanna Dohm
365 Pacific Street
Brooklyn
NY 11217.

North Carolina Herpetological Society
North Carolina State Museum of Natural History
P.O. Box 27647
Raleigh
NC 27611.

North Texas Herpetological Society
P.O. Box 470771
Fort Worth
TX 76147.

Northeast Colorado Herpetological Society
c/o Roger Klingenberg
6247 West 10th
Greeley
CO 80631.

Northern California Herpetological Society
P.O. Box 1363
Davis, CA 95617 - 1363.

Northern Nevada Herpetological Society
c/o Bill Gill
348 $^1/_2$ Wheeler Avenue
Reno, NV 89502

Northern Ohio Association of Herpetologists
Department of Biology
Case Western Reserve University
Cleveland, OH 44106.

Oklahoma Herpetological Society
c/o Patrick Mulvany
7315 East 81st Place
Tulsa, OK 74133.

Oregon Herpetological Society
c/o Steven Aveldson
8435 Derbyshire Lane
Eugene, OR 97405.

Pacific Northwest Herpetological Society
1308 North 8th Street
Tacoma, WA 98403.

Palm Beach County Herpetological Society
c/o Greg Longhurst
P.O. Box 125
Loxahatchee, FL 33470.

Philadelphia Herpetological Society
1548 Pratt Street
Philadelphia, PA 19124.

Rocky Mountain Herpetological Society
c/o Dave Baker
605 W. Colorado Avenue
Colorado Springs
CO 80905.

San Diego Herpetological Society
P.O. Box 4439
San Diego
CA 92104-0439.

San Diego Turtle and Tortoise Society
c/o 6957 Tanglewood Road
San Diego
CA 92111.

Society for the Study of Amphibians and Reptiles
c/o Douglas H. Taylor, Treasurer
Department of Zoology
Miami University
Oxford
OH 45056.

South Texas Amphibian and Reptile Society
c/o Greg Luther
P.O. Box 233
Angleton
TX 77515.

Southern Arizona Herpetological Society
c/o Tom Boyden
4521 West Mars Street
Tucson
AZ 85704.

South Mississippi Herpetological Society
P.O. Box 10047
Gulfport
MS 39505.

South Texas Herpetological Society
c/o James Maples, Jr.
927 Wilson
Alice
TX 78332.

Southern Mississippi Herpetological Society
c/o Ted Crawford
404 Ridge Drive
Biloxi, MS 39532.

Southwestern Herpetological Society
San Fernando Valley Chapter
P.O. Box 7469
Van Nuys, CA 91409.

Southwestern Herpetologists Society
c/o San Bernadino County Museum
2024 Orange Tree Lane
Redlands
CA 92373.

St. Louis Herpetological Society
P.O. Box 9216
St. Louis
MO 63117.

Susquehanna Herpetological Society
c/o Sam Burleigh
211 South Market Street
Muncy
PA 17756.

Tampa Bay Herpetological Society
3310 - A Carlton Arms Drive
Tampa, FL 33614.

Texas Herpetological Society
HC 53, Box 3225
Bulverde
TX 78163.

Toledo Herpetological Society
c/o Toledo Zoogical Society
2700 Broadway
Toledo
OH 43609.

Troup County Association of Herpetologists
c/o C. W. Dodgen
801 Grant Street
La Grange, GA 30240.

Tuscon Herpetological Society
P.O. Box 31531
Tuscon
AZ 85751 - 1531.

Turtle and Tortoise Education Adoption Media
3245 Military Avenue
Los Angeles, CA 90034.

Utah Herpetologists Society
P.O. Box 9361
Salt Lake City, UT 84109.

Virginia Herpetological Society
Route 2, Box 78
Brookneal, VA 24528.

Washington Herpetologists Society
c/o Frank Watrous III
12420 Rock Ridge Road
Herndon, VA 22070.

Wisconsin Herpetological Society
9137 West Mill Road
Milwaukee, WI 53225 - 1701.

World Congress of Herpetology
Kraig Adler, Secretary-General
Cornell University
Section of Neurobiology
and Behavior
Seeley G. Mudd Hall
Ithaca, NY 14853-0240.

Zoo Fauna Association
c/o Bern Tryon
Houston Zoological Park
Box 1562
Houston, TX 77001.

Appendix 2

Climatic Data - Thailand

The North (1951 - 1980)

2.1. Chiang Mai
2.2. Chiang Rai
2.3. Mae Hong Son
2.4. Mae Sariang
  Mae Hong Son Province
2.5. Uttaradit

Prime source:
Ministry of Communications
The Meteorology Department
Meteorological Statistics Section

Note
To change Celsius to Farenheit, use the following formula:
$$F = C \times 9/5 + 32$$

Appendix 2

THE NORTH (1951 - 1980)

2.1. Chiang Mai. 18° 47' N, 98° 59' E. Elevation 312 meters.

Temp. °C	Jan	Feb	Mar	Apr	May	Jun	Jul	Aug	Sep	Oct	Nov	Dec	Year
Mean	20.1	22.2	25.7	28.4	28.0	27.2	26.8	26.3	26.2	25.5	23.5	20.8	25.1
Mean max.	28.9	31.9	34.8	36.1	34.1	32.3	31.6	30.7	31.0	30.8	29.8	28.5	31.7
Mean min.	13.2	14.1	17.5	21.4	23.3	23.6	23.4	23.2	22.9	21.6	18.7	14.9	19.8
Ext. max.	34.7	37.3	39.6	41.5	41.4	37.9	37.5	35.4	36.1	35.3	34.5	33.5	41.5
Ext. min.	03.7	07.3	10.0	13.2	19.6	19.1	20.5	20.0	16.8	13.3	06.0	05.0	03.7
Mean daily range	15.7	17.8	17.3	14.7	10.8	08.7	08.2	07.5	08.1	09.2	11.1	13.6	11.9
Relative Humidity (%)													
Mean	74.0	65.0	59.0	60.0	73.0	79.0	81.0	83.0	83.0	81.0	79.0	77.0	74.0
Mean max.	94.2	90.5	84.3	83.5	90.0	93.1	93.4	94.3	94.5	94.5	94.2	94.4	91.7
Mean min.	43.1	33.9	31.5	37.6	51.8	60.7	62.7	66.6	64.9	60.1	53.9	48.5	51.3

Appendix 2

THE NORTH (1951 - 1980)

2.2. Chiang Rai. 19° 53' N, 99° 50' E. Elevation 394 meters.

Temp. °C	Jan	Feb	Mar	Apr	May	Jun	Jul	Aug	Sep	Oct	Nov	Dec	Year
Mean	19.5	21.7	24.6	27.5	27.5	27.1	26.7	26.2	26.1	25.0	22.5	19.6	24.5
Mean max.	27.5	30.7	33.3	34.9	33.2	31.7	30.9	30.4	30.6	30.0	28.5	26.6	30.7
Mean min.	11.8	12.7	15.7	19.6	22.0	22.8	22.8	22.6	22.1	20.3	17.0	13.0	18.5
Ext. max.	32.2	34.8	39.2	41.3	41.2	38.0	38.6	35.2	37.0	35.0	33.6	31.7	41.3
Ext. min.	01.5	06.5	09.6	11.4	17.6	18.5	19.0	18.5	18.3	11.0	05.0	02.8	01.5
Mean daily range	15.7	18.0	17.6	15.3	11.2	08.9	08.1	07.8	08.5	09.7	11.5	13.6	12.2
Relative Humidity (%)													
Mean	78.0	71.0	66.0	66.0	75.0	81.0	83.0	85.0	84.0	83.0	81.0	81.0	78.0
Mean max.	95.7	93.7	90.4	89.4	92.6	94.1	94.6	95.4	95.6	95.7	96.0	96.1	94.1
Mean min.	48.8	38.4	35.3	39.4	54.1	63.4	66.0	68.9	66.4	62.3	57.4	54.1	54.6

Appendix 2

THE NORTH (1951 - 1980)

2.3. Mae Hong Son. 19° 18' N, 97° 50' E. Elevation 267 Meters.

Temp. ° C	Jan	Feb	Mar	Apr	May	Jun	Jul	Aug	Sep	Oct	Nov	Dec	Year
Mean	20.5	22.1	26.1	29.7	28.6	27.1	26.7	26.3	26.5	26.1	24.2	21.3	25.4
Mean max.	29.7	32.7	36.1	37.7	34.9	31.9	31.0	30.8	31.6	32.0	31.1	29.4	32.4
Mean min.	14.0	14.0	17.2	22.3	24.0	23.8	23.6	23.4	23.1	22.0	19.3	15.8	20.2
Ext. max.	34.5	37.0	39.5	42.4	41.4	39.4	36.2	35.0	35.5	35.4	35.2	33.9	42.4
Ext. min.	06.0	08.2	11.0	15.6	20.4	20.5	21.2	20.5	19.7	15.0	09.8	07.2	06.0
Mean daily range	15.7	18.7	18.9	15.4	10.9	08.1	07.4	07.4	08.5	10.8	11.8	13.6	12.2
Relative Humidity (%)													
Mean	74.0	66.0	55.0	53.0	70.0	80.0	83.0	85.0	84.0	82.0	79.0	77.0	74.0
Mean max.	94.8	92.6	85.3	79.6	89.1	93.1	94.1	95.3	95.3	95.4	95.0	95.0	92.1
Mean min.	42.1	31.9	26.0	30.1	48.6	64.1	68.1	70.3	67.3	62.9	54.9	48.6	51.3

Appendix 2

THE NORTH (1951 - 1980)

2.4. Mae Sariang, Mae Hong Son Province. 18° 10' N, 97° 56' E. Elevation 212 meters.

Temp. °C	Jan	Feb	Mar	Apr	May	Jun	Jul	Aug	Sep	Oct	Nov	Dec	Year
Mean	21.7	23.6	27.6	30.7	29.3	27.2	26.6	26.3	26.9	26.8	25.2	22.4	26.2
Mean max.	30.8	33.6	36.6	37.9	34.8	31.4	30.4	30.2	31.5	32.5	31.9	30.6	32.7
Mean min.	13.1	12.9	16.9	22.1	24.0	23.5	23.1	23.0	23.1	22.1	19.2	15.3	19.9
Ext. max.	36.4	38.8	41.5	44.1	42.7	38.6	38.7	36.4	37.4	36.3	36.4	36.0	44.1
Ext. min.	03.3	06.2	08.7	13.8	19.2	20.5	20.8	20.6	19.7	13.4	06.5	05.0	03.3
Mean daily range	17.7	20.7	19.7	15.8	10.8	07.9	07.3	07.2	08.4	10.4	12.7	15.3	12.8
Relative Humidity (%)													
Mean	73.0	65.0	55.0	55.0	71.0	81.0	83.0	85.0	83.0	80.0	77.0	76.0	74.0
Mean max.	95.2	94.0	89.0	84.4	90.0	94.0	94.5	94.9	95.0	95.0	94.9	95.5	93.0
Mean min.	43.7	34.2	29.3	33.7	53.2	68.4	71.5	73.3	68.7	62.6	55.5	49.8	53.7

Appendix 2

THE NORTH (1951 - 1980)

2.5. Uttaradit. 17°37' N, 100°06' E. Elevation 63 meters.

Temp. °C	Jan	Feb	Mar	Apr	May	Jun	Jul	Aug	Sep	Oct	Nov	Dec	Year
Mean	24.0	26.2	29.2	31.3	30.2	28.8	28.3	27.8	27.9	27.7	26.0	24.1	27.6
Mean max.	32.1	34.6	37.1	38.4	35.9	33.7	33.0	32.3	32.6	33.0	32.5	31.6	33.9
Mean min.	15.7	17.5	20.6	23.6	24.6	24.3	24.0	23.9	23.7	22.6	19.9	16.8	21.4
Ext. max.	37.7	39.3	42.7	44.5	43.3	40.2	40.2	37.6	37.2	36.6	36.6	36.6	44.5
Ext. min.	04.5	10.0	13.0	16.3	20.8	20.1	20.3	20.6	16.1	15.1	10.2	07.5	04.5
Mean daily range	16.4	17.1	16.5	14.8	11.3	09.4	09.0	08.4	08.9	10.4	12.6	14.8	12.5
Relative Humidity (%)													
Mean	70.0	66.0	63.0	63.0	73.0	80.0	82.0	84.0	84.0	81.0	76.0	72.0	75.0
Mean max.	90.9	88.5	85.3	84.6	90.3	93.4	93.9	95.0	95.5	94.3	92.9	91.2	91.3
Mean min.	41.5	37.6	35.8	38.2	52.4	63.3	65.4	68.4	67.1	60.7	52.6	45.7	52.4

Appendix 3

Climatic Data - Thailand

The Central Region (1951 - 1980)

    3.1. Bangkok
    3.2. Buhmipol Dam
        Tak Province
    3.3. Nakhon Sawan
    3.4. Phitsanulok
    3.5. Suphan Buri

Prime source:
Ministry of Communications
The Meteorology Department
Meteorological Statistics Section

<u>Note</u>
To change Celsius to Farenheit, use the following formula:
$$F = C \times 9/5 + 32$$

Appendix 3

THE CENTRAL REGION (1951 - 1980)

3.1. Bangkok. 13°44' N, 100°34' E. Elevation 2 meters.

Temp. °C	Jan	Feb	Mar	Apr	May	Jun	Jul	Aug	Sep	Oct	Nov	Dec	Year
Mean	25.6	27.2	28.6	29.6	29.1	28.6	28.1	27.8	27.6	27.5	26.6	25.5	27.7
Mean max.	31.9	32.7	33.8	34.9	34.1	33.0	32.5	32.2	31.9	31.7	31.3	31.3	32.6
Mean min.	20.6	22.8	24.6	25.7	25.4	25.1	24.8	24.7	24.4	24.3	22.8	20.7	23.8
Ext. max.	36.0	36.6	39.8	40.0	39.4	37.7	37.8	36.3	36.0	35.3	35.1	35.2	40.0
Ext. min.	09.9	14.9	16.5	19.9	21.1	21.7	21.9	21.2	21.3	18.3	14.2	10.5	09.9
Mean daily range	11.3	09.9	09.2	09.2	08.7	07.9	07.7	07.5	07.5	07.4	08.5	10.6	08.8
Relative Humidity (%)													
Mean	73.0	76.0	77.0	77.0	79.0	79.0	80.0	81.0	84.0	83.0	79.0	74.0	78.0
Mean max.	91.6	92.9	92.5	91.4	93.2	92.5	92.5	93.7	95.3	95.2	93.4	91.4	93.0
Mean min. *	Not available												

Appendix 3.

THE CENTRAL REGION (1951 - 1980)

3.2. Buhmipol Dam, Tak Province. 17° 15' N, 99° 01' E. Elevation 142 meters.

Temp. °C	Jan	Feb	Mar	Apr	May	Jun	Jul	Aug	Sep	Oct	Nov	Dec	Year
Mean	24.2	27.4	30.6	32.2	30.0	28.8	28.3	28.1	27.5	27.0	25.6	23.8	27.8
Mean max.	30.5	33.8	36.6	37.8	34.6	32.6	32.1	32.0	31.4	31.1	30.3	29.6	32.7
Mean min.	16.8	19.3	22.2	24.8	25.0	24.6	24.4	24.1	23.5	22.5	20.3	17.6	22.1
Ext. max.	36.4	39.2	41.4	42.0	42.2	38.7	37.4	36.4	36.9	35.5	34.4	35.6	42.2
Ext. min.	07.0	11.0	14.4	18.4	20.4	21.8	21.3	21.8	20.3	14.4	11.0	06.4	06.4
Mean daily range	13.7	14.5	14.4	13.0	09.6	08.0	07.7	07.9	07.9	08.6	10.0	12.0	10.6
Relative Humidity (%)													
Mean	62.0	51.0	46.0	50.0	66.0	71.0	71.0	72.0	78.0	78.0	74.0	70.0	66.0
Mean max.	87.5	77.2	71.2	72.2	84.2	86.7	86.7	88.3	93.5	95.0	94.1	92.6	85.8
Mean min.	40.8	32.0	30.0	33.8	51.5	57.5	58.3	59.1	63.4	62.2	55.5	48.7	49.4

Appendix 3

THE CENTRAL REGION (1951 - 1980)

3.3. Nakhon Sawan. 15° 48' N, 100° 10 'E. Elevation 34 meters.

Temp. °C	Jan	Feb	Mar	Apr	May	Jun	Jul	Aug	Sep	Oct	Nov	Dec	Year
Mean	25.6	28.3	30.7	31.9	30.6	29.6	29.0	28.5	28.0	27.9	26.7	25.2	28.5
Mean max.	32.2	34.5	36.7	37.9	36.1	34.5	33.8	33.1	32.2	32.0	31.5	31.1	33.8
Mean min.	17.7	21.0	23.7	25.3	25.1	24.7	24.3	24.1	23.9	23.5	21.0	18.2	22.7
Ext. max.	37.0	39.8	41.2	42.5	42.7	41.0	38.9	37.8	36.3	35.9	35.7	35.8	42.7
Ext. min.	06.1	12.0	14.2	17.0	20.3	21.4	20.9	20.9	20.4	18.4	11.9	08.2	06.1
Mean daily range	14.5	13.5	13.0	12.6	11.0	09.8	09.5	09.0	08.3	08.5	10.5	12.9	11.1
Relative Humidity (%)													
Mean	63.0	62.0	61.0	61.0	70.0	74.0	75.0	78.0	82.0	80.0	73.0	67.0	70.0
Mean max.	87.3	86.9	87.3	86.5	89.1	90.5	91.5	92.9	95.5	94.7	92.4	89.9	90.4
Mean min.	41.3	40.3	39.1	40.8	51.2	56.6	58.4	62.0	66.4	63.3	53.9	45.9	51.6

Appendix 3

THE CENTRAL REGION (1951 - 1980)

3.4. Phitsanulok. 16°49' N, 100°16' E. Elevation 44 meters.

Temp. °C	Jan	Feb	Mar	Apr	May	Jun	Jul	Aug	Sep	Oct	Nov	Dec	Year
Mean	23.9	26.2	28.7	30.4	29.3	28.4	27.9	27.6	27.6	27.5	26.0	23.9	27.3
Mean max.	31.6	33.7	35.9	37.4	35.6	33.7	32.8	32.3	32.2	32.4	31.8	31.0	33.4
Mean min.	17.7	20.2	22.9	24.9	25.0	24.6	24.4	24.3	24.4	23.9	21.4	18.4	22.7
Ext. max.	36.7	38.0	40.5	42.8	42.0	38.7	38.4	36.3	36.6	35.3	36.0	35.6	42.8
Ext. min.	07.5	13.1	13.5	17.0	21.6	22.0	21.6	22.0	21.5	17.6	12.5	09.4	07.5
Mean daily range	13.9	13.5	13.0	12.5	10.6	09.1	08.4	08.0	07.8	08.5	10.4	12.6	10.7
Relative Humidity (%)													
Mean	68.0	66.0	65.0	64.0	73.0	79.0	80.4	82.0	83.0	80.0	75.0	70.0	74.0
Mean max.	87.2	84.7	82.5	81.8	88.1	91.5	92.5	93.3	93.4	92.3	90.2	88.6	88.9
Mean min.	42.7	42.3	41.8	42.3	53.2	61.0	63.7	65.7	66.1	60.9	52.9	46.0	53.2

Appendix 3

THE CENTRAL REGION (1951 - 1980)

3.5. Suphan Buri. 14°29' N, 100°08' E. Elevation 7 meters.

Temp. °C	Jan	Feb	Mar	Apr	May	Jun	Jul	Aug	Sep	Oct	Nov	Dec	Year
Mean	26.0	28.2	30.2	31.6	30.6	29.8	29.1	28.8	28.4	28.0	26.8	25.5	28.6
Mean max.	31.8	34.1	36.2	37.2	35.5	34.3	33.4	33.0	32.0	31.2	30.4	30.4	33.3
Mean min.	18.9	21.1	23.3	25.0	25.1	24.8	24.5	24.5	24.5	24.3	22.0	19.4	23.1
Ext. max.	36.7	39.8	41.0	42.2	42.6	39.8	40.0	37.1	36.0	34.5	34.9	35.0	42.6
Ext. min.	09.2	12.0	14.8	19.4	20.9	20.2	21.1	20.8	20.8	19.0	14.5	10.0	09.2
Mean daily range	12.9	13.0	12.9	12.2	10.4	09.5	08.9	08.5	07.5	06.9	08.4	11.0	10.2
Relative Humidity (%)													
Mean	66.0	66.0	65.0	64.0	70.0	71.0	73.0	75.0	79.9	79.7	74.8	68.7	71.0
Mean max.	88.4	91.6	91.4	88.7	88.9	87.9	89.1	89.7	92.6	92.8	91.2	88.3	90.0
Mean min.	44.2	43.1	41.7	42.7	52.1	55.4	58.0	60.4	66.5	66.6	59.4	50.1	53.3

Appendix 4

Climatic Data - Thailand

The Northeast (1951 - 1980)

    4.1. Mukdahan
        Nakhon Phanom Province
    4.2. Nong Khai
    4.3. Sakon Nakhon
    4.4. Surin
    4.5. Udon Thani

<u>Prime source:</u>
Ministry of Communications
The Meteorology Department
Meteorological Statistics Section

<u>Note</u>
To change Celsius to Farenheit, use the following formula:
$$F = C \times 9/5 + 32$$

Appendix 4

THE NORTHEAST (1951 - 1980)

4.1. Mukdahan, Nakhon Phanom Province. 16°32' N, 104°43' E. Elevation 138 meters.

Temp. °C	Jan	Feb	Mar	Apr	May	Jun	Jul	Aug	Sep	Oct	Nov	Dec	Year
Mean	22.7	24.9	28.1	29.7	29.1	28.5	28.1	27.4	27.3	26.6	24.6	22.6	26.6
Mean max.	29.5	31.4	34.3	35.3	33.9	32.4	31.8	31.1	31.0	30.9	29.7	28.5	31.7
Mean min.	14.6	17.4	21.1	23.6	24.4	24.4	24.1	24.0	23.5	21.6	18.4	15.5	21.1
Ext. max.	36.3	38.7	40.7	41.8	40.5	40.0	36.2	36.5	35.5	35.0	36.4	35.0	41.8
Ext. min.	03.2	09.2	10.1	15.3	19.3	18.8	20.4	20.9	17.7	12.9	09.4	05.3	03.2
Mean daily range	14.9	14.0	13.2	11.7	09.5	08.0	07.7	07.1	07.5	09.3	11.3	13.0	10.6
Relative Humidity (%)													
Mean	64.0	62.0	60.0	63.0	73.0	79.0	79.0	82.0	82.0	74.0	69.0	67.0	71.0
Mean max.	90.8	87.6	83.0	82.8	89.0	91.1	92.1	93.9	94.1	89.9	88.4	90.1	89.4
Mean min.	45.0	43.1	42.0	45.6	57.2	65.1	66.7	69.9	67.7	61.0	54.5	50.3	55.7

Appendix 4

THE NORTHEAST (1951 - 1980)

4.2. Nong Khai. 17°58' N, 102°43' E. Elevation 174 meters.

Temp. °C	Jan	Feb	Mar	Apr	May	Jun	Jul	Aug	Sep	Oct	Nov	Dec	Year
Mean	22.4	24.3	27.5	29.1	28.3	28.1	27.7	27.2	27.3	26.8	24.3	22.4	26.3
Mean max.	29.2	31.2	34.1	35.0	33.5	32.3	31.9	31.1	31.3	31.4	30.1	29.1	31.7
Mean min.	15.9	17.9	21.2	23.5	24.2	24.5	24.4	24.1	23.9	22.6	19.1	16.4	21.5
Ext. max.	35.1	37.9	40.5	41.8	40.5	37.6	37.3	36.2	35.4	35.4	35.1	34.4	41.8
Ext. min.	05.2	09.6	12.3	15.8	18.0	21.2	21.1	21.5	19.9	15.4	09.7	05.5	05.2
Mean daily range	13.3	13.3	12.9	11.5	09.3	07.8	07.5	07.0	07.4	08.8	11.0	12.7	10.2
Relative Humidity (%)													
Mean	67.0	64.0	63.0	68.0	78.0	82.0	83.0	85.0	83.0	76.0	70.0	67.0	74.0
Mean max.	90.6	87.1	86.2	88.7	93.1	94.9	95.0	96.0	95.4	92.9	90.9	90.9	91.8
Mean min.	44.4	42.8	42.1	48.8	60.3	67.3	68.0	70.8	67.0	57.9	49.6	45.2	55.4

Appendix 4

THE NORTHEAST (1951 - 1980)

4.3. Sakon Nakhon. 17° 09' N, 104°08' E. Elevation 171 meters.

Temp. °C	Jan	Feb	Mar	Apr	May	Jun	Jul	Aug	Sep	Oct	Nov	Dec	Year
Mean	22.5	24.6	27.7	29.4	28.6	28.2	28.0	27.4	27.2	26.6	24.7	22.6	26.5
Mean max.	29.1	30.9	33.7	35.0	33.4	32.2	31.9	31.1	31.0	31.0	30.1	28.9	31.5
Mean min.	14.3	17.2	20.9	23.3	24.1	24.4	24.3	24.0	23.5	21.4	17.9	14.7	20.8
Ext. max.	36.4	39.2	41.2	41.9	39.9	39.8	36.8	36.5	35.4	35.4	36.9	35.5	41.9
Ext. min.	00.5	07.6	09.6	14.0	18.8	21.0	21.2	21.0	19.2	13.3	06.9	04.0	00.5
Mean daily range	14.8	13.7	12.8	11.7	09.3	07.8	07.6	07.1	07.5	09.6	12.2	14.2	10.7
Relative Humidity (%)													
Mean	64.0	63.0	62.0	65.0	77.0	80.0	80.0	82.0	82.0	74.0	69.0	67.0	72.0
Mean max.	90.8	86.4	83.5	84.5	90.4	91.2	91.3	93.0	93.9	91.8	91.6	92.3	90.1
Mean min.	45.0	44.6	44.6	48.7	61.5	67.4	67.7	70.2	68.4	58.5	50.8	47.7	56.3

Appendix 4

THE NORTHEAST (1951 - 1980)

4.4. Surin. 14°53'N, 103°30'E. Elevation 146 meters.

Temp. °C	Jan	Feb	Mar	Apr	May	Jun	Jul	Aug	Sep	Oct	Nov	Dec	Year
Mean	24.3	26.6	29.2	30.0	29.1	28.4	28.0	27.7	27.3	26.9	25.3	24.0	27.2
Mean max.	31.2	33.3	35.5	36.0	34.5	33.2	32.6	32.2	31.5	31.0	30.4	30.1	32.6
Mean min.	16.6	19.1	22.0	23.7	24.1	23.9	23.6	23.4	23.2	22.5	19.8	17.2	21.6
Ext. max.	36.6	38.4	40.8	41.6	41.6	38.8	37.4	37.1	36.7	35.8	36.2	35.8	41.6
Ext. min.	06.4	11.0	11.0	15.2	20.0	19.8	19.6	20.0	19.0	16.3	11.9	08.2	06.4
Mean daily range	14.6	14.2	13.5	12.3	10.4	09.3	09.0	08.8	08.3	08.5	10.6	12.9	11.0
Relative Humidity (%)													
Mean	65.0	62.0	62.0	66.0	75.0	79.0	80.0	82.0	84.0	80.0	75.0	69.0	73.0
Mean max.	87.7	84.9	83.1	85.4	90.8	93.3	93.1	94.0	95.4	92.8	90.5	89.3	90.0
Mean min.	42.7	42.5	41.2	45.1	55.6	61.8	63.2	65.6	68.3	65.6	57.3	48.9	54.8

Appendix 4

THE NORTHEAST (1951 - 1980)

4.5. Udon Thani. 17°23' N, 102°48' E. Elevation 177 meters.

Temp. °C	Jan	Feb	Mar	Apr	May	Jun	Jul	Aug	Sep	Oct	Nov	Dec	Year
Mean	21.7	24.2	27.2	29.1	28.5	28.1	27.7	27.3	27.0	26.4	24.3	21.9	26.1
Mean max.	29.6	31.8	34.6	35.9	34.3	32.8	32.4	31.6	31.3	31.3	30.3	29.2	32.1
Mean min.	15.1	17.9	21.3	23.8	24.5	24.8	24.5	24.3	23.9	22.5	19.2	15.8	21.5
Ext. max.	36.5	38.5	42.2	43.9	42.6	39.7	37.2	37.1	35.8	35.9	37.2	35.1	43.9
Ext. min.	02.5	09.4	10.0	11.8	18.8	21.5	20.8	21.0	20.5	15.6	08.4	05.5	02.5
Mean daily range	14.5	13.9	13.3	12.1	09.8	08.0	07.9	07.3	07.4	08.8	11.1	13.4	10.6
Relative Humidity (%)													
Mean	69.0	66.0	63.0	66.0	77.0	80.0	80.0	83.0	83.0	77.0	72.0	70.0	74.0
Mean max.	89.6	86.4	83.6	85.4	91.0	92.0	92.2	93.4	94.4	91.8	90.6	90.3	90.1
Mean min.	44.3	42.5	40.7	44.5	56.7	63.0	63.8	66.7	66.1	57.0	49.9	45.5	53.4

Appendix 5

Climatic Data - Thailand

The Southeast (1951 - 1980)

5.1. Aranyaprathet
     Prachin Buri Province
5.2. Chanthaburi
5.3. Chon Buri
5.4. Khlong Yai
     Trat Province
5.5. Prachin Buri

Prime source:
Ministry of Communications
The Meteorology Department
Meteorological Statistics Section

Note

To change Celsius to Farenheit, use the following formula:
$$F = C \times 9/5 + 32$$

Appendix 5

THE SOUTHEAST (1951 - 1980)

5.1. Aranyaprathet, Prachin Buri Province. 13°42' N, 102°35' E. Elevation 47 meters.

Temp. °C	Jan	Feb	Mar	Apr	May	Jun	Jul	Aug	Sep	Oct	Nov	Dec	Year
Mean	25.8	27.9	29.5	29.9	28.9	28.2	27.6	27.4	27.3	27.2	26.1	25.2	27.6
Mean max.	32.2	34.2	35.9	36.3	34.7	33.3	32.3	32.0	31.8	31.8	31.3	31.1	33.1
Mean min.	18.1	21.0	23.3	24.2	24.5	24.2	23.8	23.8	23.8	23.1	20.8	18.4	22.4
Ext. max.	38.0	39.7	40.0	41.0	40.5	39.8	36.0	35.7	35.5	35.3	36.5	36.0	41.0
Ext. min.	07.6	12.5	13.7	17.0	21.5	20.3	20.8	20.7	20.5	17.0	10.2	10.0	07.6
Mean daily range	14.1	13.2	12.6	12.1	10.2	09.1	08.5	08.2	08.0	08.7	10.5	12.7	10.7
Relative Humidity (%)													
Mean	64.0	65.0	66.0	71.0	78.0	81.0	83.0	84.0	84.0	81.0	75.0	69.0	75.0
Mean max.	86.8	87.2	88.3	89.5	92.7	93.6	94.2	95.1	95.6	94.2	91.0	89.1	91.5
Mean min.	43.0	43.0	44.0	48.4	58.1	63.2	65.8	67.0	68.1	63.8	57.0	49.5	55.9

Appendix 5

THE SOUTHEAST (1951 - 1980)

5.2. Chanthaburi. 12°36' N, 102°07' E. Elevation 3 meters.

Temp. °C	Jan	Feb	Mar	Apr	May	Jun	Jul	Aug	Sep	Oct	Nov	Dec	Year
Mean	25.1	26.2	27.1	27.8	27.6	27.2	26.9	26.8	26.5	26.4	25.8	25.0	26.5
Mean max.	31.9	32.4	32.7	33.4	32.3	31.0	30.6	30.4	30.5	31.4	31.3	31.1	31.6
Mean min.	19.8	21.3	22.6	23.5	24.1	24.3	24.1	24.1	23.7	23.1	21.8	20.2	22.7
Ext. max.	37.1	37.6	37.2	37.1	36.8	35.3	34.0	34.6	35.0	35.3	35.4	36.0	37.6
Ext. min.	11.2	14.8	14.8	18.8	21.4	21.0	21.1	19.0	21.0	16.9	13.0	08.9	08.9
Mean daily range	12.1	11.1	10.1	09.9	08.2	06.7	06.5	06.3	06.8	08.3	09.5	10.9	08.9
Relative Humidity (%)													
Mean	73.0	78.0	80.0	82.0	85.0	87.0	87.0	87.0	89.0	85.0	77.0	72.0	82.0
Mean max.	88.9	93.2	94.6	95.1	96.0	96.0	96.0	96.0	97.0	94.9	89.5	86.4	93.6
Mean min.	50.4	56.2	60.4	61.2	68.8	73.4	73.9	74.3	74.5	67.6	59.6	52.6	64.4

Appendix 5.

THE SOUTHEAST (1951 - 1980)

5.3. Chon Buri. 13°22' N, 100°59' E. Elevation 1 meter.

Temp. °C	Jan	Feb	Mar	Apr	May	Jun	Jul	Aug	Sep	Oct	Nov	Dec	Year
Mean	25.9	27.4	28.8	29.6	29.3	28.9	28.6	28.3	27.8	27.4	26.6	25.8	27.9
Mean max.	31.5	32.2	33.3	34.2	33.4	32.6	32.0	31.7	31.5	31.6	31.4	31.5	32.2
Mean min.	20.2	22.5	24.3	25.5	25.5	25.5	25.1	25.0	24.5	23.8	22.1	20.4	23.7
Ext. max.	37.5	37.6	37.8	38.4	38.2	37.1	35.9	35.8	35.5	35.9	36.2	36.7	38.4
Ext. min.	09.9	16.5	17.5	20.4	21.2	21.0	20.5	20.9	20.6	17.9	14.2	12.0	09.9
Mean daily range	11.3	09.7	09.0	08.7	07.9	07.1	06.9	06.7	07.0	07.8	09.3	11.1	08.5
Relative Humidity (%)													
Mean	68.0	72.0	72.0	73.0	76.0	75.0	76.0	77.0	81.0	81.0	74.0	67.0	74.0
Mean max.	85.0	87.8	87.5	87.4	88.6	87.7	88.6	89.6	92.3	92.6	88.6	84.4	88.3
Mean min.	51.2	55.5	56.1	56.2	60.4	61.3	62.4	63.4	66.6	65.4	55.9	48.6	58.6

449

Appendix 5

THE SOUTHEAST (1951 - 1980)

5.4. Khlong Yai, Trat Province. 11°47' N, 102°53' E. Elevation 2 meters.

Temp. °C	Jan	Feb	Mar	Apr	May	Jun	Jul	Aug	Sep	Oct	Nov	Dec	Year
Mean	26.5	27.3	28.0	28.7	28.2	27.3	27.0	26.8	26.7	26.9	27.2	26.8	27.3
Mean max.	31.1	31.3	32.0	32.7	32.1	30.5	30.2	29.8	30.0	30.9	31.4	31.3	31.1
Mean min.	20.4	22.0	23.1	23.9	24.0	23.6	23.5	23.5	23.2	22.8	22.1	21.0	22.8
Ext. max.	35.7	34.9	36.2	35.4	35.0	33.8	34.2	32.6	33.0	34.9	35.8	35.3	36.2
Ext. min.	13.0	15.0	16.2	20.0	21.1	21.4	19.9	21.0	19.8	18.8	13.7	13.8	13.0
Mean daily range	10.7	09.3	08.9	08.8	08.1	06.9	06.7	06.3	06.8	08.1	09.3	10.3	08.3
Relative Humidity (%)													
Mean	70.0	75.0	77.0	78.0	82.0	86.0	86.0	87.0	87.0	83.0	75.0	69.0	79.0
Mean max.	89.0	93.0	94.3	94.2	94.7	95.7	95.7	95.4	96.1	95.3	89.4	85.1	93.2
Mean min.	56.0	60.7	63.7	64.3	68.8	75.6	76.0	77.4	76.6	70.9	60.5	54.5	67.1

Appendix 5

THE SOUTHEAST (1951 - 1980)

5.5. Prachin Buri. 14°03' N, 101°22' E. Elevation 5 meters.

Temp. °C	Jan	Feb	Mar	Apr	May	Jun	Jul	Aug	Sep	Oct	Nov	Dec	Year
Mean	26.9	28.5	30.0	30.4	29.5	28.8	28.3	28.2	28.0	28.1	27.5	26.6	28.4
Mean max.	32.4	33.9	35.4	35.9	34.3	32.8	32.1	31.8	31.6	31.9	31.8	31.6	33.0
Mean min.	19.2	21.7	23.6	24.6	24.9	24.6	24.4	24.4	24.4	24.1	22.0	19.7	23.1
Ext. max.	37.0	38.0	39.8	40.7	40.4	39.8	36.4	35.2	35.2	35.2	35.6	35.8	40.7
Ext. min.	10.2	15.0	14.6	19.8	21.4	20.8	20.6	21.6	21.4	19.0	13.8	10.8	10.2
Mean daily range	13.2	12.2	11.8	11.3	09.4	08.2	07.7	07.4	07.2	07.8	09.8	11.9	09.9
Relative Humidity (%) Mean	60.0	63.0	65.0	70.0	76.0	79.0	81.0	82.0	82.0	77.0	68.0	61.0	72.0
Mean max.	84.5	87.3	89.3	90.8	92.5	93.7	94.4	94.8	95.0	90.7	85.0	82.5	90.1
Mean min.	43.2	45.1	46.8	51.5	60.9	66.1	68.4	69.5	70.4	64.5	54.1	46.2	57.2

Appendix 6

Climatic Data - Thailand

The West (1951 - 1980)

6.1. Hua Hin
 Prachuap Khiri Khan Province
6.2. Kanchanaburi
6.3. Mae Sot
 northern Tak Province
6.4. Prachuap Khiri Khan
6.5. Tak

Prime source:
Ministry of Communications
The Meteorology Department
Meteorological Statistics Section

Note
To change Celsius to Farenheit, use the following formula:
$$F = C \times {}^9/_5 + 32$$

Appendix 6

THE WEST (1951 - 1980)

6.1. Hua Hin, Prachuap Khiri Khan Province. 12° 35' N, 99° 57' E. Elevation 5 meters.

Temp. °C	Jan	Feb	Mar	Apr	May	Jun	Jul	Aug	Sep	Oct	Nov	Dec	Year
Mean	25.4	26.7	28.2	29.4	29.1	28.7	28.2	28.1	27.7	27.1	26.3	25.3	27.5
Mean max.	29.1	30.5	31.8	33.1	33.0	32.9	32.4	32.3	31.7	30.5	29.6	28.9	31.3
Mean min.	20.8	22.1	23.5	24.8	25.2	25.0	24.6	24.6	24.1	23.7	22.7	21.4	23.5
Ext. max.	32.3	35.0	36.2	36.7	37.2	37.5	37.2	36.3	36.3	34.5	33.0	32.1	37.5
Ext. min.	13.9	15.4	18.7	21.9	22.4	22.2	22.0	21.9	21.2	19.1	17.2	13.9	13.9
Mean daily range	08.3	08.4	08.3	08.3	07.8	07.9	07.8	07.7	07.6	06.8	06.9	07.5	07.8
Relative Humidity (%)													
Mean	72.0	76.0	75.0	74.0	76.0	75.0	76.0	76.0	80.0	83.0	78.0	73.0	76.0
Mean max.	84.6	88.3	87.1	86.0	87.9	86.8	88.1	88.3	91.1	92.6	87.8	83.4	87.7
Mean min.	59.1	60.6	59.3	59.4	61.6	59.6	60.7	60.5	64.7	69.2	65.8	60.3	61.7

Appendix 6

THE WEST (1951 - 1980)

6.2. Kanchanaburi. 14°01' N, 99°32' E. Elevation 28 meters.

Temp. °C	Jan	Feb	Mar	Apr	May	Jun	Jul	Aug	Sep	Oct	Nov	Dec	Year
Mean	25.6	28.1	30.3	31.4	29.9	28.9	28.3	28.2	27.8	27.2	26.1	25.0	28.1
Mean max.	32.3	34.9	37.2	37.9	35.4	33.7	33.1	32.9	32.6	31.5	30.7	30.8	33.6
Mean min.	17.7	20.6	23.0	24.9	25.0	24.6	24.2	24.0	23.8	23.0	20.8	18.2	22.5
Ext. max.	38.1	40.3	41.9	43.5	41.6	38.4	37.8	37.8	37.6	37.3	37.5	37.2	43.5
Ext. min.	05.5	12.1	11.2	17.2	21.5	22.0	20.8	21.5	20.8	16.2	11.6	06.8	05.5
Mean daily range	14.6	14.3	14.2	13.0	10.4	09.1	08.9	08.9	08.8	08.5	09.9	12.6	11.1
Relative Humidity (%)													
Mean	62.0	60.0	57.0	59.0	69.0	72.0	73.0	74.0	76.0	79.0	73.0	67.0	68.0
Mean max.	87.1	85.0	81.9	81.6	86.6	87.8	88.4	88.9	91.5	93.0	90.9	88.5	87.6
Mean min.	41.3	39.0	35.8	38.8	52.1	57.3	58.0	58.5	61.3	64.2	57.4	48.4	51.0

Appendix 6

THE WEST (1951 - 1980)

6.3. Mae Sot, northern Tak Province. 16°40' N, 98°33' E. Elevation 196 meters.

Temp. °C	Jan	Feb	Mar	Apr	May	Jun	Jul	Aug	Sep	Oct	Nov	Dec	Year
Mean	23.2	25.5	28.4	30.2	28.6	26.7	25.9	25.8	26.3	26.6	25.2	23.2	26.3
Mean max.	31.0	33.4	35.7	36.6	33.7	30.7	29.7	29.4	30.6	31.6	31.2	30.8	32.0
Mean min.	14.6	16.3	19.9	22.8	24.0	23.5	23.1	23.0	23.1	22.2	19.1	15.5	20.6
Ext. max.	35.5	37.5	39.5	40.5	39.8	34.9	36.9	34.1	34.5	36.5	37.0	37.1	40.5
Ext. min.	04.8	09.2	12.8	16.1	20.0	21.1	20.9	20.5	19.3	14.2	09.1	04.7	04.7
Mean daily range	16.4	17.1	15.8	13.8	09.7	07.2	06.6	06.4	07.5	09.4	12.1	15.3	11.4
Relative Humidity (%)													
Mean	70.0	63.0	58.0	60.0	74.0	84.0	85.0	86.0	85.0	81.0	76.0	73.0	74.0
Mean max.	94.6	91.6	85.8	83.6	90.1	94.9	95.2	95.9	95.9	95.8	95.7	95.5	92.9
Mean min.	41.7	35.9	34.2	39.1	56.1	69.2	72.6	74.0	69.7	62.1	52.3	46.1	54.4

Appendix 6

THE WEST (1951 - 1980)

6.4. Prachuap Khiri Khan. 11°48' N, 99°48' E. Elevation 4 meters.

Temp. ° C	Jan	Feb	Mar	Apr	May	Jun	Jul	Aug	Sep	Oct	Nov	Dec	Year
Mean	25.2	26.5	27.8	29.1	28.9	28.2	27.9	27.7	27.7	27.0	26.4	25.4	27.3
Mean max.	29.8	30.9	32.1	33.5	33.4	32.5	32.2	31.8	32.1	30.8	29.9	29.6	31.6
Mean min.	19.3	20.8	22.2	23.7	24.6	24.5	24.2	24.1	23.9	23.1	22.1	20.4	22.7
Ext. max.	35.0	36.8	38.5	39.3	38.9	37.8	37.6	36.9	38.0	35.8	33.6	33.8	39.3
Ext. min.	10.5	12.2	16.1	19.5	21.2	21.2	21.4	21.1	20.0	18.6	13.0	11.4	10.5
Mean daily range	10.5	10.1	09.9	09.8	08.8	08.0	08.0	07.7	08.2	07.7	07.8	09.2	08.9
Relative Humidity (%) Mean	77.0	80.0	79.0	77.0	78.0	78.0	78.0	79.0	79.0	83.0	79.0	75.0	78.0
Mean max.	90.2	93.6	92.7	91.0	90.6	89.6	89.5	90.2	90.8	93.6	89.9	86.6	90.7
Mean min.	62.2	63.9	62.6	60.9	62.4	63.6	63.4	64.7	63.9	69.6	66.6	61.7	63.8

Appendix 6

THE WEST (1951 - 1980)

6.5. Tak. 16°53'N, 99°09'E. Elevation 121 meters.

Temp. °C	Jan	Feb	Mar	Apr	May	Jun	Jul	Aug	Sep	Oct	Nov	Dec	Year
Mean	23.6	27.2	30.6	31.8	29.8	28.5	28.1	27.8	27.4	26.6	25.1	23.3	27.5
Mean max.	31.4	34.7	37.3	38.4	35.1	32.9	32.3	32.0	31.8	31.3	30.7	30.3	33.2
Mean min.	15.4	18.8	23.3	25.8	25.5	25.0	24.8	24.5	23.9	22.6	19.8	16.3	22.1
Ext. max.	37.8	39.9	41.9	43.5	43.0	38.8	38.4	37.0	36.7	35.0	35.1	36.0	43.5
Ext. min.	04.7	10.5	11.6	17.0	20.0	22.3	21.9	18.2	18.0	14.6	09.9	05.8	04.7
Mean daily range	16.0	15.9	14.0	12.6	09.6	07.9	07.5	07.5	07.9	08.7	10.9	14.0	11.1
Relative Humidity (%)													
Mean	65.0	53.0	47.0	52.0	68.0	73.0	73.0	74.0	79.0	83.0	79.0	73.0	68.0
Mean max.	90.3	77.6	67.6	71.7	83.2	85.2	84.8	86.4	90.9	95.6	95.0	93.8	85.2
Mean min.	38.0	30.7	28.9	34.2	51.5	59.2	59.6	60.8	63.8	63.8	55.3	46.7	49.4

Appendix 7

Climatic Data - Thailand

The South (1951 - 1980)

7.1. Chumphon
7.2. Nakhon Si Thammarat
7.3. Narathiwat
7.4. Ranong
7.5. Songkhla

Prime source:
Ministry of Communications
The Meteorology Department
Meteorological Statistics Section

Note
To change Celsius to Farenheit, use the following formula:
$$F = C \times 9/5 + 32$$

Appendix 7

THE SOUTH (1951 - 1980)

7.1. Chumphon. 10° 29' N, 99° 11' E. Elevation 3 meters.

Temp. ° C	Jan	Feb	Mar	Apr	May	Jun	Jul	Aug	Sep	Oct	Nov	Dec	Year
Mean	25.2	26.4	27.5	28.5	28.0	27.4	27.0	27.0	26.9	26.6	25.8	25.1	26.8
Mean max.	29.8	31.0	32.4	33.7	33.0	31.6	31.2	30.9	31.1	30.6	29.5	29.2	31.2
Mean min.	20.1	21.1	22.0	23.2	23.7	23.7	23.5	23.6	23.5	23.2	22.2	20.5	22.5
Ext. max.	34.0	36.2	37.8	38.6	38.2	36.0	35.0	35.5	34.8	35.6	34.6	33.2	38.6
Ext. min.	12.1	14.7	16.1	19.4	20.9	21.5	19.9	20.8	21.2	19.1	15.1	12.2	12.1
Mean daily range	09.7	09.9	10.4	10.5	09.3	07.9	07.7	07.3	07.6	07.4	07.3	08.7	08.7
Relative Humidity (%) Mean	81.0	81.0	79.0	79.0	82.0	83.0	83.0	83.0	84.0	86.0	85.0	82.0	82.0
Mean max.	95.4	95.9	95.9	95.9	95.8	95.4	95.6	95.1	95.6	96.6	96.1	95.1	95.7
Mean min.	65.3	63.6	61.4	60.3	64.5	67.6	68.4	69.6	69.0	72.0	71.9	67.5	66.8

Appendix 7

THE SOUTH (1951 - 1980)

7.2. Nakhon Si Thammarat. 08°28' N, 99°58' E. Elevation 7 meters.

Temp. °C	Jan	Feb	Mar	Apr	May	Jun	Jul	Aug	Sep	Oct	Nov	Dec	Year
Mean	26.1	26.9	28.0	28.7	28.5	28.5	28.1	28.0	27.7	27.0	26.1	25.9	27.5
Mean max.	29.7	31.1	32.7	33.6	33.5	33.3	33.1	33.0	32.6	31.2	29.4	29.1	31.9
Mean min.	21.8	21.7	22.1	23.1	23.7	23.6	23.1	23.2	23.0	22.9	22.7	22.4	22.8
Ext. max.	34.6	35.4	38.0	37.1	37.3	37.7	36.5	37.1	36.6	35.0	34.2	32.6	38.0
Ext. min.	17.2	17.2	17.8	18.6	20.2	20.7	19.4	19.5	19.4	20.2	18.0	17.1	17.1
Mean daily range	07.9	09.4	10.6	10.5	09.8	09.7	10.0	09.8	09.6	08.3	06.7	06.7	09.1
Relative Humidity (%)													
Mean	83.0	80.0	78.0	79.0	80.0	77.0	77.0	77.0	80.0	84.0	86.0	85.0	81.0
Mean max.	95.6	95.4	95.1	95.1	94.4	92.8	93.3	92.8	94.6	95.9	96.0	95.5	94.7
Mean min.	66.1	60.6	57.0	57.7	59.6	57.5	57.0	57.5	59.2	67.0	73.4	71.4	62.0

Appendix 7

THE SOUTH (1951 - 1980)

7.3. Narathiwat. 06°25' N, 101°49' E. Elevation 2 meters.

Temp. ° C	Jan	Feb	Mar	Apr	May	Jun	Jul	Aug	Sep	Oct	Nov	Dec	Year
Mean	26.1	26.9	27.7	28.6	28.4	28.0	27.7	27.5	27.4	27.0	26.2	25.9	27.3
Mean max.	29.8	30.8	31.9	33.0	32.8	32.4	32.1	32.0	31.9	30.8	29.3	29.0	31.3
Mean min.	22.2	22.3	22.7	23.4	23.8	23.5	23.1	23.1	23.1	23.1	23.0	22.8	23.0
Ext. max.	33.6	35.1	35.8	36.4	36.9	36.0	35.8	36.0	36.4	35.0	33.9	32.6	36.9
Ext. min.	17.1	17.5	19.0	19.8	20.5	21.0	20.7	20.6	20.2	20.3	18.7	19.8	17.1
Mean daily range	07.6	08.5	09.2	09.6	09.0	08.9	09.0	08.9	08.8	07.7	06.3	06.2	08.3
Relative Humidity (%)													
Mean	81.0	79.0	78.0	77.0	79.0	79.0	79.0	79.0	80.0	83.0	86.0	85.0	80.0
Mean max.	94.1	94.0	94.3	94.0	94.0	94.4	94.7	94.7	94.9	95.7	96.7	95.7	94.8
Mean min.	69.6	67.4	65.2	64.7	65.2	64.9	65.2	65.1	65.9	71.2	76.0	75.3	68.0

Appendix 7

THE SOUTH (1951 - 1980)

7.4. Ranong. 09°58' N, 98°38' E. Elevation 7 meters.

Temp. °C	Jan	Feb	Mar	Apr	May	Jun	Jul	Aug	Sep	Oct	Nov	Dec	Year
Mean	25.8	26.9	28.0	28.3	27.2	26.4	26.2	26.0	25.9	25.9	25.8	25.6	26.5
Mean max.	31.7	33.2	34.1	34.1	31.8	30.3	30.1	29.8	29.8	30.5	30.6	30.8	31.4
Mean min.	20.3	20.7	21.8	23.1	23.6	23.4	23.2	23.2	23.0	22.6	22.0	21.0	22.3
Ext. max.	35.4	37.0	38.0	38.0	38.0	34.8	34.4	33.1	34.2	35.0	34.8	35.0	38.0
Ext. min.	13.7	15.0	16.8	20.0	20.6	20.1	19.6	20.1	20.2	20.0	16.0	15.1	13.7
Mean daily range	11.4	12.5	12.3	11.0	08.2	06.9	06.9	06.6	06.8	07.9	08.6	09.8	09.1
Relative Humidity (%)													
Mean	76.0	74.0	74.0	78.0	85.0	88.0	88.0	88.0	89.0	88.0	84.0	79.0	83.0
Mean max.	93.5	94.2	94.0	94.6	96.4	96.7	96.9	96.9	97.5	97.3	95.3	93.2	95.5
Mean min.	57.4	53.9	53.6	59.4	70.8	76.7	76.5	77.6	77.1	74.3	69.8	62.8	67.5

Appendix 7

THE SOUTH (1951 - 1980)

7.5. Songkhla. 07°12' N, 100°36' E. Elevation 4 meters.

Temp. °C	Jan	Feb	Mar	Apr	May	Jun	Jul	Aug	Sep	Oct	Nov	Dec	Year
Mean	26.9	27.4	28.0	28.7	28.4	28.1	27.8	27.8	27.5	27.0	26.6	26.6	27.6
Mean max.	29.6	30.4	31.4	32.7	33.1	32.8	32.6	32.6	32.3	31.1	29.6	29.2	31.4
Mean min.	23.9	24.0	23.9	24.3	24.4	24.1	23.7	23.8	23.7	23.6	23.7	23.9	23.9
Ext. max.	34.0	32.8	36.5	38.2	36.5	36.2	36.1	36.2	35.9	35.2	34.4	33.3	38.2
Ext. min.	19.1	20.3	19.8	20.5	21.8	20.9	20.6	20.8	21.0	20.7	19.9	20.5	19.1
Mean daily range	05.7	06.4	07.5	08.4	08.7	08.7	08.9	08.8	08.6	07.5	05.9	05.3	07.5
Relative Humidity (%)													
Mean	77.0	76.0	76.0	77.0	79.0	78.0	78.0	77.0	79.0	83.0	84.0	82.0	79.0
Mean max.	87.8	87.7	90.3	91.9	92.9	92.3	92.4	92.0	92.5	94.6	94.2	91.1	91.6
Mean min.	68.1	65.8	64.6	63.5	63.1	62.1	61.8	61.0	62.5	68.2	73.9	72.7	65.6

Appendix 8

Water Surface Conditions
Off the Coasts of Thailand

(Depth 0 - 5 Meters)

GULF OF THAILAND

Year	Average Temperature °C	Average Salinity (parts per 1000)
1967		
Jan - Feb	24.79	30.62
Feb - Mar	27.52	30.85
Mar - Apr	29.17	30.98
Apr - May	30.44	29.26
Jun - Jul	29.36	28.01
Jul - Aug	29.46	29.27
Aug - Sep	28.88	25.39
Sep - Oct	29.17	24.72
Nov - Dec	28.73	30.23
1967 - 1968		
Dec - Jan	26.30	30.78
Annual Av.	28.38	29.01

ANDAMAN SEA, WEST OF PHUKET

Year	No. of Readings	Min. Temp.°C	Max. Temp.°C	Av. °C
1980 15 Mar–12 May	50	29.00	30.60	29.80

Data are less complete for the Andaman Sea, but it can be assumed that the monthly figures for the Gulf of Thailand and the Andaman Sea would be essentially the same. Note that the average temperatures for the only comparable months (March, April, and May) are the same for both bodies of water.

Prime source:
Unpublished data of the Oceanographic Division of the Department of Hydrography, The Royal Thai Navy.

Appendix 9

Protected Snakes of Thailand

The Royal Government of Thailand is to be commended for enacting legislation designed to protect certain beneficial snakes by prohibiting their export. In 1982, the following snakes were banned from export:

> *Elaphe flavolineata*
> *Elaphe radiata*
> *Gonyosoma oxycephalum*
> *Ptyas carinatus*
> *Ptyas korros*
> *Ptyas mucosus*
> *Python curtus brongersmai*
> *Python molurus bivittatus*
> *Python reticulatus*
> *Xenopeltis unicolor*

Such protection is badly needed as Thailand has provided considerable quantities of animals for the international animal trade for many years. For example, over a recent four-year period, export permits were issued for 351,433 of the snakes listed above plus *Naja*. This number reflects an enormous drain from an exhaustable natural resource. If permitted to continue, wild populations would inevitably decline and ultimately disappear. The table overleaf displays the number of permits issued for certain species for the period 1980 - 1984.

Table 9.1. Permits Issued, 1980 - 1984.

Snake	Permits
Elaphe	24,121
Gonyosoma oxycephalum	990
Ptyas	105,231
Python curtus brongersmai	252
Python molurus bivittatus	7,659
Python reticulatus	26,043
Xenopeltis unicolor	328
Naja	186,809
Total	351,433

Protecting live snakes from export is an excellent beginning, but legislation controlling both the consumption and the export of snakes and their products is necessary if the conservation of beneficial and endangered snakes is to be effective. Many farming communities, where farmers are often part-time snake hunters, derive substantial income from the capture and sale of snakes for their meat and skin. There is a large market for both in Thailand and abroad. Legislation encouraging the breeding of snakes, as well as other wild animals, would surely curtail the decline of wild populations and provide legitimate economic opportunity to a portion of the population. Legislation allowing the export of live, captive bred snakes and their products, as well as the domestic consumption thereof, would diminish predation of wild populations. Domestic production of wild animals could be an effective and innovative conservation measure. I hasten to add this is not a new idea. After all, each of our domestic species were once wild animals.

Appendix 10

Analysis of the
Common Thai Snake Names

The purpose of this appendix is to help the reader to understand the Thai names of the snakes found in Thailand. Although the translation of the Thai name of a given snake is sometimes similar to that of the English name, this is not always true. Often people look at the same thing and yet see it differently.

The compilation of this appendix has proven to be a difficult task as English and Thai are very dissimilar, both in written and spoken forms. For instance, the alphabets differ. Furthermore, there are no spaces left between the words that are written in Thai. For example งูดินธรรมดา would appear as งู ดิน ธรรมดา in the English written style. Furthermore, there is no standard system of transliterating Thai into English and transliterations often vary. Thai is a tonal language and the tones must be carefully pronounced as one single word may have several different meanings, depending upon the tone. For example, the word *mai* may mean not or wood, depending upon the spoken tone used.

To simplify these complexities and to create a useful appendix, the following method will be followed. The names of the Thai snakes are listed in the order of textual appearance. Each Thai snake name begins with the word ง (ngoo ¹) = snake. This will not be repeated in each name. Furthermore, each member of a genus often, but not always, has the first two or three words of its name in common. When this happens, these words will be explained and not repeated in the discussions of individual snake names.

The Thai language uses five tones indicated by the following numbers:

1) even, 2) high, 3) low, 4) falling, 5) rising.

To the extent possible, the transliteration used by Gordon H. Allison will be followed. [1] Where this is not possible, the author has reproduced the sound by using his own system of transliteration.

Despite strenuous efforts, it has not been possible to determine or even make an educated guess at the names of three of Thailand's snakes. I suppose that this is to be expected when one considers that Thailand is an ancient civilization and has a diverse population with a language influenced and enriched by many cultures. It is likely that the meanings of the names of those three snakes have been lost in the antiquity and complexity of the language.

I am very much indebted to Jarujin Nabhitabhata of the Thailand Institute of Scientific and Technological Research and to Professor Puangkaew Latriratanakul and her colleagues in the Thai Language Department, Faculty of Arts, Silpakorn University. Without their help many more than three names would remain unknown.

---

1. Allison, Gordon H. **Jumbo English-Thai Dictionary**, 3rd ed., Odeon Store, Bangkok. 1978.

Appendix 10

ANALYSIS OF THE THAI NAME	TC	ENGLISH EQUIVALENT	Page
Both genera *Ramphotyphlops* and *Typhlops* are known as งูดิน. งู (ngoo ^{1}) = snake ดิน (din ^{1}) = earth		**Earth Snakes** (Fossorial)	85
* 1. งูดินหัวขาว หัว (hua ^{5}) = head ขาว (khao ^{5}) = white	0	White-headed Earth Snake	85
2. งูดินธรรมดา ธรรมดา (tham ^{2} ma ^{3} da ^{1}) = common	0	Common Earth Snake Common throughout Thailand	86
3. งูดินใหญ่อินโดจีน ใหญ่ (yai ^{3}) = big อินโดจีน (in ^{1} do ^{1} jeen ^{1}) = Indochina	0	Indochinese Big Earth Snake Type locality is "Cochin China"	88
4. งูดินใหญ่มลายู ใหญ่ (yai ^{3}) = big มลายู (ma ^{2} la ^{1} yoo ^{1}) = Malaya	0	Malayan Big Earth Snake Also found in Malaya Larger than average earth snake	88
5. งูดินหัวเหลือง หัว (hua ^{5}) = head เหลือง (leuang ^{5}) = yellow	0	Yellow-headed Earth Snake	89

TC = Toxicity Classification

0 = **nonvenomous**
1 = **mildly venomous**
2 = **very venomous** (see page 75 for definition)
U = **unclassified due to insufficient data**

* These serial numbers refer to the order of appearance in the text.

Appendix 10

ANALYSIS OF THE THAI NAME	TC	ENGLISH EQUIVALENT	Page
6. งูดินโคราช โคราช (kho[1] rat[2]) = Khorat	0	Khorat Earth Snake Khorat refers to the north-east where it was first seen	90
7. งูดินลายขีด ลาย (lai[1]) = pattern or stripe ขีด (kheed[3]) = stripe	0	Striped Earth Snake	90
8. งูดินเมืองตรัง เมือง (meuang[1]) = city ตรัง (trang[1]) = Trang	0	Trang Earth Snake First found in the South in Trang Province	91
9. งูผ้าขี้ริ้ว ผ้าขี้ริ้ว (pha[4] khee[4] riu[2]) = rag	0	Rag Snake Loose skin makes for a rag look when on land	94
10. งูงวงช้าง งวงช้าง (nguang[1] chang[2]) = elephant's trunk	0	Elephant Trunk Snake Resembles the trunk of an elephant	95
11. งูก้นขบ ก้น (kon[4]) = end of tail ขบ (khob[3]) = to bite	0	Tail Can Bite Snake Some Thais believe it can bite with the tail	98
งูสองหัว สอง (sawng[5]) = two หัว (hua[5]) = head	0	Two-headed Snake When disturbed the tail is raised giving a two-headed appearance	98
12. งูแสงอาทิตย์ แสง (saeng[5]) = sun rays อาทิตย์ (ah[1] thit[2]) = sun	0	Sunshine Snake Iridescent	101
13. งูหลามปากเป็ด หลาม (lahm[5]) = bloated ปาก (pak[3]) = mouth เป็ด (pet[3]) = duck	0	Bloated Duck-mouthed Snake Thick bodied - mouth resembles that of a duck	104

Appendix 10 471

ANALYSIS OF THE THAI NAME	TC	ENGLISH EQUIVALENT	Page
14. งูหลาม หลาม (lahm ^{5}) = bloated	0	Bloated Snake Thick bodied and looks as though it has overeaten	106
15. งูเหลือม เหลือม (leuam ^{5}) = shiny - a northern Thai word	0	Shiny Snake Iridescent	108
The name of each member of genus *Calamaria* and the single member of genus *Pseudorabdion* begins with งูพงอ้อ. ง (ngoo ^{1}) = snake พงอ้อ (pong ^{1} aw ^{2}) = a mass of reeds		**Reed Snakes**	111
16. งูพงอ้อหลากลาย หลาก (lock ^{3}) = many ลาย (lai ^{1}) = pattern or stripe	0	Many-patterned Reed Snake Varied pattern	113
17. งูพงอ้อท้องเหลือง ท้อง (thawng ^{2}) = belly เหลือง (leuang ^{5}) = yellow	0	Yellow-bellied Reed Snake	114
18. งูพงอ้อหัวขาว หัว (hua ^{5}) = head ขาว (khao ^{5}) = white	0	White-headed Reed Snake	114
19. งูพงอ้อเล็ก เล็ก (lek ^{2}) = small	0	Small Reed Snake	115
งูพงอ้อหัวยาว หัว (hua ^{5}) = head ยาว (yao ^{1}) = long	0	Long-headed Reed Snake	115

ANALYSIS OF THE THAI NAME	TC	ENGLISH EQUIVALENT	Page
Each of the three members of genus *Ahaetulla* are called งูเขียว. งู (ngoo [1]) = snake เขียว (kheeo [5]) = green		**Green Snakes** Members of the genus *Chrysopelea* are also called Green Snakes	117
20. งูเขียวหัวจิ้งจกมลายู หัว (hua [5]) = head จิ้งจก (jing [2] jok [3]) = house gecko (Hemidactylus sp.) มลายู (ma [2] la [1] yoo [1]) = Malaya	1	Malayan House Gecko-headed Green Snake More common in Malaya	118
21. งูเขียวปากแหนบ ปาก (pak [3]) = mouth แหนบ (naep [3]) = tweezer	1	Tweezer-mouthed Green Snake Long and narrow mouth	119
22. งูเขียวหัวจิ้งจก หัว (hua [5]) = head จิ้งจก (jing [2] jok [3]) = house gecko	1	House Gecko-headed Green Snake	120
งูง่วงกลางดง ง่วง (nguang [4]) = sleepy กลาง (klang [1]) = center ดง (dong [1]) = jungle	1	Central Jungle Sleepy Snake Some Thais believe that the bite causes sleepiness	120
Note. Thai members of *Boiga* do not have similar words in their names.			
23. งูเขียวดง เขียว (kheeo [5]) = green ดง (dong [1]) = jungle	1	Jungle Green Snake A forest snake	122

Appendix 10

ANALYSIS OF THE THAI NAME	TC	ENGLISH EQUIVALENT	Page
24. งูแส้หางม้า แส้ (sae⁴) = whip หาง (hang⁵) = tail ม้า (ma²) = horse	1	Horse-tailed Whip Snake Tail pattern similar to a horse tail	131
งูกะปิ กะปิ (ga³ phee¹) = shrimp paste	1	Shrimp Paste Snake Similar body color	131
งูกินไข่ กิน (kin¹) = eat ไข่ (khai³) = egg	1	Egg Eating Snake	131
25. งูปล้องทอง ปล้อง (plang⁴) = a segment ทอง (thawng¹) = gold	1	Golden Segment Snake Yellow body bands	133
26. งูดงคาทอง ดงคา (dong¹ kha¹) = lalang grass - tall Thai grass ทอง (thawng¹) = gold	1	Golden Lalang Snake Lives in tall grass	135
27. งูกระ กระ (gra²) = freckle	1	Freckled Snake	137
28. งูแม่ตะงาวรังนก แม่ตะงาว (mae⁴ ta³ ngaow¹) = *Vipera russelli siamensis* รังนก (rang¹ nok²) = bird nest	1	Bird Nest Russell's Viper Color and pattern resembles *Vipera russelli siamensis*. Frequents bird nests	138
29. งูต้องไฟ ต้อง (tong²) = touch ไฟ (fai¹) = fire	1	Fire Touching Snake Red body resembles embers	139

Appendix 10

ANALYSIS OF THE THAI NAME	TC	ENGLISH EQUIVALENT	Page
30. งูแส้หางม้าเทา แส้ (sae [4]) = whip หาง (hang [5]) = tail ม้า (ma [2]) = horse เทา (thao [1]) = gray	1	Gray Horse-tailed Whip Snake	141
งูแส้หางม้าเล็ก แส้ (sae [4]) = whip หาง (hang [5]) = tail ม้า (ma [2]) = horse เล็ก (lek [2]) = small	1	Small Horse-tailed Whip Snake Shorter than *Boiga cynodon*	141
31. งูเขียวดงลาย เขียว (kheeo [5]) = green ดง (dong [1]) = jungle ลาย (lai [1]) = pattern or stripe	1	Green Striped Jungle Snake More pattern than *Boiga cyanea*	143
Two of the three members of genus *Chrysopelea* are called งูเขียว.		**Green Snakes**	145
32. งูเขียวพระอินทร์ พระอินทร์ (pra [2] in [1]) = Indra	1	Indra Green Snake As the Hindu God, Indra - green	146
งูเขียวลายดอกหมาก ลาย (lai [1]) = pattern or stripe ดอกหมาก (dawk [3] mawk [3]) = Betel Palm flower	1	Pale-spotted Green Snake Resembles palm flower pattern of pale spots	146
33. งูเขียวร่อน ร่อน (rawn [4]) = glide	1	Gliding Green Snake One of the three Thai flying snakes	147
34. งูดอกหมากแดง ดอกหมาก (dawk [3] mawk [3]) = Betel Palm Flower แดง (daeng [1]) = red	1	Red Betel Palm Flower Snake	148

ANALYSIS OF THE THAI NAME	TC	ENGLISH EQUIVALENT	Page
Each member of genus *Dendrelaphis* is called งูสายม่าน. สาย (sai ^{5}) = string ม่าน (man ^{4}) = curtain		**Curtain String Snakes** Slender	149
35. งูสายม่านแดงหลังลาย แดง (daeng ^{1}) = red หลัง (lang ^{5}) = back ลาย (lai ^{1}) = pattern or stripe	0	Stripe-backed Red Curtain String Snake	150
36. งูสายม่านคอขีด คอ (khaw ^{1}) = neck ขีด (kheed ^{3}) = line or stripe	0	Stripe-necked Curtain String Snake	150
37. งูสายม่านหลังทอง หลัง (lang ^{5}) = back ทอง (thawng ^{1}) = gold	0	Gold-backed Curtain String Snake	151
38. งูสายม่านพระอินทร์ พระอินทร์ (pra ^{2} in ^{1}) = Indra	0	Indra Curtain String Snake Green - Hindu God, Indra	152
งูสายม่านธรรมดา ธรรมดา (tham ^{2} ma ^{3} da ^{1}) = common	0	Common Curtain String Snake Has a wide range	152
39. งูสายม่านลายเฉียง ลาย (lai ^{1}) = pattern or stripe เฉียง (chiang ^{5}) = oblique, diagonal	0	Diagonally-striped Curtain String Snake	153
40. งูสายม่านเกล็ดใต้ตาใหญ่ เกล็ด (klet ^{3}) = scale ใต้ (tai ^{4}) = under ตา (ta ^{1}) = eye ใหญ่ (yai ^{3}) = big	0	Big-suboculared Curtain String Snake Suboculars large and noticeable	154

ANALYSIS OF THE THAI NAME	TC	ENGLISH EQUIVALENT	Page
41. งูสายน้ำผึ้ง สายน้ำผึ้ง (sai⁴ nam² pheung⁴) = Honey creeper - used for home decoration	U	Honey Creeper Snake Close resemblance to the plant	155
งูเถา เถา (taw¹) = slender vine, stem	U	Vine Snake Very slender	155
Four of the six Thai members of genus *Elaphe* have names beginning with งูทางมะพร้าว. ง (ngoo¹) = snake ทางมะพร้าว (thang¹ ma² praow²) = central rib of a coconut leaf. The remaining two are known as งูกาบหมาก (gab¹ mawk³) as pattern resembles the sheath which covers the inflorescence of the Areca Palm.		**Central Rib of Coconut Leaf Snakes** **Areca Palm Sheath Snakes**	156 156
42. งูทางมะพร้าวดำ ดำ (dam¹) = black	0	Black Central Rib of Coconut Leaf Snake Dark in color	157
งูหลุนฉุน หลุน (lun¹) = hole ฉุน (chun¹) = crowded	0	Unknown	157
43. งูทางมะพร้าวแดงแถบดำ แดง (daeng¹) = red แถบ (taeb³) = band or bar ดำ (dam¹) = black	0	Black-banded Red Central Rib of Coconut Leaf Snake	159
44. งูทางมะพร้าวแดง แดง (daeng¹) = red	0	Red Central Rib of Coconut Leaf Snake	160
งูบ้องไฟ บ้องไฟ (bawng² fai¹) = rocket	0	Rocket Snake Bright red	160

ANALYSIS OF THE THAI NAME	TC	ENGLISH EQUIVALENT	Page
45. งูทางมะพร้าวธรรมดา ธรรมดา (tham² ma³ da¹) = common	0	Common Central Rib of Coconut Leaf Snake Has wide range	161
46. งูกาบหมากหางนิล หาง (hang⁵) = tail นิล (nin¹) = jet stone	0	Dark Black Areca Palm Sheath Snake Very black	162
งูใบ้ ใบ้ (bai²) = mute	0	Mute Snake	162
47. งูกาบหมากดำ ดำ (dam¹) = black	0	Black Areca Palm Sheath Snake	163
48. งูเขียวกาบหมาก เขียว (kheeo⁵) = green	0	Green Areca Palm Sheath Snake Closely related to genus *Elaphe*	164
49. งูหมอก หมอก (mawk³) = Thai use of the English "mock"	1	Mock Snake Mildly venomous but resembles a venomous viper	166
50. งูม่านทอง ม่าน (man⁴) = curtain ทอง (thawng¹) = gold	1	Golden Curtain Snake	167
The three Thai members of genus *Ptyas* have names beginning with งูสิง. ง (ngoo¹) = snake สิง (sing⁵) = northern Thai for "dark"		**Dark Snakes** Dark coloration	168

Appendix 10

ANALYSIS OF THE THAI NAME	TC	ENGLISH EQUIVALENT	Page
51. งูสิงหางดำ หาง (hang [5]) = tail ดำ (dam [1]) = black	0	Black-tailed Dark Snake	169
งูบองหมาบควาย บองหมาบ (bawng [1] mab [3]) = unknown ควาย (khwai [1]) = buffalo	0	Unknown	169
52. งูสิงธรรมดา ธรรมดา (tham [2] ma [3] da [1]) = common	0	Common Dark Snake Common throughout the country	170
53. งูสิงหางลาย หาง (hang [5]) = tail ลาย (lai [1]) = pattern or stripe	0	Stripe-tailed Dark Snake	179
54. งูควนขนุน ควนขนุน (khuan [1] khanun [5]) = place in southern Thailand	U	Khuan Khanun Snake First found in South - Amphoe Khuan Khnun, Phatthalung Province	180

Note.
Homalopsine snakes do not have features of their names in common.

55. งูปากกว้างท้องสัน ปาก (pak [3]) = mouth กว้าง (kwang [4]) = wide ท้อง (thawng [2]) = belly สัน (sen [5]) = a keel	U	Keel-bellied Wide-mouthed Snake	183
งูเปี้ยว (pheeo [2]) = Thai children's game - running	U	Fast Snake	183

ANALYSIS OF THE THAI NAME	TC	ENGLISH EQUIVALENT	Page
56. งูปลาหลังม่วง ปลา (pla ^{1}) = fish หลัง (lang ^{5}) = back ม่วง (muang ^{4}) = purple	U	Purple-backed Fish Snake	185
57. งูปากกว้างน้ำเค็ม ปาก (pak ^{3}) = mouth กว้าง (kwang ^{4}) = wide น้ำเค็ม (nam ^{2} khem ^{1}) = salt water	1	Salt Water Wide-mouthed Snake Habitat near brackish water	186
58. งูไซ ไซ (sai ^{1}) = fish trap	1	Fish Trap Snake Often found in fish traps	188
59. งูสายรุ้งธรรมดา สายรุ้ง (sai ^{5} roong ^{2}) = rainbow ธรรมดา (tham ^{2} ma ^{3} da ^{1}) = common	1	Common Rainbow Snake	189
60. งูสายรุ้งลาย สายรุ้ง (sai ^{5} roong ^{2}) = rainbow ลาย (lai ^{1}) = pattern or stripe	1	Striped Rainbow Snake	190
61. งูปลิง ปลิง (pling ^{1}) = leech	1	Leach Snake Body shape and color resembles a leach	190
62. งูสายรุ้งดำ สายรุ้ง (sai ^{5} roong ^{2}) = rainbow ดำ (dam ^{1}) = black	1	Black Rainbow Snake	191
63. งูกระด้าง กระด้าง (gra ^{3} dang ^{2}) = stiff	1	Stiff Snake Stiffens when held	194

ANALYSIS OF THE THAI NAME	TC	ENGLISH EQUIVALENT	Page
64. งูปลาหลังเทา ปลา (pla ^{1}) = fish หลัง (lang ^{5}) = back เทา (thao ^{1}) = gray	1	Gray-backed Fish Snake	196
65. งูปลาตาแมว ปลา (pla ^{1}) = fish ตา (ta ^{1}) = eye แมว (maeo ^{1}) = cat	U	Cat-eyed Fish Snake	197
งูปากกว้างตาเล็ก ปาก (pak ^{3}) = mouth กว้าง (kwang ^{4}) = wide ตา (ta ^{1}) = eye เล็ก (lek ^{2}) = small	U	Small-eyed Wide-mouthed Snake	197
66. งูหัวกระโหลก หัวกระโหลก (hua ^{5} gra ^{3} lok ^{3}) = skull	1	Skull Snake	198
งูเหลียมอ้อ เหลียม (leuam ^{5}) = python อ้อ (aw ^{2}) = tall reeds	1	Reed Python Snake	198
Subfamily *Lycodintinae* is diverse and most, but not all Thai members have the words งูปล้องฉนวน or งูปี่แก้ว within their name. งูปล้องฉนวน งู (ngoo ^{1}) = snake ปล้องฉนวน (plang ^{4} chanuan ^{5}) = segment or band		**Banded Snakes**	199
ปี่แก้ว (pee ^{3} kaeo ^{4}) = classical Thai song or flute		**Flute Snakes** Some Thai believe the snake makes musical sounds	199

Appendix 10 481

	ANALYSIS OF THE THAI NAME	TC	ENGLISH EQUIVALENT	Page
67.	งูปล้องฉนวนภูเขา ภูเขา (phoo [1] khao [5]) = mountain	1	Mountain Banded Snake Habitat - high elevations	200
68.	งูปล้องฉนวนธรรมดา ธรรมดา (tham [2] ma [3] da [1]) = common	0	Common Banded Snake Common and widespread	201
69.	งูปล้องฉนวนมลายู มลายู (ma [2] la [1] yoo [1]) = Malaya	0	Malayan Banded Snake	202
70.	งูปล้องฉนวนบอร์เนียว บอร์เนียว (ba [1] neeo [1]) = Borneo	U	Bornean Banded Snake Type locality - Sarawak, Borneo	204
71.	งูสายทองลายแถบ สายทอง (sai [5] thawng [1]) = golden line ลาย (lai [1]) = pattern or stripe แถบ (taeb [3]) = band or bar	0	Banded Golden-lined Snake Small and slender snake	206
72.	งูสายทองคอแหวน สายทอง (sai [5] thawng [1]) = golden line คอ (khaw [1]) = neck แหวน (waen [5]) = ring	0	Ring-necked Golden-lined Snake	207
73.	งูสายทองมลายู สายทอง (sai [5] thawng [1]) = golden line มลายู (ma [2] la [1] yoo [1]) = Malaya	0	Malayan Golden-lined Snake	208
74.	งูสร้อยเหลือง สร้อย (soi [4]) = necklace เหลือง (leuang [5]) = yellow	1	Yellow Necklace Snake Has a yellow neckband	209

ANALYSIS OF THE THAI NAME	TC	ENGLISH EQUIVALENT	Page
75. งูปล้องฉนวนเมืองเหนือ เมืองเหนือ (meuang ^{1} neua ^{5}) = northern	1	Northern Banded Snake Locality northern Thailand	210
76. งูปล้องฉนวนลายเหลือง ลาย (lai ^{1}) = pattern or stripe เหลือง (leuang ^{5}) = yellow	1	Yellow-striped Banded Snake	211
งูปล้องฉนวนลาว ลาว (lao ^{1}) = Laos	1	Laotian Banded Snake Type locality - Laos	211
77. งูปล้องฉนวนบ้าน บ้าน (ban ^{4}) = house	1	House Banded Snake Often found near housing	211
78. งูปี่แก้วหัวลายหัวใจ หัว (hua ^{5}) = head ลาย (lai ^{1}) = pattern or stripe หัวใจ (hua ^{5} jai ^{1}) = heart	1	Heart Pattern-headed Flute Snake Heart marking on head	214
79. งูปี่แก้วลายกระธรรมดา ลายกระ (lai ^{1} gra ^{2}) = flecked ธรรมดา (tham ^{2} ma ^{3} da ^{1}) = common	1	Common Flecked Musical Snake	215
80. งูปี่แก้วลายกระเมืองจีน ลายกระ (lai ^{1} gra ^{2}) = flecked เมืองจีน (meuang ^{1} jeen ^{1}) = Chinese	1	Chinese Flecked Flute Snake Type locality - Amoy, China	216
81. งูปี่แก้วธรรมดาลายจาง ธรรมดา (tham ^{2} ma ^{3} da ^{1}) = common ลาย (lai ^{1}) = pattern or stripe จาง (jang ^{1}) = pale	1	Common Pale-patterned Flute Snake Color is faded	217

Appendix 10 483

	ANALYSIS OF THE THAI NAME	TC	ENGLISH EQUIVALENT	Page
82.	งูปี่แก้วธรรมดาลายเข้ม ธรรมดา (tham² ma³ da¹) = common ลาย (lai¹) = pattern or stripe เข้ม (khem⁴) = dark	1	Common Dark Flute Snake	218
83.	งูปี่แก้วภูหลวง ภูหลวง (phu¹ luang⁵) = Phu Luang	1	Phu Luang Flute Snake First found at Phu Luang, Loei Province	227
84.	งูปี่แก้วหลังจุดวงแหวน หลัง (lang⁵) = back จุด (joot³) = spot วงแหวน (wong¹ waen⁵) = a ring or circle	1	Circular Spot-backed Flute Snake	228
85.	งูปี่แก้วสีจาง สี (see⁵) = color จาง (jang¹) = pale	1	Pale Colored Flute Snake Snake Dull color	229
86.	งูปี่แก้วใหญ่ ใหญ่ (yai³) = big	1	Big Flute Snake	229
87.	งูงอดเขมร งอด (ngawt¹) = spotted เขมร (khmen⁵) = Cambodia - Cambodian language	1	Spotted Cambodian Snake Type locality: Cambodia	230
88.	งูคุด คุด (khot³) = to turn inwards	1	Inward Turning Snake When discovered tries to conceal head under body	231
89.	งูงอด งอด (ngawt¹) = spotted	1	Spotted Snake	232

Appendix 10

ANALYSIS OF THE THAI NAME	TC	ENGLISH EQUIVALENT	Page
Of Subfamily *Natricinae*, genera *Amphiesma* and *Rhabdophis* share the words งูลายสาบ in their names. ลาย (lai [1]) = pattern or stripe สาบ (sab [3]) = smelly Emit a foul odor when handled.		**Smelly Striped Snakes**	233
Genera *Opisthotropis, Parahelicops, Sinonatrix* and *Xenochrophis* share the words งูลายสอ. ลายสอ (lai [1] saw [5]) = white pattern This is not easily understood as white is not dominant.		**White-patterned Snakes**	233
90. งูลายสาบท้องสามขีด ท้อง (thawng [2]) = belly สาม (sam [5]) = three ขีด (kheed [3]) = stripe	0	Stripe-bellied Smelly Striped Snake	234
91. งูลายสาบท่าสาร ท่าสาร (ta [4] san [5]) = Ta San	0	Ta San Smelly Striped Snake Type locality -Ta San, Ranong Province	235
92. งูลายสาบมลายู มลายู (ma [2] la [1] yoo [1]) = Malaya	0	Malayan Smelly Striped Snake Type locality - Gunong Ines, Perak, Malaysia	235
93. งูลายสาบดอกหญ้า ดอกหญ้า (dawk [3] ya [4]) = a grass flower	0	Grass Flower Smelly Striped Snake	237
94. งูลายสอลาวเหนือ ลาว (lao [1]) = Laos เหนือ (neua [5]) = north	0	Northern Laos White-patterned Snake First found northern Laos	238

Appendix 10

ANALYSIS OF THE THAI NAME	TC	ENGLISH EQUIVALENT	Page
95. งูลายสอสองสี สอง (sawng ^{5}) = two สี (see ^{5}) = color	0	Bicolor White-patterned Snake	239
96. งูลายสอหมอบุญส่ง หมอ (maw ^{5}) = doctor บุญส่ง (boon ^{3} song ^{5}) = a family name. Dr. Boonsong is a famous Thai naturalist	U	Dr. Boonsong's White-patterned Snake	240
97. งูลายสาบจุดดำขาว จุด (joot ^{3}) = spot ดำ (dam ^{1}) = black ขาว (khao ^{5}) = white	1	Black and White Spotted Smelly Striped Snake	242
98. งูลายสาบเขียวขั้นดำ เขียว (kheeo ^{5}) = green ขั้น (khwen ^{4}) = girdle ดำ (dam ^{1}) = black	1	Black-girdled Green Smelly Striped Snake	243
99. งูลายสาบสีจาง สี (see ^{5}) = color จาง (jang ^{1}) = pale	1	Pale Colored Smelly Striped Snake	244
100. งูลายสาบคอแดง คอ (khaw ^{1}) = neck แดง (daeng ^{1}) = red	1	Red-necked Smelly Striped Snake	245
101. งูลายสอเมืองจีน เมืองจีน (meuang ^{1} jeen ^{1}) = Chinese	0	Chinese White-patterned Snake Type locality - Kuatun Mts., N.W. Fukien Prov., China	248
งูลายสอเกล็ดใต้ตาสอง เกล็ด (klet ^{3}) = scale ใต้ (tai ^{4}) = under ตา (ta ^{1}) = eye สอง (sawng ^{5}) = two	0	Two-subocular White-patterned Snake	248

ANALYSIS OF THE THAI NAME	TC	ENGLISH EQUIVALENT	Page
102. งูลายสอลายสามเหลี่ยม ลาย (lai ^{1}) = pattern or stripe สามเหลี่ยม (sam ^{5} liam ^{3}) = triangle	0	Triangularly-patterned White-patterned Snake	249
103. งูลายสอธรรมดา ธรรมดา (tham ^{2} ma ^{3} da ^{1}) = common	0	Common White-patterned Snake Extensive range	250
104. งูลายสอใหญ่ ใหญ่ (yai ^{3}) = big	0	Big White-patterned Snake	252
The Thai names of members of Subfamily *Pareatinae* have the words งูกินทาก in common. กิน (kin ^{1}) = eat ทาก (tak ^{1}) = slug		**Slug Eating Snakes**	254
105. งูกินทากหัวโหนก หัวโหนก (hua ^{5} nook ^{3}) = prominent forehead	0	Prominent-foreheaded Slug Eating Snake	256
งูบอ บอ (ba ^{1}) = boa - corrupt form of the Latin word "boa"	0	Boa Snake	256
106. งูกินทากเกล็ดสัน เกล็ด (klet ^{3}) = scale สัน (sen ^{5}) = keel	0	Keel-scaled Slug Eating Snake	257
107. งูกินทากลายขวั้น ลาย (lai ^{1}) = pattern or stripe ขวั้น (khwen ^{4}) = girdle	0	Girdle-striped Slug Eating Snake	258

Appendix 10 487

ANALYSIS OF THE THAI NAME	TC	ENGLISH EQUIVALENT	Page
108. งูกินทากเกล็ดเรียบ เกล็ด (klet ^{3}) = scale เรียบ (riap ^{4}) = smooth	0	Smooth-scaled Slug Eating Snake	259
งูกินทากสีน้ำตาล สี (see ^{5}) = color น้ำตาล (nam ^{2} tan ^{1}) = brown	0	Brown Slug Eating Snake	259
109. งูกินทากจุดดำ จุด (joot ^{3}) = spot ดำ (dam ^{1}) = black	0	Black-spotted Slug Eating Snake	260
110. งูกินทากมลายู มลายู (ma ^{2} la ^{1} yoo ^{1}) = Malaya	0	Malayan Slug Eating Snake Type locality - Malacca	260
111. งูกินทากจุดขาว จุด (joot ^{3}) = spot ขาว (khao ^{5}) = white	0	White-spotted Slug Eating Snake	261
Note. The four Thai members of Subfamily *Pseudoxenodontinae* do not have many identical words in their names.			
112. งูรังแหหัวแดง รังแห (rang ^{1} hae ^{5}) = net, network or reticulation หัว (hua ^{5}) = head แดง (daeng ^{1}) = red	1	Red-headed Reticulated Snake	263
113. งูรังแหหลังศร รังแห (rang ^{1} hae ^{5}) = net, network or reticulation หลัง (lang ^{5}) = back ศร (sawn ^{5}) = arrow	1	Arrow-backed Reticulated Snake Dorsal pattern of arrow shapes	264

Appendix 10

ANALYSIS OF THE THAI NAME	TC	ENGLISH EQUIVALENT	Page
114. งูหัวศร หัว (hua ^{5}) = head ศร (sawn ^{5}) = arrow	0	Arrowhead Patterned Snake	265
งูหัวลายลูกศร หัว (hua ^{5}) = head ลาย (lai ^{1}) = stripe or pattern ลูกศร (look ^{4} sawn ^{5}) = arrow	0	Arrow Pattern-headed Snake	265
115. งูลายสาบตาโต ลาย (lai ^{1}) = pattern or stripe สาบ (sab ^{3}) = smelly ตา (ta ^{1}) = eye โต (toh ^{1}) = big	1	Big-eyed Smelly Striped Snake	266
งูลายสาบคอบั้ง ลาย (lai ^{1}) = pattern or stripe สาบ (sab ^{3}) = smelly คอ (khaw ^{1}) = neck บั้ง (bung ^{2}) = chevron	1	Chevron-necked Smelly Striped Snake Emits foul odor when disturbed	266
Thailand's three members of Subfamily *Sibynopheinae* share the following words in their names งูคอขวั้น. ง (ngoo ^{1}) = snake คอ (khaw ^{1}) = neck ขวั้น (khwen ^{4}) = girdle		**Girdle-necked Snakes**	276
116. งูคอขวั้นหัวดำ หัว (hua ^{5}) = head ดำ (dam ^{1}) = black	0	Black-headed Girdle-necked Snake	277
117. งูคอขวั้นปลายหัวดำ ปลาย (plai ^{1}) = end, tip หัว (hua ^{5}) = head ดำ (dam ^{1}) = black	0	Black-snouted Girdle-necked Snake	278

Appendix 10   489

ANALYSIS OF THE THAI NAME	TC	ENGLISH EQUIVALENT	Page
118. งูคอขวั้นหัวลายสามเหลี่ยม หัว (hua⁵) = head ลาย (lai¹) = pattern or stripe สามเหลี่ยม (sam⁵ liam³) = triangle	0	Triangular Pattern-headed Girdle-necked Snake	279
119. งูขอนไม้ ขอนไม้ (khan⁵ mai²) = log	0	Log Snake First found by Dr. Taylor inside a log	280
งูท้องขาว ท้อง (thawng²) = belly ขาว (khao⁵) = white	0	White-bellied Snake	280
The names of two of the three Thai members of Genus *Bungarus* contain the words งูสามเหลี่ยม. งู (ngoo¹) = snake สามเหลี่ยม (sam⁵ liam³) = triangle		**Triangle Snakes** In cross section the body is triangular	282
120. งูทับสมิงคลา ทับ (tub²) = dwelling สมิง (saming⁵) = tiger คลา (khla¹) = abundant	2	Abundant Tiger House Snake. Exact meaning is unknown. As the snake is abundant it often enters homes causing venomous bites. The banded stripes and ferocious nature possibly create the "tiger".	284
121. งูสามเหลี่ยม สามเหลี่ยม (sam⁵ liam³) = triangle	2	Triangle Snake	285
122. งูสามเหลี่ยมหัวหางแดง หัว (hua⁵) = head หาง (hang⁵) = tail แดง (daeng¹) = red	2	Red-headed Red-tailed Triangle Snake	287

Appendix 10

ANALYSIS OF THE THAI NAME	TC	ENGLISH EQUIVALENT	Page
All Thai members of Genus *Naja* are called งูเห่า. ง (ngoo [1]) = snake เห่า (how [3]) = to hiss		**Hissing Snakes**	288
123. งูเห่าหม้อ หม้อ (maw [4]) = earthen pot blackened by charcoal fire	2	Black Hissing Snake	294
124. งูเห่าสีนวล สี (see [5]) = color นวล (nuan [4]) = cream	2	Cream-colored Hissing Snake	295
125. งูเห่าด่างพ่นพิษ ด่าง (dang [2]) = piebald พ่น (phon [4]) = to spray พิษ (phit [2]) = poison	2	Poison Spraying Piebald Hissing Snake	297
126. งูเห่าดำพ่นพิษ ดำ (dam [1]) = black พ่น (phon [4]) = to spray พิษ (phit [2]) = poison	2	Poison Spraying Black Hissing Snake	298
127. งูเห่าอิสานพ่นพิษ อิสาน (isan [3]) = N.E. Thailand พ่น (phon [4]) = to spray พิษ (phit [2]) = poison	2	Poison Spraying North-eastern Hissing Snake	299
128. งูเห่าทองพ่นพิษ ทอง (thawng [1]) = gold พ่น (phon [4]) = to spray พิษ (phit [2]) = poison	2	Poison Spraying Golden Hissing Snake	300
129. งูจงอาง จงอาง (jong [1] ang [1]) = unknown	2	Unknown	303

ANALYSIS OF THE THAI NAME	TC	ENGLISH EQUIVALENT	Page
Subfamily *Hydropheinae* contains a number of diverse snakes and this is reflected in the names of Thai members. The most common are: งูทากลาย. ทาก (tak ^{1}) = slug ลาย (lai ^{1}) = pattern or stripe		**Slug-patterned Snakes**	305
งูทะเล ทะเล (tha ^{2} leh ^{1}) = sea		**Sea Snakes**	305
งูคออ่อน คอ (khaw ^{1}) = neck อ่อน (awn ^{3}) = weak Thin necks		**Weak-necked Snakes**	305
งูแสมรัง แสมรัง (samae ^{5} rang ^{1}) = a mangrove forest tree		**Mangrove Snakes**	305
งูชายธง ชายธง (chai ^{1} thong ^{1}) = triangular end of pennant		**Triangle Snakes** Triangular markings on the sides	305
130. งูทากลายท้องขาว ท้อง (thawng ^{2}) = belly ขาว (khao ^{5}) = white	2	White-bellied Slug-patterned Snake	315
131. งูทะเลจุดขาว จุด (joot ^{3}) = spot ขาว (khao ^{5}) = white	2	White-spotted Sea Snake	317
132. งูทากลาย ทาก (tak ^{1}) = slug ลาย (lai ^{1}) = pattern or stripe	2	Slug-patterned Snake	318

Appendix 10

ANALYSIS OF THE THAI NAME	TC	ENGLISH EQUIVALENT	Page
133. งูคอย่อนหัวโต หัว (hua ^{5}) = head โต (toh ^{1}) = big	2	Big-headed Weak-necked Snake	319
งูคอย่อนปากจะงอย ปาก (pak ^{3}) = mouth จะงอย (ja ^{3} ngai ^{1}) = bird's beak	2	Bird-beaked Weak-necked Snake Rostral extends over lower lip	319
134. งูแสมรังท้องเหลือง ท้อง (thawng ^{2}) = belly เหลือง (leuang ^{5}) = yellow	2	Yellow-bellied Mangrove Snake	321
135. งูแสมรังลายเยื้อง ลายเยื้อง (lai ^{1} euang ^{2}) = diagonal	2	Diagonally-striped Mangrove Snake	322
งูแสมรังเกล็ดหยาบ เกล็ด (klet ^{3}) = scale หยาบ (yeeab ^{3}) = rough	2	Rough-scaled Mangrove Snake	322
136. งูแสมรังเหลืองลายคราม เหลือง (leuang ^{5}) = yellow ลาย (lai ^{1}) = pattern or stripe คราม (khram ^{1}) = indigo	2	Indigo-striped Yellow Mangrove Snake	323
งูแสมรังลายฟ้า ลาย (lai ^{1}) = pattern or stripe ฟ้า (fa ^{2}) = blue	2	Blue Mangrove Snake	323
137. งูแสมรังปล้องลายแถบ ปล้อง (plang ^{2}) = segment ลายแถบ (lai ^{1} taeb ^{2}) = banded pattern	2	Banded Segmented Mangrove Snake	324

Appendix 10   493

ANALYSIS OF THE THAI NAME	TC	ENGLISH EQUIVALENT	Page
138. งูแสมรังลายแถบ ลายแถบ (lai [1] taeb [2]) = banded pattern	2	Banded Mangrove Snake	325
139. งูฝักมะรุม ฝัก (fak [2]) = pod มะรุม (ma [3] room [1]) = Horseradish tree	2	Horseradish Pod Snake Body shape resembles the pod of the Horseradish tree	326
140. งูทะเลอ่าวเปอร์เซีย อ่าว (ao [3]) = gulf เปอร์เซีย (per [1] sia [1]) = Persia	2	Persian Gulf Sea Snake New to Thailand	327
141. งูแสมรังหางขาว หาง (hang [5]) = tail ขาว (khao [5]) = white	2	White-tailed Mangrove Snake	329
142. งูแสมรังเหลือง เหลือง (leuang [5]) = yellow	2	Yellow Mangrove Snake	330
143. งูแสมรังเทาอ่าวไทยตอนล่าง เทา (thao [1]) = gray อ่าว (ao [3]) = gulf ไทย (thai [1]) = Thailand ตอน (tawn [1]) = part ล่าง (long [4]) = lower	2	Lower Gulf of Thailand Gray Mangrove Snake	331
144. งูแสมรังเทาอ่าวไทยตอนบน เทา (thao [1]) = gray อ่าว (ao [3]) = gulf ไทย (thai [1]) = Thailand ตอน (tawn [1]) = part บน (bon [1]) = upper	2	Upper Gulf of Thailand Gray Mangrove Snake	332

ANALYSIS OF THE THAI NAME	TC	ENGLISH EQUIVALENT	Page
145. งูแสมรังเทาหัวดำเหลือง เทา (thao ^{1}) = gray หัว (hua ^{5}) = head ดำ (dam ^{1}) = black เหลือง (leuang ^{5}) = yellow	2	Black and Yellow-headed Gray Mangrove Snake	333
146. งูแสมรัง แสมรัง (samae ^{5} rang ^{1}) = Mangrove forest tree	2	Mangrove Snake	335
งูสามเหลี่ยมลายข้าวหลามตัด ลายข้าวหลามตัด (lai ^{1} khao ^{4} lahm ^{5} tat ^{3}) = diamond	2	Diamond Triangle Snake	335
147. งูสามเหลี่ยมหัวโต หัว (hua ^{5}) = head โต (toh ^{1}) = big	2	Big-headed Triangle Snake	336
งูกะรังหัวโต กะรัง (ga ^{3} rang ^{1}) = coral หัว (hua ^{5}) = head โต (toh ^{1}) = big	2	Big-headed Coral Snake	336
148. งูไอ้งั่ว ไอ้งั่ว (ai ^{2} ngua ^{3}) = a variety of Thai bird	2	Bird Snake Color similarity	338
149. งูคอย่อนหัวเข็ม หัว (hua ^{5}) = head เข็ม (khem ^{5}) = needle	2	Needle-headed Weak-necked Snake	339
150. งูสามเหลี่ยมหลังดำ หลัง (lang ^{5}) = back ดำ (dam ^{1}) = black	2	Black-backed Triangle Snake	341

Appendix 10   495

ANALYSIS OF THE THAI NAME	TC	ENGLISH EQUIVALENT	Page
151. งูชายธงท้องขาว ท้อง (thawng ²) = belly ขาว (khao ⁵) = white	2	White-bellied Triangle Snake	343
152. งูเสมียนรังหัวสั้น เสมียน (samian ⁵) = clerk Probably a play on words as it is similar to แสม (samae ⁵). Therefore, เสมียนรัง is possibly equivalent to แสมรัง (samae ⁵ rang ¹) = a tree found in mangrove forests. หัว (hua ⁵) = head สั้น (san ⁴) = short	2	Short-headed Mangrove Snake Head relatively short	344
The two Thai members of Subfamily *Laticaudinae* are named งูสมิงทะเล. งู (ngoo ¹) = snake, สมิง (saming ⁵) = tiger ทะเล (tha ² leh ¹) = sea		**Sea Tiger Snakes** Striped like a Tiger	345
153. งูสมิงทะเลปากเหลือง ปาก (pak ³) = mouth เหลือง (leuang ⁵) = yellow	2	Yellow-mouthed Sea Tiger Snake	346
154. งูสมิงทะเล สมิงทะเล (saming ⁵ tha ² leh ¹) = sea tiger	2	Sea Tiger Snake	348
Thai members of Genus *Calliophis* are called งูปล้องหวาย. ปล้อง (plang ⁴) = segment of plant stem หวาย (hwai ⁵) = rattan		**Rattan Segment Snakes**	349

Appendix 10

ANALYSIS OF THE THAI NAME	TC	ENGLISH EQUIVALENT	Page
155. งูปล้องหวายเทา เทา (thao [1]) = gray	2	Gray Rattan Segment Snake	350
156. งูปล้องหวายลายขวั้นดำ ลาย (lai [1]) = pattern or stripe ขวั้น (khwen [4]) = girdle ดำ (dam [1]) = black	2	Black-girdled Rattan Segment Snake	351
งูธิดาพระอาทิตย์ ธิดาพระอาทิตย์ (ti [1] daa [1] phra [2] ah [1] thit [1]) = daughter of the sun	2	Daughter of the Sun Snake Bright red coloration	351
157. งูปล้องหวายหางแหวน หาง (hang [5]) = tail แหวน (waen [5]) = ring	2	Ring-tailed Rattan Segment Snake	352
158. งูปล้องหวายหัวดำ หัว (hua [5]) = head ดำ (dam [1]) = black	2	Black-headed Rattan Segment Snake	353
159. งูปล้องหวายหลังเส้น หลัง (lang [5]) = back เส้น (sen [4]) = line	2	Line-backed Rattan Segment Snake	354
Both Thai members of genus *Maticora* are called งูพริก. งู (ngoo [1]) = snake พริก (phrik [2]) = red chili		**Chili Snakes** Color similarity	363
160. งูพริกท้องแดง ท้อง (thawng [2]) = belly แดง (daeng [1]) = red	2	Red-bellied Chili Snake	363

Appendix 10

ANALYSIS OF THE THAI NAME	TC	ENGLISH EQUIVALENT	Page
161. งูพริกสีน้ำตาล สี (see⁵) = color น้ำตาล (nam² tan¹) = brown	2	Brown Chili Snake	365
162. งูกะปะ กะปะ (ga² pha³) = abbreviated Malayan name, Kapak Bodoh = ax-shaped	2	Ax-headed Snake The head is shaped like an ax	368
Thai members of the Genus *Trimeresurus* are called งูเขียวหางไหม้. งู (ngoo¹) = snake เขียว (kheeo⁵) = green หางไหม้ (hang⁵ mai⁴) = burnt-tailed		**Burnt-tailed Green Snakes** Tails are red	371
163. งูเขียวหางไหม้ท้องเหลือง ท้อง (thawng²) = belly เหลือง (leuang⁵) = yellow	2	Yellow-bellied Burnt-tailed Green Snake	373
164. งูเขียวหางไหม้สุมาตราหัวเขียว สุมาตรา (su³ ma¹ tra¹) = Sumatra หัว (hua⁵) = head เขียว (kheeo⁵) = green	2	Green-headed Sumatran Burnt-tailed Green Snake	375
165. งูหางแม่บกาญจน์ หางแม่บ (hang⁵ haem⁴) = burnt-tailed - a northern Thai dialect กาญจน์ (gan³) = formal name for gold or short for Kanchanburi Province. See following note, page 498.	2	Kanchanaburi Burnt-tailed Snake	376

ANALYSIS OF THE THAI NAME	TC	ENGLISH EQUIVALENT	Page
165. (continued from page 497). This name is confusing as it contains words of a northern Thai dialect, but it is not a northern snake. The last word in the name is equally inappropriate as the pattern does not contain gold nor is the snake from Kanchanaburi Province, it is from the South. Possibly this is due to the original misidentification of the snake (see page 377).		Kanchanaburi Burnt-tailed Snake	376
166. งูเขียวหางไหม้ตาโต ตา (ta¹) = eye โต (toh¹) = big	2	Big-eyed Burnt-tailed Green Snake	378
167. งหางแบ่มภูเขา หางแบ่ม (hang⁵ haem⁴) = burnt-tailed - northern dialect ภูเขา (phoo¹ khao⁵) = mountain	2	Mountain Burnt-tailed Snake Habitat - high elevations, but the tail is not red	381
168. งูเขียวหางไหม้ท้องเขียว ท้อง (thawng²) = belly เขียว (kheeo⁵) = green	2	Green-bellied Burnt-tailed Green Snake	384
169. งูปาล์มแดง ปาล์ม (pam¹) = palm แดง (daeng¹) = red	2	Red Palm Snake	386
170. งูปาล์ม ปาล์ม (pam¹) = palm	2	Palm Snake Originally found in a palm tree	387

Appendix 10   499

ANALYSIS OF THE THAI NAME	TC	ENGLISH EQUIVALENT	Page
171. งูพังกา พังกา (peng² ka¹) = common mangrove forest tree	2	Mangrove Tree Snake	390
งูเขียวม่วงหางไหม้ ม่วง (muang⁴) = purple หางไหม้ (hang⁵ mai⁴) = burnt-tailed	2	Purple Burnt-tailed Green Snake Color is purplish-brown	390
172. งูเขียวไผ่ เขียว (kheeo⁵) = green ไผ่ (pai³) = bamboo	2	Bamboo Green Snake	393
173. งูเขียวหางไหม้สุมาตราหัวดำ สุมาตรา (su³ ma¹ tra¹) = Sumatra หัว (hua⁵) = head ดำ (dam¹) = black	2	Black-headed Sumatran Burnt-tailed Green Snake Type locality - Sumatra	394
174. งูเขียวตุ๊กแก เขียว (kheeo⁵) = green ตุ๊กแก (took² kae¹) = Tokay gecko	2	Tokay Gecko Green Snake	404
งูกะปะเสือ กะปะ (ga² pha³) = Malay - Ax-shape เสือ (seua⁵) = Tiger	2	Tiger Ax-heaed Snake Slightly striped as a tiger Head triangular	404
175. งูแมวเซา แมว (maeo¹) = cat เซา (sao¹) = sleepy	2	Sleepy Cat Snake It is locally believed that the bite causes motionless sleep	407

Bibliography

Bogert, Charles Mitchell.
1943   Dentitional Phenomena in Cobras and Other Elaphids With Notes on Adaptive Modification of Fangs. *Bulletin of the American Museum of Natural History*: New York. Vol. LXXXI (193) : 285-360 and plates.

Boswall, Jeffrey and Frith, Clifford B.
1978   Cantor's Water Snake, Cantoria violacea (Girard) A Vertebrate New to the Fauna of Thailand. *Natural History Bulletin of the Siam Society*: Bangkok. Vol. 27: 187-188.

Breen, John F.
1974   *Encyclopedia of Reptiles and Amphibians.* New Jersey: T. F. H. Publications.

Brongersma, L. O.
1933   V. The Herpetological Fauna of Pulu Weh. *Zoologische Mededeelingen*. Leiden: Vol. 16: 1-2.

Bunuan Tumwipat and Wirot Nutphand.
1982   *The Treatment of Poisonous Snakebite Victims and The Poisonous Snakes of Thailand.* In Thai. Bangkok: Peekhanet.

Campden-Main, Simon M.
1968   The Subspecies of Calliophis maculiceps (Gunther). *British Journal of Herpetology.* British Herpetological Society: London. Vol. 4 (3): 49-50.

1969   The Status of Oligodon taeniatus (Gunther), 1861, and Oligodon mouhoti (Boulenger), 1914 (Serpentes, Colubridae). *Herpetologica*: Pittsburgh. Vol. 25 (4): 295-299.

1970   *A Field Guide to the Snakes of South Vietnam.* Washington D.C: Division of Reptiles and Amphibians, Smithsonian Institute.

1970   The Identity of Oligodon cyclurus (Cantor, 1839) and Revalidation of Oligodon brevicauda (Steindachner, 1876) Serpents: Colubridae. *Proceedings of the Biological Society of Washington.* Washington: Biological Society of Washington. Vol. 82 (58): 763-766.

Chote, Suvatti.
1967   *Fauna of Thailand.* 2d ed. Bangkok: Applied Scientific Research Corporation of Thailand. Reptilia. 64-521.

Cochran, Doris M.
1927   New Reptiles and Batrachians Collected By Dr. Hugh M. Smith in Siam. *Proceedings of the Biological Society of Washington.* Washington: Biological Society of Washington. Vol. 40: 179-191.

Cogger, Harold G.
1986   *Reptiles and Amphibians of Australia.* Rev. ed. New South Wales: Reed.

Cox, Merel J.
1988   Serious Effects From the Bite of the Red Cat-Eye Snake, Boiga nigriceps. *Bulletin of the Chicago Herpetological Society.* Chicago: Chicago Herpetological Society. 23 (10): 162.

Daniel, J. C.
1983   *The Book of Indian Reptiles.* Bombay: Bombay Natural History Society.

Deranlyagala, P. E. P.
1960   The Taxonomy of the Cobras of South Eastern Asia. *Spolia Zeylanica.* Sri Lanka: National Museums of Ceylon. 29: 41-63.

1960   The Taxonomy of the Cobras of South Eastern Asia - Part 2. *Spolia Zeylanica.* Sri Lanka: National Museums of Ceylon. 29: 205-231.

De Rooij, Nelly.
1917   *The Reptiles of the Indo-Australian Archipelago.* Vol. 2. Ophidia. Leiden: Brill.

Deoras, P. J.
1970   *Snakes of India.* 2d rev. ed. New Delhi: National Book Trust.

Deuve, J.
1970   *Serpents du Laos.* In French. Paris: Office de la Récherche Scientifique.

Dowling, Herndon G.
1958   A Taxonomic Study of the Ratsnakes VI. Validation of the Genera Gonysoma (Wagler) and Elaphe (Fitzinger). *Copeia.* Florida: University of Florida. Vol. 1: 29-41.

1959   Classification of the Serpents: A Critical Review. *Copeia.* Florida: University of Florida. Vol. 1: 38-52.

Engelmann, Wolf-Eberhard and Obst., Fritz Junger.
1984 *Snakes: Biology, Behavior, and Relationship to Man.* Beckenham, U.K: Croom Helm.

Fitch, Henry S.
1981 *Sexual Size Differences in Reptiles.* Miscl. 70. Lawrence: University of Kansas.

1982 *Reproduction Cycles in Tropical Reptiles.* Occasional Paper 96. Lawrence: University of Kansas.

Frith, Clifford B.
1977 A Survey of the Snakes of Phuket Island and the Adjacent Mainland Areas of Peninsular Thailand. *Natural History Bulletin of the Siam Society.* Bangkok: The Siam Society. Vol. 26 (3-4): 263-316.

1977 The Sea Snake Hydrophis spiralis (Shaw): A New Species of the Fauna of Thailand. *Natural History Bulletin of the Siam Society.* Bangkok: The Siam Society. Vol. 26: 339-344.

Frith, Clifford B. and Maciver, Donaldo.
1978 The Crab-Eating Water Snake, Fordonia leucobalia (Schleg.), Another Snake New to Thailand. *Natural History Bulletin of the Siam Society.* Bangkok: The Siam Society. Vol. 27: 189-191.

Gow, Graeme F.
1983 *Snakes of Australia.* Sydney: Angus & Robertson.

Grandison, A. G. C.
1972 The Gunong Benom Expedition 1967, 5. Reptiles and Amphibians of Gunong Benom With A Description of a New Species of Macrocalamus. *Bulletin of the British Museum (Natural History) Zoology.* London: The British Museum.Vol. 23 (4): 45-101.

Hoge, A. R. and Romano Hoge, S. A. R. W. L.
1978-1979   *Poisonous Snakes of the World.* Part I. Checklist of the Pit Vipers Viperoidea, Viperidae, Crotalinae. Sao Paolo: Institute Butantan. Vol. 42/43: 179-310.

Hoge, A. R. and Romano Hoge, S. Alma.
1980/81   Notes on Micro and Ultrastructure of "Oberhautschen". *Viperoidea.* Sao Paolo: Institute Butantan. Vol. 44/45: 81-118.

Inger, Robert F. and Colwell, Robert K.
1977   Organization of Contiguous Communities of Amphibians and Reptiles in Thailand. *Ecological Monographs.* Illinois: Field Museum of Natural History. Vol. 47 (3): 229-253.

Inger, Robert F. and Marx, Hymen.
1965   The Systematics and Evolution of the Oriental Colubrid Snakes of the Genus Calamaria. *Fieldiana: Zoology.* Vol. 49. Illinois: Chicago Natural History Museum.

Institute for Herpetological Research.
1979   *The Python Breeding Manual.* 2d print. Stanford, U.S.A: Institute for Herpetological Research.

Ko Ko Gyi.
1970   A Revision of Colubrid Snakes of the Subfamily Homalopsinae. Lawrence: University of Kansas. Vol. 20 (2): 47-223.

Kramer, Eugen.
1977   Zur Schlangenfauna Nepals. In German. *Revue Suisse Zoologie.* Geneva: Swiss Zoological Society. Vol. 84 (3): 721-761.

Kramer, Eugen and Renenass, Urs.
1981   Zur Systematik der Grunen Grubenottern der Gattung Trimeresurus (Serpentes, Crotalidae). *Revue Suisse Zoologie.* In German. Geneva: Swiss Zoological Society. Vol. 88 (1): 163-205.

Kroon, Charles.
1973   A New Colubrid Snake (Boiga) from Southeast Asia. *Copeia.* Florida: Univeristy of Florida. Vol. 3: 580-586.

Leviton, Alan E.
1963   Contributions to a Review of Philippine Snakes II. The Snakes of Genera Liopeltis and Sibynophis. *Philippine Journal of Science.* Manila Vol. 92 (3): 367-381.

1963   Contributions to a Review of Philippine Snakes III. The Genera Maticora and Calliophis. *Philippine Journal of Science.* Manila. Vol. 92 (4): 523-550.

1964   Contributions to a Review of Philippine Snakes IV. The Genera Chrysopelea and Dryophiops. *Philippine Journal of Science.* Manila. Vol. 93 (1): 131-145.

1964   Contributions to a Review of Philippine Snakes VII. The Snakes of Genera Naja and Ophiophagus. *Philippine Journal of Science.* Manila. Vol. 93 (4): 531-550.

1965   Contributions to a Review of Philippine Snakes VIII. The Snakes of Genus Lycodon H. Boie. *Philippine Journal of Science.* Manila. Vol. 94 (1): 117-140.

Lillywhite, Harvey B., Smits, Allan W., and Feder, Martin E.
1988   Body Fluid Volumes in the Aquatic Snake Acrochordus granulatus. *Journal of Herpetology.* Athens: Ohio University. Vol. 22 (4): 434-438.

Lim Boo Liat.
1979 *Poisonous Snakes of Peninsular Malaysia.* Kuala Lumpur: Malaysian Nature Society.

Loveridge, A.
1945 *Reptiles of the Pacific World.* New York: MacMillan.

Malnate, Edmond V.
1960 Systematic Division and Evolution of the Colubrid Snake Genus Natrix, With Comments on the Subfamily Natricinae. *Proceedings of the Academy of Natural Sciences of Philadelphia.* Philadelphia: Academy of Natural Sciences of Philadelphia. Vol. 112 (3): 41-71.

Martinez, Joe and Behler, John L.
1988 Erpeton Tentaculatum (Tentacled Snake), Reproduction. *Herpetological Review.* Oxford: Miami University. Vol. 19 (2): 35.

Mattison, Christopher.
1982 *The Care of Reptiles and Amphibians in Captivity.* Poole, U.K.: Blandford.

McCoy, Michael.
1980 *Reptiles of the Solomon Islands.* Handbook 7. Papua New Guinea: Wau Ecology Institute.

McDowell, S. B.
1972 The genera of sea-snakes of the Hydrophis group (Serpentes: Elapidae). London: Trans. Zoological Society of London. Vol. 32: 189-247.

1974 A Catalogue of the Snakes of New Guinea and the Solomons, With Special Reference to Those in The Bernice P. Bishop Museum, Part I, Scolecophidia. *Journal of Herpetology.* Athens: Ohio University. Vol. 8 (1): 1-57.

1975 A Catalogue of the Snakes of New Guinea and the Solomons, With Special Reference to Those in The Bernice P. Bishop Museum, Part II, Anilioidea and Pythoninea. *Journal of Herpetology.* Athens: Ohio University. Vol. 9 (1): 1-79.

1977 A Catalogue of the Snakes of New Guinea and the Solomons, With Special Reference to Those in The Bernice P. Bishop Museum, Part III, Boinae and Acrochordoidea (Reptilia, Serpentes). *Journal of Herpetology.* Athens: Ohio University. Vol. 13 (1): 1-92.

Mehrtens, John M.
1987 *Living Snakes of the World in Color.* New York: Sterling.

Ministry of Communications.
1982 *Climatological Data of Thailand, 30-year Period.* (1951-1980). Bangkok: Ministry of Communications.

Ministry of Public Health.
1985 *Annual Summary.* In Thai. Bangkok: Ministry of Public Health.

Murphy, James B. and Collins, Joseph T., ed.
1980 *Reproductive Biology and Diseases of Captive Reptiles.* Lawrence: Meseraull.

Nop Hoodarapawaanoon.
1964 *An Account of Sea Turtles and Sea Snakes.* In Thai. Bangkok: Naval Print Shop.
The author is a Captain in the Royal Thai Navy.

Phelps, Tony.
1981 *Poisonous Snakes.* Poole, U.K: Blandford.

Rasmussen, Arne Redsted.
1987   Persian Gulf Sea Snake Hydrophis lapemoides (Gray): New Records From Phuket Island, Andaman Sea, and the Southern Part of the Straits of Malacca. *Natural History Bulletin of the Siam Society.* Bangkok: The Siam Society. Vol. 35 (1-2): 57-58.

Rasmussen, Arne Redsted.
1989   Sea Snakes Thalassophina viperina (Schmidt) and Laticauda laticaudata (Linne) New Records From Phuket Island, Andaman Sea, with Remarks on Subspecies of L. laticaudata. *Natural History Bulletin of the Siam Society.* Bangkok: The Siam Society. Vol. 37 (1): 99-103.

Rasmussen, Jens Bodtker.
1982   A New Record of the Rare Opisthotropis praemaxillaris (Serpentes: Colubridae). *Amphibia - Reptilia.* Copenhagen. Vol. 3: 279-280.

Reitinger, Frank F.
1978   *Common Snakes of South East Asia and Hong Kong.* Hong Kong: Heineman.

Romer, Alfred S.
1956   *Osteology of the Reptiles.* Chicago and London: University of Chicago Press.

Ross, Richard A.
1984   *The Bacterial Diseases of Reptiles.* Stanford: The Institute for Herpetological Research.

Rossman, Douglas A. and Eberle W. Gary.
1977   Partition of the Genus Natrix, With Preliminary Observations on Evolutionary Trends in Natricine Snakes. *Herpetologica.* Pittsburgh: Vol. 33: 34-43.

Saint Girons, H.
1972   Les Serpents du Cambodge. *Memoires du Museum National D'Histoire Naturelle.* In French. Serie A; Toma LXXIV. Paris.

Schmidt, Karl P. and Inger, Robert F.
1957   *Living Reptiles of the World.* London: Hamish Hamilton.

Seigel, Richard A., Collins, Joseph T., and Novak, Susan S., ed.
1987   *Snakes: Ecology and Evolutionary Biology.* New York: Macmillan.

Smith, Hobart M., Smith, Rozella B., and Sawin. H. Lewis.
1977   A Summary of Snake Classification (Reptilia, Serpentes). *Journal of Herpetology.* Athens: Ohio University. Vol. 11 (2): 115-121.

Smith, Malcolm.
1943   *The Fauna of British India, Ceylon, and Burma, including the Whole of the Indo-Chinese Sub-Region, Reptilia and Amphibia.* Vol. III. Serpentes. London: Taylor and Francis.

Staubing, Rob.
1988   Island Romance: The Biology of the Yellow-lipped Sea Krait. *Malayan Naturalist.* Kuala Lumpur: Malayan Nature Society. Vol. 41 (3 & 4): 9-11.

Steward, J. W.
1971   *The Snakes of Europe.* Cranbury: Associated University Press.

Taylor, Edward H.
1965   The Serpents of Thailand and Adjacent Waters. *The University of Kansas Science Bulletin.* Vol. XLV (9). Lawrence: University of Kansas.

Taylor, Edward H. and Elbel, Robert E.
1958   Contributions to the Herpetology of Thailand. *The University of Kansas Science Bulletin.* Vol. XXVIII (13) Pr. II. Lawrence: University of Kansas.

Trutnau, Ludwig.
1979 *Schlangen im Terrarium.* Band I. In German. Stuttgart: Eugen Ulmer.

1979 *Schlangen im Terrarium.* Band 2. In German. Stuttgart: Eugen Ulmer.

Tweedie, M. W. F.
1983 *The Snakes of Malaya.* 3d ed. Singapore: National Printers.

Ubertazzi Tanara, Milly.
1978 *The World of Amphibians and Reptiles.* English ed. New York: Gallery.

Van Hoesal, J. K. P.
1959 *Ophidia Javanica.* In Eng. and Indon. Bogor: Museum Zooligicum Bogoriense.

Voris, Harold K. and Glodck, Garrett S.
1980 Habitat, Diet, and Reproduction of the File Snake Acrochordus granulatus, in The Straits of Malacca. *Journal of Herpetology.* Athens: Ohio University. Vol. 14 (1): 108-111.

Wall, Frank., comp. Simon M. Camden-Main.
1969 *Bibliography of the Herpetological Papers of Frank Wall (1865-1950).* Washington D.C.: U.S. National Museum.

Wirot Nutphand.
1985a *Subfamily Boiginae.* In Thai. Bklt. 1. Bangkok: Thai Zoological Center.
- Subsequently superceded by: 1985b.

1985b *Subfamily Boiginae.* In Thai. Bklt. 1. Bangkok: Thai Zoological Center.

1986 *Cobras.* In Thai. Bklt. 12. Bangkok: Thai Zoological Center.

Wirot Nutphand.
1986   *Python molurus bivittatus in Thailand*. In Thai. Bklt. 3. Bangkok: Thai Zoological Center.

1986   *Tribe Trimeresurus*. In Thai. Bklt. 5. Bangkok: Thai Zoological Center.

1986   *Wolf Snakes*. In Thai. Bklt. 9. Bangkok: Thai Zoological Center.

Wüster, Wolfgang and Thorpe, Roger S.
1989   Population affinities of the asiatic cobra (Naja naja) species complex in south-east Asia: reliability and random resampling. *Biological Journal of the Linnean Society*. London: 36: 391-409.

York, Daniel S. and Burghardt, Gordon M.
1988   Brooding in the Malayan Pit Viper, Calloselasma rhodostoma: Temperature, Relative Humidity, and Defensive Behavior. *The Herpetological Journal*. London: British Herpetological Society. Vol. 1 (6): 210-214.

Zimmerman, Elke.
1986   *Breeding Terrarium Animals*. New Jersey: T. F. H. Publications.

Index

(**Bold-face** = major reference: [ ] = plates)

A
aagaardi, Hydrophis
   torquatus, **331**, 332
atriceps, Hydrophis fasciatus, **324**
Acalyptophis, 315
   peronii, [314] **315**
Acrochordoidea, 4, 93, 97
Acrochordidae, 93
Acrochordus, 24, 93
   arafura, 32
   granulatus, 32, **94** [123]
   javanicus, 94, **95**, 96 [124]
Afronatrix, 247
Agkistrodon, 367
   piscivorus, 302
   rhodostoma, 77
Ahaetulla, 35, 117, 301
   ahaetulla, 77
   caudolineata, 77
   formosa, 77
   mycterizans, 77, **118** [129]
   nasuta, xvii, 77, **119**, 120 [129]

Ahaetulla
   prasina, 77, 118, **120**, 121 [129] [130]
Aipysurus, 316
   eydouxii, **317** [355]
albiceps, Ramphotyphlops, 77, **85**
   Typhlops, 77
albofuscus, Lycodon, 204
albolabris albolabris,
   Trimeresurus, 18, 371, **373**, 374, 379, 380 [397] [398]
albolabris Trimeresurus,
   albolabris, 18, 371, **373**, 374, 379, 380 [397] [398]
Alethinophidia, 92, 96
amikacin, 66, 67, 68, 69, 70, 71
Amphibious, Sea Snake, 396
Amphiesma, 233
   deschauenseei, **234** [268]
   groundwateri, **235**
   inas, 77, **235**, 236
   stolata chinensis, **237** [268]

Index

ampicillin, 66, 71
Anaconda, 108
andersoni, Trimeresurus
  purpureomaculatus, 392
Angel's Mountain Keelback, 238
Anilioidea, 4, 97
annamensis, Calloselasma, 367
annandalei, Kolpophis, **336**
Anomalepididae, 82
Anomalous Sea Snake, 344
anomalous, Thalassophis, **344**
Apolopeltura, 255
  boa, 254, **256** [271]
arafura, Acrochordus, 32
Ashy Kukri Snake, 215
Asian Coral Snakes, 349, 363
Assamese Mountain
  Snake, 265 [273]
Astrotia, 318
  stokesii, **318**, 319 [355]
Asuntol, 56
atra, Naja, 298, 300
atra, Naja naja, 298, 300
atriceps, Hydrophis fasciatus, **324**

B

baliodeira cochranae, Liopeltis, **206**
Bamboo Pit Viper, 393
Banded Green Cat Snake, 143 [173]
Banded Krait, [274] 285, 286
Banded Small-headed Sea
  Snake, 324
Banded Wolf Snake, 210
Barred Tree Snake, 148 [174]
barroni, Oligodon, **214** [226]
Barron's Kukri Snake, 214 [226]
Beaked Sea Snake, 319 [356]
Big-eyed Mountain
  Keelback, 266, 275
Big-eyed Pit Viper, 378 [399]
Big-headed Sea Snake, 336, 337
Bitia, 182
  hydroides, **183**

bivirgata flaviceps, Maticora,
  30, 36, **363**, 365 [395]
bivittatus, Python molurus, 103,
  106, 107, 108, 109 [126] [127]
Black and White Spitting
  Cobra, xvii, 293, 296, **297**, 298
  [311] [312]
Black-banded Sea Krait, 348
Black Dog-toothed Cat Snake, 132
Black Spitting Cobra, 293, 296,
  298, 299 [312]
Black-banded Sea Krait, 348
Black Dog-toothed Cat Snake, 132
Black Spitting Cobra, 293, 296,
  298, 299 [312]
Black-striped Mountain Racer, 159
Blind Snakes, 76, 81, 84, 88
Blood Python, 104, 105
Blue-banded Sea
  Snake, 323, 324 [357]
Blue Krait, [274] 284
Blue Long-glanded Coral
  Snake, 363, 364, 365 [395]
Blue-necked Keelback, 264 [272]
Blunt-headed Slug Snake, 256 [271]
boa, Aplopeltura, 254, **256** [271]
boa constrictor, 30
boas, 38
bocourti, Enhydris, **188** [221]
Bocourt's Water Snake, 188 [221]
Boidae, 102
Boiga, 24, 121, 122, 142
  cyanea, xvii, **122** [130] 144
  cynodon, [130] **131**, 132, 141
    [171]
  dendrophila melanota, xvii, **133**,
    134, 140 [171] 301
  drapiezii, **135**, 136 [172]
  jaspidea, xvii, **137** [172]
  mahasomi, 143
  multimaculata, 138
  multomaculata, **138**, 139 [173]

Boiga nigriceps, **139**, 140 [173]
  ocellata, **141**, 142
  saengsomi, xviii, **143** [173] 144
Booidea, 4, 28, 31, 92, 102
boonsongi, Parahelicops, **240**
Boonsong's Keelback, 240
borneensis, Lepturophis, xvii, **204**, 205 [224]
braminus,
  Ramphotyphlops, 77, **86** [123] [361]
  Typhlops, 77
brongersmai, Python curtus, **104** [126]
brookii, Hydrophis, **321** [356]
Brooke's Sea Snake, 321 [356]
Brown Flat-nosed Pit Viper, 386 [400]
Brown Kukri Snake, 231
Brown Long-glanded Coral Snake, 365 [395]
buccata,
  Homalopsis, 31, 187, **198** [223]
Bungarinae, 282, 305
Bungarus, 19, 282
  candidus, 35, 212 [274] **284**
  fasciatus, 35 [274] **285**, 286, 287
  flaviceps, **287**, 288 [307]
Burmese Python, 106, 107 [126] [127]

C

caerulescens,
  Hydrophis, **322**, 323 [356]
Calamaria, 25, 111
  leucocephala, xix, 77
  lumbricoidea, xix, 77, **113**
  pavimentata, xix, **114**
  schlegeli schlegeli, 77, 113, **114**
  siamensis, xix
  uniformis, xix
  vermiformis, 77
Calamariinae, 92, 111

Calliophis, 349, 350
  gracilis, **350** [362]
  macclellandii macclellandii, **351**, 352
  maculiceps hughi, **352**
  maculiceps maculiceps, **353**, 354 [362]
  maculiceps malcolmi, xix, 354
  maculiceps smithi, xix, **354**
Calloselasma, 18, 367, 368
  annamensis, 367
Calloselasma rhodostoma, 18, 19, 77, 367, **368**, 369, 370 [396] [397]
Calotes, 302
Cambodian Kukri Snake, 230
candidus,
  Bungarus, 35, 212 [274] **284**
canker, 66
Cantoria, 184
  violacea, xvii, **185**
Cantor's Water Snake, 185
capucinus, Lycodon, **209** [225]
carinatus, Pareas, **257** [271]
  Ptyas, 22, **169**
  Zaocys, 169
Cat-eyed Fishing Snake, 197
caudolineata, Ahaetulla, 77
caudolineatus,
  Dendrelaphis, 77, **150** [175]
Cave Dwelling Snake, 162 [178]
Cerberus, 186, 301
  rynchops, 31, 32, **186**, 187 [220]
Checkered Keelback, 252 [270]
chinensis, Amphiesma stolata, **237** [268]
Chinese Green Tree Viper, 393
Chinese Keelback, 248
chloramphenicol, 66, 67, 68, 71
chrysargus, Rhabdophis, **242** [268]
Chrysopelea, 26, 145, 148, 182, 204
  ornata, 145, **146** [174]

Index

Chrysopelea
  paradisi, 145, **147** [174]
  pelias, 145, **148** [174]
cinereus multifasciatus,
  Oligodon, **215**
  swinhonis, Oligodon, **216**
CITIES, 39
Cobras, viii, 19, 27, 29, 34, 36, 38,
  288, 289, 290, 291, 292, 293, 406
cochranae, Liopeltis baliodeira, **206**
Cochran's Coral Snake, 352
Cohn's Bronzeback, 153 [176]
Collared Reed Snake, 114
collaris, Sibynophis, **277**, 278
Coluber naja, 291
Colubrid, 27
Colubridae, 110, 282
colubrina,
  Laticauda, **346**, 347, 348 [362]
Colubrinae, 117
Colubroidea, 4, 75, 76, 110
Common Blackhead, 277
Common Blind Snake, 8 [123]
Common Bridle Snake, 201
Common Bronzeback, 152 [175]
Common Keelback, 250 [270]
Common Kukri Snake, 217 [226]
Common Malayan Racer, 157 [177]
Common Mock Viper, 166 [219]
Common Rat Snake, 179 [220]
Common Ringneck, 207
Common Wolf Snake, 209 [225]
condanarus indochinensis,
  Psammophis, **167** [219]
constrictor, boa, 30
Copperhead Racer, 161 [177]
Corneo-spectacular Swelling, 67
Crab-eating Water Snake, 196
Cream Colored Cobra, **295** [311]
Crotalinae, 28, 366, 367, 406
curtus brongersmai,
  Python, **104** [126]
cyanea, Boiga, xvii, **122** [130] 144

cyanochloris,
  Dendrelaphis, xvii, **150**
cyanocinctus,
  Hydrophis, **323**, 324 [357]
cyclurus smithi,
  Oligodon, **217** [226]
  superfluens, Oligodon, **218**
Cylindrophis, 98, 100
  rufus rufus, 36, **98**, 99 [124] [125]
cynodon,
  Boiga, [130] **131**, 132, 141 [171]

D

Dark Blue-banded Sea
  Snake, 322 [356]
davisonii, Dryocalamus, **201**
  tangsongensis, Dryocalamus, 202
Dendrelaphis, 22, 24, 150, 182
  caudolineatus, 77, **150** [175]
  cyanochloris, xvii, **150**
  formosus, xvii, 77, **151** [175]
  pictus pictus, 23, 77, **152** [175]
  striatus, xvii, **153** [176]
  subocularis, **154** [176]
dendrophila melanota,
  Boiga, xvii, **133**, 134, 140
  [171] 301
deschauenseei,
  Amphiesma, **234** [268]
diadema, Hydrophis torquatus, **332**
diardi diardi, Typhlops, **87**
  muelleri, Typhlops, xix, **88**
  Typhlops diardi, **87**
Dinodon, 199
  septentrionalis, **200**
Dog-faced Water Snake, 186 [220]
Dog-toothed Cat Snake, [130] 131
dorsalis, Oligodon, xvii, **227**
dorsolateralis, Oligodon, **228**
drapiezii, Boiga, **135**, 136 [172]
Dryocalamus, 200, 201
  davisonii, **201**
  davsonii tungsongensis, 202

Index

D

Dryocalamus
  subannulatus, xvii, 201, **202** [224]
Dryophiops, 154, 204
  rubescens, xvii, 154, **155** [176]
Dryophis mycterizans, 77
  nasuta, 77
  prasinus, 77
Duvernoy's gland, 75, 110, 111, 112, 117, 257, 276
Dwarf Reed Snake, 115, 116

E

Elaphe, 24, 156
  flavolineata, xvii, 21, **157**, 158 [177]
  floweri, xix
  porphyracea
    nigrofasciata, xvii, **159**
  porphyracea
    porphyracea, 158, **160** [177]
  radiata, 158, **161** [177]
  taeniura ridleyi, 156, **162** [178]
  taeniura taeniura, **163** [178]
Elapidae, 4, 282, 345, 366
elegans, Gonyosoma jansenii, xix
Elegant Bronzeback, 151 [175]
Elephant Trunk Snake, 95, 96 [124]
Enhydrina, 319
  schistosa, 306, **319**, 320 [355]
Enhydris, 188, 301
  bocourti, **188** [221]
  enhydris, **189** [221]
enhydris, Enhydris, **189** [221]
Enhydris jagori, **190** [221]
  plumbea, 187, **190**, **191** [222]
  smithi, **191**
Equatorial Spitting Cobra, 300 [313]
Erpeton, 192, 193
  tentaculatum, 192, **194** [222]
erythrurus, Trimeresurus, xix, 372
eydouxii, Aipysurus, **317** [355]
Eunectes murinas, 108

F

fasciatus atriceps, Hydrophis, **324**
  Bungarus, 35 [274] **285**, 286, 287
fasciatus, Hydrophis, xviii, **325**, 326 [357]
  Lycodon, **210**
File Snake, 94, 95 [123]
Flat-nosed Pit Viper, 386 [400]
flaviceps, Bungarus, **287**, 288 [307]
  Macropisthodon, 263 [272]
  Maticora bivirgata, 30, 36, **363**, 365 [395]
flavolineata, Elaphe, xvii, 21, **157**, 158 [177]
flavipunctata,
  Xenochrophis, **250** [270]
Flower Pot Snake, 86, 87 [123]
floweri, Elaphe, xix
  Typhlops, **89**
Flower's Blind Snake, 89
Flying Snakes, 145
force feeding, 60
Fordonia, 195, 301
  leucobalia, xvii, **196**
formosa, Ahaetulla, 77
formosus,
  Dendrelaphis, xvii, 77, **151** [175]

G

Gekko, 302
Gentian Violet, 65
gentamicin, 66, 67, 68, 71
Gerarda, 196
  prevostiana, **197**
Gerard's Water Snake, 197
Gliding Snakes, 145
Golden Spitting
  Cobra, 298, 300 [313]
Golden Thai Python, 107
Golden Tree Snake, 146 [174]

518 Index

Gonyosoma, 164
  jansenii, xix
  jansenii elegans, xix
  oxycephalum, **164** [178]
gracilis, Calliophis, **350** [362]
gracilis gracilis, Hydrophis, 339
gracilis gracilis,
  Microcephalophis, **339** [360]
gracilis Hydrophis, gracilis, 339
gracilis Microcephalophis,
  gracilis, **339** [360]
gramineus, Trimeresurus, 384
Granular Snake, 94 [123]
granulatus,
  Acrochordus, 32, **94** [123]
Gray Cat Snake, 141, 142
Gray Coral Snake, 350, 351 [362]
Gray's Kukri Snake, 227
Green Cat Snake, 122 [130]
Green Cat-eye Snake, 122 [130]
Green Keelback, 243
Grey Kukri Snake, 229 [267]
Gro-lux, 42
groundwateri, Amphiesma, **235**
Groundwater's Keelback, 235

H

hageni
  Trimeresurus, xvii, 371, **375**, 376,
  [398] 403
Hagen's Pit Viper, 375 [398]
hamptoni, Pareas, **258**
Hampton's Slug Snake, 258
hannah, Ophiophagus, 19, 25, 34,
  289, 301, **303**, 304 [314]
Harder's gland, 282
Hardwicke's Sea Snake, 338 [359]
hardwickii,
  Lapemis, 306, 337, **338** [359]
helleri, Rhabdophis
  subminiatus, 244
hexagonotus, Xenelaphis, xvii, **180**
Hill Wolf Snake, 200

Homalopsinae, 181
Homalopsis, 198, 301
  buccata, 31, 187, **198** [223]
Horse-tail Whip Snake, [130] 131
House Snake, 209, 210 [225]
hughi, Calliophis maculiceps, **352**
hydroides, Bitia, **183**
Hydropheinae, 305, 306, 315, 345,
  366
Hydrophis, 24, 321
Hydrophis brookii, **321** [356]
  caerulescens, **322**, 323 [356]
  cyanocinctus, **323**, 324 [357]
  fasciatus atriceps, **324**
  fasciatus fasciatus, xviii, **325**,
    326 [357]
  klossi, **326**, 327
  lapemoides, xvii, **327**
  mamillaris, xix
  ornatus ornatus, **329** [358]
  spiralis, xvii, **330** [358]
  torquatus, xviii, 331, 333
  torquatus aagaardi, **331**, 332
  torquatus diadema, **332**
  torquatus torquatus, xviii, 332,
    333, 334 [358]

I

inas, Amphiesma, 77, **235**, 236
inas, Natrix, 77
Indian Banded Wolf Snake, 210
indochinensis, Psammophis
  condanarus, **167** [219]
Indochinese Blind Snake, 88
Indochinese Rat Snake, 170 [220]
Indochinese Sand
  Snake, 167, 168 [219]
Indochinese Wolf Snake, 211 [225]
Inornate Kukri Snake, 229 [267]
inornatus, Oligodon, **229** [267]
intestinalis lineata,
  Maticora, **365** [395]
iodine, 65

Iridescent Earth Snake, 101 [125]
Isan Spitting Cobra, 293, 296, 298, **299** [313]
isanensis, Naja naja, **299**

J

Jacobson's organ, 27, 33
jagori, Enhydris, **190** [221]
jansenii, Gonyosoma, xix
    elegans, Gonyosoma, xix
Jasper Cat Snake, 137
jaspidea, Boiga, xvii, **137** [172]
javanicus,
    Acrochordus, 94, **95**, 96 [123]
    Xenodermus, **280**
jerdonii siamensis, Kerilia, **335**
Jerdon's Sea Snake, 335
joynsoni, Oligodon, **229** [267]

K

Kanburee Pit Viper, 376, 378 [398]
kanburiensis,
    Trimeresurus, **376**, 392, [398]
kaouthia, Naja, 293, **294**, 295, 296, 301, [307] [308] [309] [310] [311]
    suphanensis Naja, 295
    Naja naja, 291, 294
Keel-bellied Water Snake, 183, 184
Keeled Rat Snake, 169
Keeled Slug Snake, 257 [271]
Kerilia, 335
    jerdonii siamensis, **335**
Khorat Blind Snake, 90
khoratensis, Typhlops, **89**
King Cobra, 35, 301, 302, 303, 304, 305 [314]
klossi, Hydrophis, **326**, 327
Kloss's Sea Snake, 326, 327
Kolpophis, 24, 336
    annandalei, **336**
korros, Ptyas, 170 [220]
krait, 35, 282, 283, 286

L

laevis, Pareas, **259**
laoensis, Lycodon, 211 [225]
Laotian Wolf Snake, 211 [225]
Lapemis, 337
    hardwickii, 306, 337, **338** [359]
lapemoides, Hydrophis, xvii, **327**
Laticauda, 346
    colubrina, **346**, 347, 348 [362]
    laticaudata, **348**
laticaudata, Laticauda, **348**
Laticaudinae, 315, 345
Leptotyphlopidae, 82
Lepturophis, 24, 204
    borneensis, xvii, **204**, 205 [224]
leucobalia, Fordonia, xvii, **196**
leucocephala, Calamaria, xix, 77
lineata, Maticora
    intestinalis, **365** [395]
lineatus, Typhlops, **90** [123]
Liopeltis, 206
    baliodeira cochranae, **206**
    scriptus, **207**
    tricolor, **208**
longiceps, Pseudorabdion, **115**, 116
Long-glanded Coral Snakes, 363
Long-nosed Whip Snake, 119 [129]
Lower Gulf Black-headed Sea Snake, 331
lumbricoidea,
    Calamaria, xix, 77, **113**
Lycodon, 204, 209, 211
    albofuscus, 204
    capucinus, **209** [225]
    fasciatus, **210**
    laoensis, **211** [225]
    subcinctus, 35, **211**, 212 [225] [226]
Lycondontinae, 199

M

macclellandii, Calliophis
macclellandii, **351**, 352
macclellandii Calliophis, **351**, 352
Macropisthodon, 262
flaviceps, **263** [272]
rhodomelas, **264** [272]
macrops Pseudoxenodon, **266**
Trimeresurus, xvii, 18, 371, **378**, 379, 380 [399]
macularius, Pareas, **260**
maculiceps, Calliophis
maculiceps, **353**, 354 [362]
maculiceps hughi, Calliophis, **352**
maculiceps, Calliophis, **353**, 354
malcolmi, Calliophis, xix, 354
smithi, Calliophis, xix, **354**
mahasomi, Boiga, 143
malaccanus, Pareas, **260** [271]
Malayan Banded Wolf Snake, 211 [225] [226]
Malayan Blackhead, [273] 278
Malayan Blind Snake, 88
Malayan Bridle Snake, 202 [224]
Malayan Brown Snake, 180
Malayan Green Whip Snake, 118 [129]
Malayan Mountain Keelback, 235
Malayan Pit Viper, 368, 370 [396] [397]
Malayan Ringneck, 208
Malayan Slug Snake, 260, 261 [271]
Malayan Stripe-tailed Racer, 162
malcolmi, Calliophis
maculiceps, xix, 354
mamillaris, Hydrophis, xix
Mangrove Pit Viper, 390, 392 [401]
Mangrove Snake, 133, 134
Many Spotted Cat Snake, 138 [173]
Marble Cat Snake, 138 [173]
margaritophorus, Pareas, **261** [272]
Masked Water Snake, 198 [223]
Maticora, 349, 363

Maticora, bivirgata
flaviceps, 30, 36, **363**, 365 [395]
intestinalis lineata, 365 [395]
Maticorinae, 349
McClelland's Coral Snake, 351
melanocephalus,
Sibynophis, [273] **278**
melanota, Boiga dendrophila, xvii, 133, 134, 140 [171] 301, 371, 404
meridionalis, Trimeresurus
(Ovophis) monticola, **381**, 382 [399]
Microcephalophis, 339
gracilis gracilis, **339** [360]
mites, 55, 56
molurus bivittatus, Python, 103, **106**, 107, 108, 109 [126] [127]
molurus, Python, 106
Monocled Cobra, 294, 295 [307] [308] [309] [310] [311]
monticola meridionalis,
Trimeresurus (Ovophis) 371, **381**, 382 [399] 404
Trimeresurus, 381
Mottled Spitting
Cobra, **297** [311] [312]
mouhoti, Oligodon, **230**
Mountain Bronzeback, 154 [176]
Mountain Pit Viper, 381, 382 [399]
Mountain Slug Snake, 261 [272]
mouth rot, 66
mucosus, Ptyas, 168, **179** [220] 301
muelleri, Typhlops, xix
Typhlops diardi, xix, **88**
multifasciatus, Oligodon
cinereus, **215**
multimaculata, Boiga, 138
multomaculata,
Boiga, **138**, 139 [173]
murinus, Eunectes, 108
mycterizans,
Ahaetulla, 77, **118** [129]
Dryophis, 77

N

Naja, xvi, xix, xx, xxiii, 18, 19, 288, 290, 292, 293, 301, 304, 305
Naja atra, 298, 300
naja atra, Naja, 298, 300
naja, Coluber, 291
naja isanensis, Naja, 299
Naja kaouthia, 291, 293, **294**, 296, 301 [307] [308] [309] [310] [311]
naja kaouthia, Naja, 291, 293, 294, 295, 296
Naja kaouthia suphanensis, 295
Naja naja atra, 298, 300
   isanensis, 299
   kaouthia, 291, 293, 294, 295, 296
   sputatrix, 297
   sumatrana, 291, 300
Naja sp. 291
Naja sp. Black and White Spitting Cobra, 293, **297** [311] [312]
Naja sp. Black Spitting Cobra, 293, **298** [312]
Naja sp. Cream Colored Cobra, **295**, [311]
Naja sp. Isan Spitting Cobra, 293, **299** [313]
Naja sp. Mottled Spitting Cobra, **297** [311] [312]
Naja sp. Suphan Cobra, 293, **295** [311]
naja sputatrix, Naja, 297
Naja sumatrana, xvii, 291, 293, **300**, 301 [313]
naja sumatrana, Naja, 291, 300
nasuta, Ahaetulla, xvii, 77, **119**, 120 [129]
   Dryophis, 77
Natricinae, 233, 262
Natrix, 246, 247
   inas, 77
   percarinata, 77
   piscator, 77

Natrix,
   trianguligera, 77
necrotizing dermatitis, 67
   stomatitis, 66
Needle-headed Sea Snake, 339 [360]
Nerodia, 247
nigriceps, Boiga, **139**, 140 [173]
nigrocinctus, Rhabdophis, xvii, **243**
nigrofasciata, Elaphe porphyracea, xvii, **159**
Northern Keelback, 234
nuchalis, Plagiopholis, 265 [273]

O

ocellata, Boiga, **141**, 142
Oligodon, 111, 213
   barroni, **214** [226]
   cinereus multifasciatus, **215**
   cinereus swinhonis, **216**
   cyclurus smithi, **217** [226]
   cyclurus superfluens, **218**
   dorsalis, xvii, **227**
   dorsolateralis, **228**
   inornatus, **229** [267]
   joynsoni, **229** [267]
   mouhoti, **230**
   purpurascens purpurascens, **231**
   quadrilineatus, xix
   taeniatus, xix, 230, **232** [267]
Ophiophagus, 301
   hannah, 19, 25, 34, 289, 301, **303**, 304 [314]
Opisthotropis, 238
   praemaxillaris, **238**
   spenceri, **239**
Opistoglyph, 111, 117, 122
Orange-necked Keelback, 263 [272]
Oriental Rat Snake, 179 [220]
Oriental Whip Snake, 120 [129] [130]
ornata, Chrysopelea, 145, **146** [174]

ornatus, Hydrophis
  ornatus, **329** [358]
  ornatus, Hydrophis, **329** [358]
Ovophis, 381 [399]
oxycephalum Gonysoma, **164** [178]

P

panophthalmitis, 67
paradisi,
  Chrysopelea, 145, **147** [174]
Paradise Tree Snake, 147 [174]
Parahelicops, 240
  boonsongi, **240**
Parasibynophis, 276
Pareas, 254, 255, 257
  carinatus, **257** [271]
  hamptoni, **258**
  laevis, **259**
  macularius, **260**
  malaccanus, **260** [271]
  margaritophorus, **261** [272]
Pareatinae, 254, 255
pavimentata, Calamaria, xix, **114**
Pelamis, 340
  platurus, **341** [360] [361]
pelias, Chrysopelea, **148** [174]
percarinata, Natrix, 77
  Sinonatrix, 77, **248**
peronii, Acalyptophis, [314] **315**
Peron's Sea Snake, [314] **315**
Persian Gulf Sea Snake, 327
pictus, Dendrelaphis pictus, 23, 77, **152** [175]
  pictus, Dendrelaphis, 23, 77, **152** [175]
piscator, Natrix, 77
  Xenochrophis, 77, **252** [270]
piscavorus, Agkistrodon, 302
Pit Vipers, 367
Plagiopholis, 264
  nuchalis, **265** [273]
platurus, Pelamis, **341** [360] [361]

plumbea, Enhydris, 187, **190**, **191** [222]
Plumbeous Water Snake, 190 [222]
Polysporin, 66, 67, 68
Pope's Pit Viper, 384, 385 [400]
popeorum
  Trimeresurus popeorum, 371, 379, **384**, 385, 394 [400]
  popeorum, Trimeresurus, 371, 379, **384**, 385, 394 [400]
porphyracea, Elaphe
  porphyracea, 158, 159, **160** [177]
  nigrofasciata, Elaphe, xvii, **159**
  porphyracea, Elaphe, 158, 159, **160** [177]
praemaxillaris, Opisthotropis, **238**
Praescutata, 342
  viperina, **343** [361]
prasina, Ahaetulla, 77, 118, **120**, 121 [129] [130]
prasinus, Dryophis, 77
prevostiana, Gerarda, **197**
Psammodynastes, 165
  pictus, 165
  pulverulentus, 165, **166**, 167 [219]
Psammophis, 167
  condanarus
    indochinensis, **167** [219]
Pseudorabdion, 111, 115
  longiceps, **115**, 116
Pseudorhabdion, 115
Pseudoxenodon, 266
  macrops, **266**
Pseudoxenodontinae, 262
Ptyas, 22, 168
  carinatus, 22, **169**
  korros, **170** [220]
  mucosus, 168, **179** [220] 301
Puff-faced Water Snake, 198 [223]
pulverulentus, Psammodynastes, 165, **166**, 167 [219]

puniceus puniceus,
  Trimeresurus, **386**, 388, 389, 390 [400]
  Trimeresurus puniceus, **386**, 388, 389, 390 [400]
  wiroti, Trimeresurus, xviii, xix, **387**, 390 [400]
purpurascens, Oligodon
  purpurascens, **231**
purpureomaculatus andersoni, Trimeresurus, 392
purpureomaculatus purpureomaculatus, Trimeresurus, 377, **390** [401]
purpureomaculatus Trimeresurus purpureomaculatus, 377, **390** [401]
Python, 28, 33, 50, 104
  curtus brongersmai, **104** [126]
  molurus bivittatus, 103, **106**, 107, 108, 109 [126] [127]
  molurus molurus, 106
  reticulatus, 76, 103, 106, **108**, 109 [123] 301, 302
Pythonidae, 102, 103, 105

Q
quadrilineatus, Oligodon, xix

R
radiata, Elaphe, 158, **161** [177]
Rainbow Water Snake, 189 [221]
Ramphotyphlops, 27, 85, 86
  albiceps, 77, **85**
  braminus, 77, **86** [123] [361]
Red Cat-eye Snake, 139 [173]
Red-headed Krait, 287, 288 [307]
Red Mountain Racer, 160 [177]
Red-necked
  Keelback, 245, 246 [269]
Red-tailed Pipe
  Snake, 98 [124] [125]
Red-tailed Racer, 164 [178]

Red Whip Snake, 155 [176]
Reef Sea Snake, 329 [358]
respiratory infections, xvi, 65, 67
Reticulated Python, 108, 109 [128]
reticulatus, Python, 76, 103, 106, **108**, 109 [128] 301, 302
Rhabdophis, 241
  chrysargus, **242** [268]
  nigrocinctus, xvii, **243**
  subminiatus helleri, **244**
  subminiatus
    subminiatus, **245** [269]
rhodomelas,
  Macropisthodon, **264** [272]
rhodostoma, Agkistrodon, 77
  Calloselasma, 18, 19, 77, 367, **368**, 369, 370 [395] [397]
ridleyi, Elaphe
  taeniura, 156, **162** [178]
rostral, rubbed, 40, 41, 64
roundworms, 57
rubescens,
  Dryophiops, xvii, 154, **155** [176]
rufus, Cylindrophis
  rufus, 36, **98**, 99 [124] [125]
  rufus, Cylindrophis, 36, **98**, 99 [124] [125]
russelli russelli, Vipera, [402] 408
  siamensis, Vipera, 18, 19, 35, 36 [402] 406, **407**, 408
  Vipera russelli, [402] 408
rynchops, Cerberus, 31, 32, **186**, 187 [220]

S
saengsomi, Boiga, xviii, **143** [173] 144
scale rot, 67
schistosa,
  Enhydrina, 306, **319**, 320 [356]
schlegeli, Calamaria
  schlegeli, 77, 113, **114**

schlegeli,
  schlegeli, Calamaria, 77, 113, **114**
Schmidt's Red-necked
  Keelback, 244
Scolecophidia, 4, 81, 92
scriptus, Liopeltis, **207**
septentrionalis, Dinodon, **200**
serous cells, 110, **111**
shedding, 54
Shell pest strip, 55, 56
Shore Pit Viper, 390 [401]
Short-tailed Python, 104 [126]
siamensis, Calamaria, xix
  Kerilia jerdonii, **335**
  Typhlops, xix
  Vipera russelli, 18, 19, 35, 36 [402] 406, **407**, 408
Siamese Cobra, 294, 295 [307] [308] [309] [310] [311]
Siamese Palm Viper, 387, 388, 389, 390 [400]
Siamese Russell's Viper, [402] 407
Sibynodontophis, 276
Sibynopheinae, 276
Sibynophis, 111, 276
  collaris, **277**, 278
  melanocephalus, [273] **278**
  triangularis, [273] **279**
Sinonatrix, 246, 247
  percarinata, 77, **248**
  trianguligera, 77, **249** [269] [270]
Slender Wolf Snake, 204 [224]
Small-spotted Coral Snake, 353
smithi, Enhydris, **191**
  Calliophis maculiceps, xix, **354**
  Oligodon cyclurus, **217** [226]
Smith's Mountain Keelback, 239
Smith's Water Snake, 191
Smooth Slug Snake, 259
South Chinese Kukri Snake, 216
Speckle-bellied Keelback, 242
spenceri, Opisthotropis, **239**

spiralis, Hydrophis, xvii, **330** [358]
spitting cobras, 288, 290, 292
Spotted Cat Snake, 138 [173]
Spotted Slug Snake, 260
sputatrix, Naja naja, 297
stejnegeri, stejnegeri,
  Trimeresurus, 371, 384, 385, **393**
stejnegeri, Trimeresurus
  stejnegeri, 371, 384, 385, **393**
Stokes' Sea Snake, 318 [355]
stokesii, Astrotia, **318**, 319 [355]
stolata chinensis,
  Amphiesma, **237** [268]
stress, 68
striatus,
  Dendrelaphis, xvii, **153** [176]
Stripe-backed Coral Snake, 354
Stripe-tailed Racer, 163
Striped Blind Snake, 90 [123]
Striped Bronzeback, 150
Striped Kukri Snake, 232 [267]
Striped Ringneck, 206
Striped Sea Snake, 325 [357]
Striped Water Snake, 190 [221]
subannulatus, Dryocalamus, xvii, 201, **202** [224]
subcinctus, Lycodon, 35, **211**, 212 [225] [226]
subminiatus helleri,
  Rhabdophis, **244**
  Rhabdophis
    subminiatus, **245** [269]
subminiatus,
  Rhabdophis, **245** [269]
subocularis Dendrelaphis, **154** [176]
Sumatran Pit Viper, 394, 403
sumatrana, Naja, xvii, 291, 293, **300**, 301 [313]
  Naja naja, 291, 300
sumatranus
  Trimeresurus, 371, 375, 376, **394**

Sunbeam Snake, 101 [125]
superfluens, Oligodon
 cyclurus, **218**
superior labial gland, 110
Suphan
 Cobra, 293, 294, **295**, 296 [311]
suphanensis, Naja kaouthia, 295
swinhonis, Oligodon cinereus, **216**

T

taeniatus,
 Oligodon, xix, 230, **232** [267]
taeniura, Elaphe
 taeniura, 156, **163** [178]
 ridleyi, Elaphe, **162** [178]
 taeniura, Elaphe, **163** [178]
tapeworms, 57
Taylor's Kukri Snake, 218
Temple Pit Viper, 404 [401]
Tentacled Snake, 194, 195 [222]
tentaculatum,
 Erpeton, 192, **194** [222]
Tetracycline, 66, 71
Thalassophis, 344
 anomalus, **344**
Thamnophis, 288
The Convention on International
 Trade in Endangered Species of
 Flora & Fauna, 39
Thermal sensitivity, 28
thermotheraphy, 57, 58, 65, 66
ticks, 55, 56
torquatus aagaardi,
 Hydrophis, **331**, 332
 diadema, Hydrophis, **332**
 Hydrophis, xviii, 331, 333
torquatus torquatus, Hydrophis,
 xviii, 332, **333**, 334 [358]
Trang Blind Snake, 91
trangensis, Typhlops, **91**
Triangle Blackhead, [273] 279
Triangle Keelback, 249 [269] [270]

triangularis, Sibynophis, [273] **279**
trianguligera, Natrix, 77
 Sinonatrix, 77, **249** [269] [270]
tricolor, Liopeltis, 208
Trimeresurus, xvi, xix, xx, xxiii, 17,
 18, 19, 22, 25, 34, 35, 288, 301,
 371, 381
 albolabris albolabris, 18, **373**, 374,
  379, 380 [397] [398]
 erythrurus, xix, 372
 gramineus, 384
 hageni, xvii, **375**, 376 [398] 403
 kanburiensis, **376**, 392
 macrops, xvii, 18, **378**, 379, 380
  [399]
 monticola, 381
 (Ovophis) monticola meridionalis,
  381, 382 [399]
 popeorum popeorum, 379, **384**,
  385, 394 [400]
 puniceus puniceus, **386**, 388, 389,
  390 [400]
 puniceus wiroti, xviii, xix, **387**,
  388, 389, 390 [400]
 purpureomaculatus andersoni, 392
 purpureomaculatus
  purpureomaculatus, 377, **390**
  [401]
 stejnegeri stejnegeri, 384, 385, **393**
 sumatranus, 375, 376, **394**
 (Tropidolaemus)
  wagleri, xvii [401] [402] **404**
 wagleri, 402
 wiroti, xviii, 388, 389
Tropidolaemus, xix, **404**
True-lite, 42
True Vipers, 406
tungsongensis, Dryocalamus
 davisonii, 202
Two-headed
 Snake, 98, 99 [124] [125]
tylosin, 66, 68, 71

Typhlina, 85
Typhlopidae, 82, 84, 97
Typhlops, 24, 27, 85, 87, 89, 98, 353
  albiceps, 77
  braminus, 77
  diardi diardi, **88**
  diardi muelleri, xix, **88**
  floweri, **89**
  khoratensis, **90**
  lineatus, **90** [123]
  muelleri, xix
  siamensis, xix
  trangensis, **91**

U

unicolor, Xenopeltis, **101** [125]
uniformis, Calamaria, xix
Upper Gulf Black-headed Sea Snake, 332
Uropeltidae, 97

V

Varanus, 302
Variable Reed Snake, 113
vermiformis, Calamaria, 77
violacea, Cantoria, xvii, **185**
Vipera, 406
  russelli russelli, [402] 408
  russelli siamensis, 18, 19, 35, 36 [402] 406, **407**, 408
Viperidae, 4, 64, 366
viperina, Praescutata, **343**, 361
Viperinae, 366, 406
Viperine Sea Snake, 343 [361]
Vipers, 29
Vita-lite, 42

W

wagleri Trimeresurus, 371, 402
  Trimeresurus (Tropidolaemus), xvii, **404** [401] [402]
Wagler's Pit Viper, [401] [402] 404, 405
Wall's Bronzeback, 150
Wall's Kukri Snake, 228
Wart Snake, 95 [123]
Weak-jawed Blind Snakes, 82, 83
West Coast Black-headed Sea Snake, 333 [358]
White-headed Blind Snake, 85
White-headed Reed Snake, 114
White-lipped Pit Viper, **373**, 374 [397] [398]
White-spotted Cat Snake, 135
White-spotted Sea Snake, 317 [355]
White-striped Keelback, 237
wiroti, Trimeresurus, xviii, 388, 389
  Trimeresurus puniceus, xviii, **387**, 388, 389, 390 [400]
Wirot's Pit Viper, 387 [400]
Worms, 57

X

Xenelaphis, 180
  hexagonotus, xvii, **180**
Xenochrophis, 250
  flavipunctata, **250** [270]
  piscator, 77, **252** [270]
Xenoderminae, 280
Xenodermine Snake, 280
Xenodermus, 110, 280
  javanicus, **280**
Xenopeltidae, 97, 100
Xenopeltis, 100
  unicolor, **101** [125]

Y

Yellow-bellied Sea Snake, 341, 342 [360] [361]
Yellow-lipped Sea Krait, 346 [362]
Yellow Sea Snake, 330 [358]

Z

Zaocys carinatus, 169